Formation and Decay of Disinfection By-Products in the Distribution System

Awwa Research Foundation

About the Awwa Research Foundation

The Awwa Research Foundation (AwwaRF) is a member-supported, international, nonprofit organization that sponsors research to enable water utilities, public health agencies, and other professionals to provide safe and affordable drinking water to consumers.

The Foundation's mission is to advance the science of water to improve the quality of life. To achieve this mission, the Foundation sponsors studies on all aspects of drinking water, including supply and resources, treatment, monitoring and analysis, distribution, management, and health effects. Funding for research is provided primarily by subscription payments from approximately 1,000 utilities, consulting firms, and manufacturers in North America and abroad. Additional funding comes from collaborative partnerships with other national and international organizations, allowing for resources to be leveraged, expertise to be shared, and broad-based knowledge to be developed and disseminated. Government funding serves as a third source of research dollars.

From its headquarters in Denver, Colorado, the Foundation's staff directs and supports the efforts of more than 800 volunteers who serve on the board of trustees and various committees. These volunteers represent many facets of the water industry, and contribute their expertise to select and monitor research studies that benefit the entire drinking water community.

The results of research are disseminated through a number of channels, including reports, the Web site, conferences, and periodicals.

For subscribers, the Foundation serves as a cooperative program in which water suppliers unite to pool their resources. By applying Foundation research findings, these water suppliers can save substantial costs and stay on the leading edge of drinking water science and technology. Since its inception, AwwaRF has supplied the water community with more than $300 million in applied research.

More information about the Foundation and how to become a subscriber is available on the Web at **www.awwarf.org.**

Formation and Decay of Disinfection By-Products in the Distribution System

Prepared by:
Hélène Baribeau, **Lina Boulos**, **Hiriti Haileselassie**, and **Gil Crozes**
Carollo Engineers, P.C. Fountain Valley, CA 92708

Philip C. Singer, **Cora Nichols**, and **Savita A. Schlesinger**
University of North Carolina, Chapel Hill, NC 27599

Richard W. Gullick
American Water, Voorhees, NJ 08043

Sherry L. Williams
Newport News Waterworks, Yorktown, VA 23693

Roy L. Williams and **Laura Fountleroy**
Old Dominion University, Norfolk, VA 23529

and

Susan A. Andrews and **Erin Moffat**
University of Waterloo, Waterloo, Ont., Canada N2L 3G1

Jointly sponsored by:
Awwa Research Foundation
6666 West Quincy Avenue, Denver, CO 80235-3098

and

U.S. Environmental Protection Agency
Washington D.C.

Published by:

Awwa Research Foundation

American Water Works Association

IWA Publishing

DISCLAIMER

This study was jointly funded by the Awwa Research Foundation (AwwaRF) and the U.S. Environmental Protection Agency (USEPA) under Cooperative Agreement No. CR828216-01. AwwaRF and USEPA assume no responsibility for the content of the research study reported in this publication or for the opinions or statements of fact expressed in the report. The mention of trade names for commercial products does not represent or imply the approval or endorsement of either AwwaRF or USEPA. This report is presented solely for informational purposes.

CONTENTS

LIST OF TABLES ix

LIST OF FIGURES xiii

FOREWORD xvii

ACKNOWLEDGMENTS xix

EXECUTIVE SUMMARY xxi

CHAPTER 1: BACKGROUND AND INTRODUCTION 1
Project Goals and Objectives 2
Project Tasks and Report Organization 3

CHAPTER 2: LITERATURE REVIEW 5
Definition, Occurrence, and Health Effects of Disinfection By-Products 5
Trihalomethanes 6
Haloacetic Acids 8
N-nitrosodimethylamine 11
Regulatory Background 12
N-nitrosodimethylamine 13
Chlorination and Chloramination 14
Chlorine Reactions 14
Chloramine Reactions 14
Formation and Stability of Disinfection By-Products 17
Effect of Natural Organic Matter on Disinfection By-Products 18
Effect of Bromide Ion on Disinfection By-Products 20
Effect of Water Temperature on Disinfection By-Products 23
Effect of pH on Disinfection By-Products 26
Effect of Disinfectant Concentration on Disinfection By-Products 27
Stability of Disinfection By-Products 28
Change in Water Quality in the Distribution System 28
Change in Disinfectant Concentrations in the Distribution System 29
Changes in Disinfection By-Product Concentrations and Speciation in the Distribution System 31
Influence of Seasons and Water Temperature on Disinfection By-Products in the Distribution System 37
Influence of pH on Disinfection By-Products in the Distribution System 38
Influence of Pipe Material on Disinfection By-Products 39
Influence of Pipe Diameter on Disinfection By-Products 41
Influence of Hydraulic Conditions on Disinfection By-Products in the Distribution System 41

Correlations Between Disinfection By-Products and
Water Quality Characteristics 42
Conclusions 45

CHAPTER 3: MATERIALS AND METHODS 47
Selection of Water Utilities 47
Utility A – Free Chlorine 49
Utility B – Free Chlorine 52
Utility C – Chloramines 52
Utility D – Chloramines 53
Utility E – Chloramines Periodically Switching to Free Chlorine 53
Selection of Sampling Locations 54
Utility A – Free Chlorine 54
Utility B – Free Chlorine 57
Utility C – Chloramines 57
Utility D – Chloramines 62
Utility E – Chloramines Periodically Switching to Free Chlorine 66
Sampling Strategy 68
Analytical parameters 71
Disinfection By-Products 74
Organic Matter 80
General Water Quality Parameters 83
Controls 84
Inter-Laboratory Comparison Experiments 84

CHAPTER 4: CHANGES IN DISINFECTION BY-PRODUCTS AND WATER
QUALITY IN FULL-SCALE DISTRIBUTION SYSTEMS 85
Trihalomethanes and Haloacetic Acids 85
Free Chlorinated Systems 86
Chloraminated Systems 101
Chloraminated System Periodically Switching to Free Chlorine 114
Effect of Seasons on THMs and HAAs 131
N-Nitrosodimethylamine 131
Chloraminated System 131
Free Chlorinated System 136
Chloraminated System Periodically Switching to Free Chlorine 136
System Comparison 140
Other Water Quality Parameters 141
Chlorine Residuals and Heterotrophic Plate Counts 141
Natural Organic Matter 148
Nitrification in the Chloraminated Systems 151
Summary and Conclusions 155
Trihalomethanes and Haloacetic Acids 155
N-Nitrosodimethylamine 160

CHAPTER 5 GUIDELINES TO SELECT DISTRIBUTION SYSTEM SAMPLING LOCATIONS FOR DISINFECTION BY-PRODUCT MONITORING 161
Stage 2 Disinfectants / Disinfection By-Products Rule 161
Initial Distribution System Evaluation........ 161
IDSE Schedule and System Classification 162
Initial Distribution System Evaluation Options........ 162
Findings from this Study........ 174
Case Study 177
Standard Monitoring Plan Preliminary Steps 177
Standard Monitoring Program Results........ 181
Proposed Stage 2 DBPR Compliance Monitoring Sites........ 181
Summary and Conclusions 182

CHAPTER 6: SUMMARY, CONCLUSIONS AND RECOMMENDATIONS 185
Recommendations........ 188

APPENDIX A: MICROBIAL DEGRADATION OF HAAs AND OTHER DBPs........ 191

APPENDIX B: BATCH INCUBATIONS IN BOTTLES........ 231

APPENDIX C: CORRELATIONS BETWEEN DBP CONCENTRATIONS WATER AGE, AND WATER QUALITY PARAMETERS........ 237

APPENDIX D: EFFECTS OF PIPE MATERIAL, PIPE DIAMETER, BOOSTER CHLORINATION STATION, AND STORAGE RESERVOIRS........ 249

APPENDIX E: INTER-LABORATORY COMPARISON EXPERIMENTS........ 287

REFERENCES 301

ABBREVIATIONS 319

TABLES

2.1	DBPs grouped into subclasses by common mode of action (MOA)	6
2.2	Haloacetic acid species	8
2.3	Toxicological effects of various HAAs	10
2.4	NDMA concentrations measured in California during the first round of sampling	12
3.1	Water utilities selected for the full-scale distribution system monitoring	48
3.2	Water quality at the participating utilities	50
3.3	Distribution storage information for Utility A	51
3.4	Summary of plant production during the samplings conducted at Utility A	55
3.5	Sample site characteristics for Utility A	56
3.6	Summary of plant production during the samplings conducted at Utility B	58
3.7	Sample site characteristics for Utility B	58
3.8	Summary of plant production during the samplings conducted at Utility C	59
3.9	Sample site characteristics for Utility C	61
3.10	Hydraulic scenarios for the samplings conducted at Utility D	63
3.11	Average residence time for all hydraulic scenarios for Utility D	64
3.12	Flushing information during the samplings conducted in Utility D's distribution system	65
3.13	Summary of plant production during the samplings conducted at Utility E	67
3.14	Characteristics of sample collection sites in Utility E	67
3.15	Monitoring program to address the seasonal variability of DBPs in distribution systems	69
3.16	Monitoring program to examine the effect of a switch in final disinfectant (Utility E)	70

3.17 Monitoring parameters, location of analysis and methods used 72

4.1 THM4 concentrations measured during each sampling in Utility A, with water temperature at the POE 88

4.2 THM4 concentrations measured during each sampling in Utility B, with water temperature at the POE 89

4.3 THM4 LRAA values calculated in Utilities A and B 90

4.4 Mass ratios of $CHCl_3$ to THM4 and chlorine-containing THMs to THM4 calculated in the free chlorinated systems 93

4.5 HAA5 and HAA9 concentrations measured during each sampling in Utility A 96

4.6 HAA5 and HAA9 concentrations measured during each sampling in Utility B 97

4.7 HAA5 LRAA values calculated in Utilities A and B 98

4.8 Mass ratios of chlorine-containing HAAs to HAA9 and HAA5 to HAA9 calculated in the free chlorinated systems 99

4.9 THM4 concentrations measured during each sampling in Utility C 103

4.10 THM4 concentrations measured during each sampling in Utility D 104

4.11 THM4 LRAA values calculated in Utilities C and D 106

4.12 Mass ratios of $CHCl_3$ to THM4, chlorine-containing THMs to THM4, and bromine-containing THMs to THM4 calculated in the chloraminated systems 108

4.13 HAA5 and HAA9 concentrations measured during each sampling in Utility C 111

4.14 HAA5 and HAA9 concentrations measured during each sampling in Utility D 112

4.15 HAA5 LRAA values calculated in Utilities C and D 113

4.16 Mass ratios of chlorine-containing HAAs to HAA9, bromine-containing HAAs to HAA9 and HAA5 to HAA9 calculated in the chloraminated systems 115

4.17 Mass ratios of $CHCl_3$ to THM4 calculated in Utility E ... 121

4.18 Mass ratios of chlorine-containing HAAs to HAA9 and HAA5 to HAA9 measured in Utility E ... 127

4.19 Comparison of NDMA results across utilities ... 140

4.20 Occurrence of NDMA concentrations greater than the MCL ... 141

4.21 TOC concentration, UV absorbance at 254 nm and AOC concentration measured at the POE of Utilities A and B during each sampling ... 149

4.22 TOC concentration, UV absorbance at 254 nm and AOC concentration measured at the POE of Utilities C and D during each sampling ... 150

4.23 TOC concentration, UV absorbance at 254 nm and AOC concentration measured at the POEs of Utility E during each sampling ... 151

4.24 Concentrations of total ammonia, nitrite and nitrate measured in absence of nitrification in the distribution systems of Utilities C and D ... 152

4.25 Fate of THM4 and HAA9 in full-scale distribution systems ... 156

5.1 SMP sampling and Stage 2 DBPR compliance monitoring requirements for 100-percent purchasing systems ... 166

5.2 SMP sampling and Stage 2 DBPR compliance monitoring requirements for producing systems ... 167

5.3 Water treatment plant flow rates for both POE locations in Utility E ... 176

5.4 Utility D water ages at each sampling location and sampling date ... 180

5.5 Disinfectant residual data at each location for each sampling date and average disinfectant residual data ... 180

FIGURES

2.1 Distributions of ICR plant-quarter THM4 and HAA5 concentration ranges 16

2.2 HAA5 (A) and THM4 (B) masses from February 1998 to January 1999 19

2.3 Averaged ICR THM4 data showing the relationship between influent bromide level, average THM4 concentration and total percent of brominated THM4 species 22

2.4 Averaged ICR HAA6 data showing the relationship between influent bromide level, average HAA6 concentration, and total percent of brominated HAA6 species 23

2.5 Effect of incubation temperature on THM4 formation.. 25

2.6 Effect of incubation temperature on HAA6 formation.. 25

2.7 Free chlorine residual as a function of water age in a full-scale distribution system.... 30

2.8 Comparison of THM4 concentrations leaving the treatment plant and in the distribution system 33

2.9 Example of HAA results obtained in warm water in the chlorinated train................... 36

2.10 Chlorine decay curve, THM4 production curve and HAA6 production curve............. 40

2.11 Correlations between THMs and DOC (Graph A), and HAA5 and UV absorbance at 254 nm (Graph B) using simple linear regression............... 44

2.12 Distribution of HAA5 concentrations as a function of influent TOC level.................. 45

3.1 Distribution of estimated water age obtained at Utility A sampling locations 56

3.2 Distribution of estimated water age obtained at Utility B sampling locations 59

3.3 Distribution of estimated water age obtained at Utility C sampling locations 61

3.4 Distribution of estimated water age obtained at Utility D sampling locations 65

3.5 Distribution of estimated water age obtained at Utility E sampling locations 68

4.1 THM4 concentration, HAA9 concentration and free chlorine residual as a function of water age measured during the September 23, 2003 sampling in Utility A.. 87

4.2 THM4 concentration, HAA9 concentration and free chlorine residual as a function of water age measured during the August 12, 2003 sampling in Utility B........ 87

4.3 THM4 speciation as a function of water age measured during the June 10, 2003 sampling in Utility A 92

4.4 THM4 speciation as a function of water age measured during the June 10, 2003 sampling in Utility B 92

4.5 HAA9 speciation as a function of water age measured during the September 23, 2003 sampling in Utility A 95

4.6 HAA9 speciation as a function of water age measured during the January 7, 2003 sampling in Utility B 95

4.7 HAA5 and HAA9 concentrations as a function of water age measured during the September 23, 2003 sampling at Utility A 100

4.8 HAA5 and HAA9 concentrations as a function of water age measured during the April 29, 2003 sampling at Utility B 100

4.9 THM4, HAA9 and total chlorine residual as a function of water age measured during the June 11, 2003 sampling in Utility C........ 101

4.10 THM4, HAA9 and total chlorine residual as a function of water age measured in Utility D 102

4.11 THM speciation as a function of water age measured in Utility C........ 107

4.12 THM4 speciation as a function of water age measured during the September 23, 2003 sampling in Utility D 109

4.13 HAA9 speciation as a function of water age measured during the January 7, 2003 sampling in Utility C 110

4.14 HAA9 speciation as a function of water age measured during the April 8, 2003 sampling in Utility D........ 110

4.15 HAA5 and HAA9 concentrations as a function of water age measured during the January 7, 2003 sampling at Utility C........ 116

4.16 HAA5 and HAA9 concentrations as a function of water age measured during the June 10, 2003 sampling at Utility D 116

4.17 THM4 concentration as a function of water age for all samplings sorted by POE water in Utility E 117

4.18 HAA9 concentrations as a function of water ages for all samplings sorted by POE water at Utility E 117

4.19 THM4 concentration as a function of water age measured before, during and after the switch to free chlorine in Utility E 119

4.20 pH as a function of water age measured in Utility E 120

4.21 THM4 speciation as a function of water age measured in Utility E 122

4.22 HAA9 concentration as a function of water age measured before, during and after the switch to free chlorine in Utility E 124

4.23 HAA5 and HAA9 concentrations as a function of water age in Utility E 128

4.24 HAA9 speciation as a function of water age measured in Utility E 129

4.25 DBP concentrations as a function of water age measured in Utility A 132

4.26 DBP concentrations as a function of water age measured in Utility B 133

4.27 DBP concentrations as a function of water age measured in Utility C 134

4.28 NDMA concentration and total chlorine residual as a function of water age measured during the June 11, 2003 sampling in Utility C 135

4.29 NDMA concentrations as a function of water age measured in Utility C 135

4.30 NDMA concentration and free chlorine residual as a function of water age measured during the February 25, 2003 sampling in Utility A 137

4.31 NDMA concentrations as a function of water age measured in Utility A 137

4.32 NDMA concentrations as a function of water age for all samplings sorted by POE water in Utility E 138

4.33 NDMA concentration and chlorine residual as a function of water age measured in Utility E 138

4.34 Free chlorine residual and HPC as a function of water age measured during the December 3, 2002 sampling in Utility A 142

4.35 Free chlorine residual and HPC as a function of water age measured during the August 12, 2003 sampling in Utility B 142

4.36 Total chlorine residual and HPC as a function of water age measured during the October 15, 2002 sampling in Utility C 143

4.37 Total chlorine residual and HPC as a function of water age measured during the sampling in Utility D 144

4.38 HPC as a function of water age for all samplings sorted by POE water in Utility E 145

4.39 Chlorine residual as a function of water age for all samplings sorted by POE water in Utility E 146

4.40 HPC and chlorine residual as a function of water age measured in Utility E 147

4.41 Nitrogen species as a function of water age measured during the August 12, 2003 sampling in Utility C 152

4.42 Nitrogen species as a function of water age measured in Utility D 154

4.43 Free ammonia concentration as a function of water age for all samplings sorted by POE water in Utility E 155

5.1 System classification and IDSE schedule flow chart 163

5.2 Guidelines for IDSE options 164

5.3 Selection of preliminary SMP sites 168

5.4 Selection of final SMP sites 169

5.5 Stage 2 DBPR compliance monitoring site selection 171

5.6 Recommended limits of sampling period for historical data for systems on the early and late schedules 172

FOREWORD

The Awwa Research Foundation (AwwaRF) is a nonprofit corporation that is dedicated to the implementation of a research effort to help utilities respond to regulatory requirements and traditional high-priority concerns of the industry. The research agenda is developed through a process of consultation with subscribers and drinking water professionals. Under the umbrella of a Strategic Research Plan, the Research Advisory Council prioritizes the suggested projects based upon current and future needs, applicability, and past work; the recommendations are forwarded to the Board of Trustees for final selection. The foundation also sponsors research projects through the unsolicited proposal process; the Collaborative Research, Research Applications, and Tailored Collaboration programs; and various joint research efforts with organizations such as the U.S. Environmental Protection Agency, the U.S. Bureau of Reclamation, and the Association of California Water Agencies.

This publication is a result of one of these sponsored studies, and it is hoped that its findings will be applied in communities throughout the world. The following report serves not only as a means of communicating the results of the water industry's centralized research program but also as a tool to enlist the further support of the nonmember utilities and individuals.

Projects are managed closely from their inception to the final report by the foundation's staff and large cadre of volunteers who willingly contribute their time and expertise. The foundation serves a planning and management function and awards contracts to other institutions such as water utilities, universities, and engineering firms. The funding for this research effort comes primarily from the Subscription Program, through which water utilities subscribe to the research program and make an annual payment proportionate to the volume of water they deliver and consultants and manufacturers subscribe based on their annual billings. The program offers a cost-effective and fair method for funding research in the public interest.

A broad spectrum of water supply issues is addressed by the foundation's research agenda: resources, treatment and operations, distribution and storage, water quality and analysis, toxicology, economics, and management. The ultimate purpose of the coordinated effort is to assist water suppliers to provide the highest possible quality of water economically and reliably. The true benefits are realized when the results are implemented at the utility level. The foundation's trustees are pleased to offer this publication as a contribution toward that end.

Walter J. Bishop
Chair, Board of Trustees
Awwa Research Foundation

Robert C. Renner, P.E.
Executive Director
Awwa Research Foundation

ACKNOWLEDGMENTS

The authors of this report wish to thank the following individuals and organizations for their contribution to this project:

- Tony Akel, Carollo Engineers
- Erika Depaz and Dan Gatti, University of North Carolina at Chapel Hill, NC
- Mark Elliott, Old Dominion University, VA
- Mike Hotaling, Debbie Rindfliesch, Betty Wrightson, Claudia Melgaard, Jessica Kivett, Jacqueline Cypress, and Michelle Harrell, Newport News Waterworks, VA
- Don Mead, Pete Haddix and Nancy Shaw, American Water.

The participating utilities are not listed here for confidentiality purposes. The support and participation of all utility staff members involved in this study was greatly appreciated.

The support from the Project Manager – Dr. Djanette Khiari, AwwaRF, and Project Advisory Committee – Dr. Hiba M. Shukairy, USEPA, Dr. David A. Reckhow, University of Massachusetts, and Mr. Russell Chinn, Metropolitan Water District of Southern California, was truly appreciated.

The authors also acknowledge the technical assistance of Stacy Fuller for the preparation of this manuscript.

EXECUTIVE SUMMARY

Disinfection by-products (DBPs) in drinking water have been a major subject of study over the past 30 years. The majority of these studies have focused on the formation and control of DBPs, primarily trihalomethanes (THMs) and more recently haloacetic acids (HAAs), during water disinfection. Most of these studies were conducted in the laboratory under controlled, uniform conditions. Some of them attempted to model the behavior of DBPs in simulated distribution systems, using mostly batch incubations in bottles. However, little is known about the changes in DBP concentrations and speciation in full-scale distribution systems. In addition, information to date suggests that laboratory batch incubations may not accurately represent the fate of DBPs in full-scale systems because of their inability to include the effect of pipe walls and biofilms, and the dynamic nature of the distribution system.

THMs and HAAs in water distribution systems are known to exhibit appreciable spatial and temporal variability. Causes of temporal variability include changes in raw water quality, flow rate, hydraulic residence time (water age), and chemical doses in the treatment plant. THMs and HAAs continue to form in distribution systems where free chlorine is employed as a secondary disinfectant, resulting in significant spatial differences in DBP concentrations. One source of this variability is the dynamic nature of water distribution systems, i.e., as demands change throughout the system and storage tanks fill and empty, water ages in the distribution system and thus, affects chlorine residuals and DBP concentrations. Another source of spatial variability is associated with DBP reactions, such as biodegradation of some HAA species (primarily the dihalogenated HAAs, or X_2AAs). Because of this, locations with maximum THM levels (maximum water ages) are likely to be different than locations with maximum HAA concentrations. Also, within the HAA class, different HAA species are expected to peak at different locations because of their different chemical properties. Furthermore, studies indicated that the locations of these maximum concentration points vary from one distribution system to another (close to the point-of-entry in some cases, or at "maximum" residence time in others). This is the conceptual basis for the design of the initial distribution system evaluation (IDSE), which is part of the Stage 2 Disinfectants/Disinfection By-Products Rule (DBPR).

The lack of data on the fate of THMs and HAAs in full-scale distribution systems is also true for other DBPs. *N*-nitrosodimethylamine (NDMA) was found in drinking water, and studies suggest that its presence might be related to disinfection processes. However, technologies for NDMA removal or control of its formation are limited. Furthermore, little information is available about the fate of NDMA in distribution systems, although preliminary results suggest that NDMA concentrations may increase with increasing water age.

PROJECT GOAL AND OBJECTIVES

The goal of this project was to understand the formation and decay of selected DBPs in full-scale distribution systems focusing on the four THMs and the nine HAAs, as well as individual THM and HAA species. Because of its critical nature, NDMA was also included. The project objectives were as follows:

- Evaluate the critical factors that affect THM and HAA behavior in distribution systems.
- Determine the fate and behavior of NDMA in distribution systems.
- Evaluate the effect of pipe material and diameter on the fate of DBPs in distribution systems.
- Examine the effect of storage reservoirs/tanks and booster chlorination stations on THM, HAA and NDMA concentrations.
- Evaluate the changes in DBP concentrations and speciation when a system seasonally switches from chloramines to free chlorine to limit potential nitrification episodes.
- Identify the compounds that result from the breakdown of degradable DBPs.
- Identify bacteria responsible for the degradation of HAA and other DBPs, and determine the degree and mechanisms of degradation.
- Determine a means of identifying the distribution system locations and times of maximum THM and HAA concentrations in order to address the Stage 2 DBPR requirements.

With the Stage 2 DBPR anticipated to regulate the sum of four THMs and five HAAs at specific distribution system locations, information is critically needed to determine and understand the formation and decay of these by-products, as well as others, within full-scale distribution systems.

RESEARCH APPROACH

To accomplish these objectives, this project was separated into four tasks.

Task 1 – Literature Review

Task 1 consisted of a literature review focusing on the factors that impact the fate and behavior of DBPs in distribution systems. The literature review also gathered information about microbial degradation of some DBPs, the mechanisms involved, and possible end products of DBP degradation. This aspect was further examined at the bench scale in Task 4.

Task 2 – Full-Scale Distribution System Monitoring

Task 2 focused on the examination of five full-scale distribution systems. First, a preliminary assessment of numerous systems was conducted to select five systems that would best meet the goal and objectives of this project. To limit the number of factors investigated and focus this project on the distribution system, utilities were selected based on specific finished water characteristics. All utilities possessed a distribution system hydraulic model available for the determination of water age throughout the system. Systems were also selected based on their simplicity to maximize system control and limit the number of factors affecting the changes in water quality. As an example, utilities that blend source waters of difference quality in the distribution system (e.g., groundwater mixed with treated surface water) were not selected to reduce complexity. Finally, minimum DBP levels were required for participation: THM4 needed to be greater than 40 μg/L and HAA5 greater than 20 μg/L.

Specific inter-system variables were also considered when selecting the participating utilities: secondary disinfectant type (free chlorine or monochloramine), presence of brominated DBPs in some sites, variation in geographical locations to include different seasonal patterns and water temperatures, presence of storage reservoirs and booster chlorination stations in one system at least, and inclusion of a chloraminated system that periodically switches to free chlorine in the distribution system. This selection process led to the following participating utilities:

- Utility A uses free chlorine as the final disinfectant and includes a storage reservoir and a booster chlorination station
- Utility B uses free chlorine as the final disinfectant
- Utilities C and D use chloramines as the final disinfectant
- Utility E uses chloramines as the final disinfectant, with a one-month switch to free chlorine once per year to limit nitrification occurrence.

Within each distribution system studied, 12 to 16 sampling locations were carefully selected to provide as complete an understanding as possible of the fate of DBPs considering the complexity of full-scale distribution systems. The sampling locations were based on several intra-system variables, including: water age (from close to the treatment plant to dead-ends), pipe material and diameter, pressure zone, disinfectant residuals (spatial and temporal variations), presence of booster chlorination stations, presence of storage reservoirs, and potential major interferences such as autoflushers.

To address the spatial and seasonal variability of DBPs in distribution systems, 6 to 8 samplings were conducted over a one year period in each distribution system studied. In Utility E, the chloraminated system that periodically switches to free chlorine, two samplings were conducted just before the switch to free chlorine. Two other samplings were conducted during the free chlorination period, and the last two samplings occurred following the return to chloramines.

The DBPs monitored throughout the study are listed below. Several general water quality parameters were also monitored, including free and total chlorine residuals, temperature, pH, UV absorbance at 254 nm, TOC, AOC, HPCs, bromide, and ammonia, nitrate, nitrite at the chloraminating utilities:

- All four THMs: chloroform ($CHCl_3$), bromodichloromethane ($CHBrCl_2$), dibromochloromethane ($CHBr_2Cl$), and bromoform ($CHBr_3$)
- All nine HAAs: monochloroacetic acid (ClAA), dichloroacetic acid (Cl_2AA), trichloroacetic acid (Cl_3AA), bromochloroacetic acid (BrClAA), bromodichloroacetic acid ($BrCl_2AA$), dibromochloroacetic acid (Br_2ClAA), monobromoacetic acid (BrAA), dibromoacetic acid (Br_2AA), and tribromoacetic acid (Br_3AA)
- NDMA in the distribution systems of Utilities A, C and E.

Task 3 – Desktop Studies

This task included a desktop evaluation aimed at developing guidelines to help water utilities determine the locations of maximum THM and HAA concentrations in their distribution systems. Such a tool is needed by water suppliers to help them meet the requirements of the Stage 2 DBPR, and particularly the IDSE.

Task 4 – Microbial Degradation of HAAs and Other DBPs

This task included bench-scale work to examine the microbial degradation of HAAs and other DBPs more closely. The aim was to identify the bacteria responsible for the degradation of HAAs and other DBPs, and to determine the degree and mechanisms of degradation.

CHANGES IN DISINFECTION BY-PRODUCTS IN FULL-SCALE DISTRIBUTION SYSTEMS

In five full-scale distribution systems, the changes in DBP concentration, DBP speciation and water quality with increasing water age were examined, as well as the effects of seasons, pipe material, pipe diameter, booster chlorination station and storage reservoir.

Trends with Increasing Water Age

Trihalomethanes

In the free chlorinated systems, THM4 concentrations increased as water age increased and chlorine residual decreased, a common trend that was expected and observed for all samplings in both systems. Increases in THM4 concentrations from the point of entry (POE) to the MRTL were much lower in Utility B (average of 58 percent) than in Utility A (average of 452 percent), consistent with a greater stability of the free chlorine residuals and lower water ages in Utility B than in Utility A.

In one of the chloraminated systems (Utility C), THM4 concentrations were relatively stable as water age increased, as expected. Surprisingly, occasional decreases in concentration were observed at the MRTL (up to 23 percent) in warmer water and when the total chlorine residual was lower (<2.0 mg/L). These decreases were consistent with the occurrence of nitrification at these locations, as shown by increases in nitrite, nitrate and HPC concentrations, and decreases in total chlorine residual and total ammonia concentration.

Over the course of the project, the studied portion of Utility D's distribution system (chloraminated system) had to supplement their typical POE with a second POE. Both POEs pulled water of similar quality from the same transmission main. The addition of a second POE changed the flow patterns in the distribution system, which changed the water age at each sampling location. All samplings conducted were grouped into four distinct scenarios, and the water ages were calculated for each hydraulic scenario. When both POEs were used, which resulted in lower water ages throughout the distribution system, THM4 concentrations remained constant with increasing water age. Concurrently, HPCs remained low and total chlorine residual remained stable. Also, nitrification did not occur, except in rare cases. However when only one POE was used, resulting in higher water ages throughout the system, THM concentrations decreased (46 to 62 percent) at the sampling locations with higher water ages and lower chlorine residuals. During these samplings, nitrification was observed at several locations in Utility D. Recent studies by Wahman et al. (2003) have reported THM biodegradation via cometabolism by nitrifying bacteria, including bacteria that have been observed in drinking water distribution systems. Although microorganism identification was not conducted during this study, it is possible that nitrifying bacteria may have degraded THMs at MRTLs. These results also suggest that high THM4 concentrations may not necessarily be found at MRTLs in chloraminated systems, which is not commonly expected.

Utility E used chloramines as the final disinfectant, but switches to free chlorine during a one-month period every spring to limit nitrification occurrences. In this system, THM4 concentrations increased 20 to 50 percent from POEs to MRTLs with increasing water age during the chloramination periods, and 135 to 161 percent during the free chlorination period. These results confirmed that final disinfection with chloramines is an effective technology to limit THM4 formation in the distribution system and help meet regulatory requirements. THM4 concentrations did not decrease in Utility E's distribution system, as observed in the other two chloraminated systems studied. Concurrently, the total chlorine residuals remained constant or decreased slightly with increasing water age during the chloramination periods (down to 2.2-2.9 mg/L at the MRTL), whereas the free chlorine residuals decreased significantly during the free chlorination period (down to <0.1-0.5 mg/L). The relatively high chlorine residuals maintained in Utility E's distribution system during the chloramination periods may have limited microbial regrowth, as shown by the relatively low HPCs and absence of signs of nitrification. These results suggest that chloraminated systems that periodically switch to free chlorine may experience peak THM concentrations during the free chlorination period that may result in non-compliance.

Haloacetic Acids

In the free chlorinated systems, HAA9 concentrations generally increased or remained constant as water age increased and chlorine residual decreased, except at the MRTLs, where a decrease in HAA9 concentrations was measured. The decrease in HAA concentrations at the MRTLs (41 percent in Utility A) was observed for the majority of the samplings, but was most noticeable during the warmer months and when the free chlorine residual was low, which are the preferred conditions for biodegradation to occur. HAA biodegradation in Utility B was not as pronounced as in Utility A, with 11 to 26 percent decreases in HAA9 concentrations in Utility B. The free chlorine residual was also more stable and water age was lower in Utility B than in Utility A, which may have prevented HAA biodegradation in Utility B.

HAA9 concentrations were relatively constant throughout the chloraminated distribution systems, with the exception of the MRTLs, where decreases in HAA concentrations occurred, particularly in Utility C (23 to 85 percent decreases). The presence of elevated HPCs and nitrification, as well as the loss in disinfectant residual at Utility C's MRTL suggests that HAA biodegradation was occurring. In Utility D, HAA9 concentrations followed either one of two trends with increasing water age, depending on the hydraulic scenario as explained above:

1. HAA9 concentrations remained constant when both POEs were used, consistent with lower HPCs, stable total chlorine residual and the absence of nitrification.
2. HAA9 concentrations remained constant and then decreased at MRTLs (59-78 percent from POE to MRTL) when only one POE was used, consistent with lower THM4 concentrations. Lower total chlorine residuals, fluctuating HPCs, and signs of nitrification were also present at the locations where HAA9 decreases occurred.

As expected, HAA9 concentrations remained relatively stable with increasing water age during both chloramination periods in Utility E; only minor increases in HAA9 concentrations were noticed. During the free chlorination period however, HAA9 concentrations increased significantly with water age (44 to 53 percent). Again, these results confirmed that final disinfection with chloramines is an effective technology to limit HAA9 formation in the distribution system. No clear decrease in HAA concentration was observed at the MRTLs, neither during the free chlorination or chloramination periods, even though the maximum water age was relatively high (92 hours). These results suggest that HAA biodegradation did not occur in Utility E over the course of this study, which was consistent with the relatively low HPCs and the absence of signs of nitrification. Also during the chloramination periods, the total chlorine residuals remained constant or decreased slightly with increasing water age. During the free chlorination period however, the chlorine residuals decreased significantly. Nonetheless, microbial regrowth was minimal in Utility E during the sampling periods.

These results suggest that a thorough examination of the data is warranted since the low disinfectant residual limits presented in the USEPA guidelines (at least 0.2 mg/L free chlorine or 0.5 mg/L as chlorine for the chloraminated systems) may allow biodegradation to occur and may not represent the location with the highest HAA concentration. Such phenomena need to be considered when selecting distribution system sampling locations for Stage 2 DBPR compliance purposes.

The HAA5-to-HAA9 weight ratios were similar in both free chlorinated systems studied, averaging at 72 and 79 percent in Utility A and Utility B, respectively. Likewise, this ratio averaged at 77 percent in Utility C (chloraminated system), but was much lower in Utility D (average of 39 percent). The opposite was observed in Utility E, where HAA5 represented almost all of the HAA9 measured (>84 percent). These results suggest that the non-regulated HAAs comprise a significant proportion of the HAA9 in most of the systems studied. Such results are concerning considering that the brominated HAAs, which are not well represented by HAA5, are those that may present the greatest health risks. Nonetheless, results suggest that the trends observed and presented above were similar for both HAA5 and HAA9, with the exception of the locations where HAA biodegradation occurred. At these locations, decreases in HAA5-to-HAA9 weight ratio were observed, as the XAAs and X_2AAs are the HAA species that are generally biodegraded. These species are included in the HAA5 calculation, whereas the less biodegradable X_3AAs constitute the majority of the remaining non-regulated HAAs.

N-Nitrosodimethylamine

Surprisingly high NDMA concentrations were measured in all systems studied. Although most sampling locations in the chloraminated system studied (Utility C) had concentrations below 20 ng/L, with the majority of samples having concentrations below 10 ng/L, very high concentrations (40 and 46 ng/L) were measured on two occasions. Even in the free chlorinated system (Utility A), NDMA concentrations over 20 ng/L were measured during two of the five samplings conducted. In the chloraminated system periodically switching to free chlorine (Utility E), NDMA concentrations were high, with 8 samples showing NDMA concentrations over 40 ng/L, one of which was the POE. There were no obvious differences between the concentrations measured in the free chlorinated and the chloraminated systems. It is likely that there were different levels of NDMA precursors in each system, regardless of their sources (i.e., present in the raw water, or introduced during treatment, such as through the use of polymers).

No particular trend in NDMA concentrations could be identified with increasing water age in a chloraminated system (Utility C) or a free chlorinated system (Utility A). Trends could not be found also in a chloraminated system that periodically switches to free chlorine (Utility E). As a result, it can be concluded that water age does not appear to affect NDMA concentration in the distribution systems studied.

Disinfection By-Product Speciation

Trihalomethanes

The chlorine-containing THMs (i.e., $CHCl_3$ and $CHCl_2Br$) comprised most of the THMs in the free chlorinated systems, representing 80 to 98 percent of the THM4, because of the lower bromide concentrations in the source water of these systems.

Utility C, one of the chloraminated systems studied, was experiencing drought conditions when sampling began. As a result, raw water bromide concentrations were higher during the first two samplings and the treated water contained mainly the bromine-containing THM species (i.e., $CHBr_2Cl$ and $CHBr_3$). Subsequent samplings conducted under lower source water bromide concentrations were dominated by $CHCl_3$ and $CHCl_2Br$, representing 92 to 100 percent of the THM4. Raw water bromide concentrations were relatively high in Utility D's source water, resulting in high proportion of bromine-containing THMs (49 to 65 percent of the THM4).

The $CHCl_3$ fraction comprised most of the THM4 (representing 83 to 93 percent of the THM4) measured in the system periodically switching final disinfectant, Utility E. These results were expected considering the low bromide concentration in the raw water. The remaining THMs consisted of $CHBrCl_2$ only. The speciation of THM4 was not affected by disinfectant type.

Haloacetic Acids

During most of the samplings conducted in the free chlorinated and chloraminated systems, the dominant HAA species were X_2AAs and X_3AAs. In Utility E, X_3AAs were slightly more dominant than X_2AAs for both disinfectant residuals studied. When decreases in HAA concentrations were observed, they were generally associated with losses in XAAs and X_2AAs (the more biodegradable HAA classes); X_3AAs remained relatively conserved with increasing water age, consistent with the fact that this classes is less biodegradable.

In Utility A, the fraction of chlorine-containing HAAs (i.e., ClAA, Cl_2AA, Cl_3AA and Cl_2BrAA) accounted for 63 to 100 percent of the HAA9 measured. In Utility B, this ratio was slightly lower, representing 57 to 93 percent of the HAA9 measured. These results are consistent with the lower bromide concentrations of the source waters of these systems.

As a result of the high raw water bromide concentration occurring in Utility C's source water at the beginning of the study, samples collected during the first two samplings contained significant concentrations of bromine-containing HAAs. Subsequent samplings conducted under lower source water bromide concentrations were dominated by the chlorine-containing HAAs, which represented 70 to 91 percent of the HAA9. Because of the relatively high bromide concentration normally present in Utility D, the fractions of bromine-containing HAAs were similar to the fractions of chlorine-containing HAAs.

When chloramine was used as the final disinfectant in Utility E, Cl_2AA and Cl_3AA were the predominant HAA species, and were approximately equal in concentration. A slightly higher concentration of Cl_3AA was measured while using free chlorine compared to Cl_2AA, which was expected.

Effect of Seasons

Because the kinetics of reaction between disinfectants and NOM are slower in colder water, THM4 and HAA9 concentrations were lowest in the colder months and highest in the warmer months in all systems studied. Also in all systems studied, the changes in THM4 concentrations with increasing water age did not follow any seasonal trends. In the case of HAA9 however, biodegradation observed at MRTLs was more pronounced in warmer water, leading to different seasonal trends with increasing water age. These results suggest that the time of sampling should be carefully considered when planning monitoring programs, or when using historical data to select sampling locations for Stage 2 DBPR compliance. To a larger scale, climatic changes such as high precipitation and runoff events are likely to distort DBP data as well. Thus data obtained in a non-typical year should be carefully assessed before selecting DBP monitoring sites.

In the case of NDMA, results show slightly greater concentrations in warm water than in colder water in all three systems studied, although this trend was not evident at each sampling location. This trend was particularly noticeable in Utility E where concentrations seemed lowest during the first two samplings conducted before the switch to free chlorine, greater during the free chlorination period, and greatest following the return to chloramination, concurrently with increases in water temperature (from 6 to 18°C, 43 to 68°F). In the case of NDMA, results show slightly greater concentrations in warm water than in colder water in the systems studied, although this trend was not evident at each sampling location.

Correlations Between Disinfection By-Products and Other Parameters

Attempts were made to find a surrogate parameter for THM4 and HAA9 to help predict the location(s) of maximum DBP concentrations in distribution systems. Results of linear correlations between DBP concentrations and several parameters including water age, chlorine residual, TOC concentration, UV absorbance, HPC, nitrite, nitrate and ammonia concentrations, showed weak and inconsistent correlations in general (Appendix C).

Effect of Pipe Material and Pipe Diameter

Within each system, consideration was made, as much as possible, to select locations with similar water age and pipe diameter but different pipe material, as well as locations with similar water age and pipe material but different pipe diameter. The study of the effect of pipe material and diameter that resulted from this effort led to the following observations (Appendix D):

- Pipe material and diameter did not have an appreciable effect on THM4 nor HAA9 concentrations in the distribution systems studied, with variations of 20 percent or less most of the time. Some important differences were observed at times, but these differences most likely resulted from variations in water age, or differences between modeled and actual water ages.
- No specific trends could be drawn regarding the effect of pipe material on NDMA concentration in the distribution systems studied. NDMA concentrations were greater in the small-diameter pipes at times, whereas at other times, concentrations were greater in the large-diameter pipes for samples with similar water ages. The majority of the samples with high NDMA concentrations occurred in small-diameter pipes in two of the systems studied, but not in the third system. Thus, pipe diameter did not seem to affect NDMA concentration.
- The type of distribution system residual disinfectant (free chlorine or chloramines) did not affect the trends observed.

Effect of a Booster Chlorination Station and a Storage Reservoir

In Utility A (free chlorinated system), samples were collected from distribution system pipes, as well as before and after a storage reservoir and a booster chlorination station to allow examination of the effect of these facilities on DBP concentrations (Appendix D). When conducting such evaluation, it is important to consider the hydraulic scenario present at the moment of sampling. In the case of a reservoir, depending on its piping configuration, aged water from the reservoir may flow back to the reservoir influent when the reservoir is emptying. It this study, it appeared that the status of the reservoirs did not affect the change in concentration of any of the DBPs studied, although further investigation is warranted to confirm this observation.

The hydraulic scenario also deserves attention around booster chlorination stations. In this study, greater free chlorine residuals were observed at times upstream of the booster chlorination station, which suggest unexpected hydraulic conditions at this location and the possibility of data distortion. Nonetheless, the booster chlorination station increased THM4 concentrations by approximately 16 to 101 percent. The fate of HAAs following booster chlorination was related to the free chlorine residual present. Decreases in concentrations were observed when the free chlorine residual decreased to low levels (<0.1 mg/L Cl_2) downstream of the chlorination station; increases were observed when the free chlorine residual remained high (1.3-1.4 mg/L); and HAA9 concentrations did not change when the free chlorine residual decreased but remained detectable (0.3-0.5 mg/L). Changes in NDMA concentrations were observed during booster chlorination, but no particular trend could be distinguished.

MICROBIAL DEGRADATION OF HALOACETIC ACIDS AND OTHER DISINFECTION BY-PRODUCTS

The aim of this task was to identify the bacteria responsible for the degradation of HAAs and other DBPs, and to determine the degree and mechanisms of degradation. Activities conducted during this task also identified the compounds that result from the breakdown of degradable DBPs. This part of the study was divided into two specific tasks, which were pursued concurrently (Appendix A):

1. Identification of the two HAA degrading bacteria previously isolated by Williams et al. (1994, 1998; "bacterial study").
2. Examination of possible mechanisms of action of the specific dehalogenase enzymes with various DBPs.

Bacterial Study

The specific objectives of the bacterial study were as follows:

1. Isolate HAA degrading bacteria from MRTL sites.
2. Culture and identify the bacteria isolated from these samples.
3. Isolate and purify the dehalogenase enzymes from these bacteria.
4. Determine the physical character (e.g., molecular weight) of the enzymes.
5. Determine the scope of activity and optimum conditions for the bacterial degradation of certain HAAs and other DBPs.

From the many different bacterial isolates that were cultured from sites that exhibited unexpected low levels of HAAs, two specific bacteria were shown to be quite effective in the degradation of various HAAs, and have been identified as a *Burkholderia* species and a *Sphingomonas* species. Another bacterial species, *Xantherobacter autotrophicus*, known as GJ-10, was used as a control in this study and was shown to effectively degrade various HAAs and other halogenated compounds.

The following conclusions regarding the relative effectiveness of the different bacteria to degrade certain halogenated DBPs were drawn:

- All three bacterial isolates were capable of degrading ClAA, Cl_2AA, BrAA, Br_2AA and BrClAA. None of the bacterial isolates were capable of effectively degrading Cl_3AA or Br_3AA, which may demonstrate a distinct stearic effect that inhibits the dehalogenase enzyme activity. There appeared to be a selectivity for the X_2AAs over the XAAs by the bacteria studied. This may be due to the basic nature of the enzymes affinity for the active site. The nucleophilic substitution center would be inductively activated by two halogens.
- The *Sphingomonas* species was more effective than the *Burkholderia* species in degrading HAAs. However GJ-10 appeared to be much more effective than either of the other two bacterial species in the degradation of these same HAAs.
- It appeared that the *Sphingomonas* species preferred the brominated substrates over the chlorinated substrates. The substrate specificity for the *Burkholderia* species

appeared to be more diverse and less selective. GJ-10 exhibited a greater affinity for Br_2AA over Cl_2AA, which may be explained by the greater leaving-group potential of the bromide over the chloride in the rate determining step of the dehalogenation mechanism.

- All three bacterial isolates appeared capable of degrading monochloroacetonitrile (MCAN) and dichloroacetonitrile (DCAN). The order of reactivity was the following: MCAN > DCAN > trichloroacetonitrile (TCAN). The *Sphingomonas* species was more effective than the *Burkholderia* species with regard to HAN degradation, however GJ-10 appeared to be more effective than either of the other two isolates.

Haloacid Degradation Mechanism

In experiments with GJ-10 enzyme extracts and Cl_2AA, two expected degradation products (glyoxylic and oxalic acids) were observed. There were two additional peaks that may have been glycolic and ClAA. These preliminary data suggest that the GJ-10 bacteria may be capable of some form of reductive dehalogenation as well as the expected hydrolytic mechanism. The glycolic acid peak could have arised from the hydrolysis of the ClAA produced in the reductive dehalogenation step.

The *Burkholderia* isolate only producted detectable levels of degradation products at very high Cl_2AA concentrations (1,000 mg/L). The degradation products observed were hydroxychloroacetic acid intermediate, glyoxylic acid and oxalic acid. No degradation products were observed in experiments using the *Sphingomonas* isolate.

RECOMMENDATIONS

The following are issues raised during this study that could be addressed in future work:

- This study addressed temporal variations in DBP concentrations and speciation in distribution systems by examining the effect of seasons. Although some information is available regarding short-term (e.g., diurnal or hourly) variations in DBP concentrations in full-scale systems, studies are required to further address this issue. Temporal studies to evaluate the effects of diurnal and hourly variations in distribution systems on DBP concentrations would be useful to confirm the conclusions drawn during this study, better understand the formation and decay of DBPs in distribution systems, determine the impact of changes in flow and demand on water quality, and assess the sensitivity of THM and HAA concentrations to variations in water age at a sampling location.
- While many utilities use chloramines to control the formation of DBPs downstream of the treatment plant, chloramines are limited in their ability to truly limit THM and HAA formation. As observed during this study, high THM and HAA concentrations were measured in all of the chloraminated systems. As such, tools other than conversion to chloramines are needed to help utilities limit DBP concentrations in their distribution system and meet the Stage 2 DBPR requirements.
- This study also showed that the use of chloramines is not without problems. Two of the utilities that utilized chloramines experienced nitrification (historical data indicate that the third utility also experienced nitrification, although nitrification was not

observed during this study because samplings were conducted in the winter and spring periods only). Investigations into the causes of nitrification, as well as nitrification prevention and control practices would be highly useful for those utilities that have nitrification problems in their distribution system.

- Interestingly, this and other studies suggest that nitrification may be "beneficial" as some nitrifying bacteria can degrade THM4 via co-metabolism. Further studies are required to better address this phenomenon, and to determine whether other microorganisms have the same capability.
- This study supports the proposed mechanism for HAA degradation via a hydrolytic process, with the formation of the glycolic, glyxoylic and oxalic acids (Appendix A). However additional studies are warranted to confirm this mechanism using analytical methods that would possess sufficient sensitivity to detect the degradation products.
- The degradation of HANs by species of *Sphingomonas*, *Burkholderia* as well as *Xanthobacter autotrophicus* GJ-10 is a relatively new and novel observation (Appendix A), and warrants further study in view of the potential release of cyanide ion as a possible degradation by-product.
- To accommodate for the high numbers of THM and HAA samples that had to be analyzed during this study, two laboratories were used. Inter-laboratory comparison tests were conducted to ensure comparability of results. Results showed some differences, mainly when the DBP concentrations were low. The reasons for these differences are unknown and the study was not designed to address them. Nonetheless, these data emphasize the need to further examine the analytical methods that are suggested to comply with the Stage 2 DBPR. As observed during this study, even reputable laboratories that use strict quality control (QC) procedures may show differences in THM or HAA concentrations, particularly when the concentrations are low. The source(s) of these differences need to be identified and examined.
- A statistical analysis of the parameters analyzed during this study using the analysis of variance (ANOVA) procedure is recommended. This would allow identification of the statistically significant variables that impact DBP concentrations in the distribution systems. This study was not designed to allow for such an analysis. To achieve accurate results from an ANOVA analysis, the sample size would need to be increased, and additional samplings would be required to broaden the range of conditions under which the distribution systems were analyzed.
- When examining water quality around a storage reservoir or a booster chlorination station, it is important to consider the hydraulic scenario present at the moment of sampling. In the case of a reservoir, depending on its piping configuration, aged water from the reservoir may flow back to the reservoir influent when the reservoir is emptying. It this study, it appeared that the status of the reservoirs did not affect the change in concentration of any of the DBPs studied, although further investigation is warranted to confirm this observation. The hydraulic scenario also deserves attention around booster chlorination stations. In this study, greater free chlorine residuals were observed at times upstream of the booster chlorination station, which suggest unexpected hydraulic conditions at this location and the possibility of data distortion. The zone of influence of a reservoir or a booster chlorination station also needs to be addressed.

CHAPTER 1
BACKGROUND AND INTRODUCTION

Disinfection by-products (DBPs) in drinking water have been a major subject of study over the past 30 years. The majority of these studies have focused on the formation and control of DBPs, primarily trihalomethanes (THMs) and more recently haloacetic acids (HAAs), in treatment plants. These studies have included investigations of water quality and treatment factors influencing DBP formation and the evaluation and implementation of methodologies for controlling DBP formation, including DBP precursor removal. Most of these studies were conducted in the laboratory under controlled, uniform conditions. Some of them attempted to model the behavior of DBPs in simulated distribution systems, using mostly batch incubations in bottles. Little is known, however, about the changes in DBP concentrations and speciation in full-scale distribution systems. In addition, information to date suggests that laboratory batch incubations may not accurately represent the fate of DBPs in full-scale systems because of their inability to include the effects of pipe walls and biofilms, and the dynamic nature of the distribution system.

Because THMs have been regulated since 1979, a significant database on THM occurrence in tap water has been developed. Most of this information, however, relates to quarterly measurements of THMs made at only selected locations in the distribution system (grab samples at three monitoring locations with average residence times and one site at a distant location). The Information Collection Rule (ICR) database contains concentrations of several DBPs (USEPA 2000a); although, monitoring criteria were similar. Accordingly, the database on THM, and particularly HAA, occurrence is sparse from both a temporal (quarterly grab samples only) and a spatial (only four locations per system) viewpoint.

This lack of data is problematic because THM and HAA concentrations in water distribution systems are known to exhibit appreciable spatial and temporal variability. One cause of temporal variability is that raw water quality changes seasonally. For example, organic carbon concentration, the nature of the organic material, bromide concentration, pH, and temperature can all exhibit significant seasonal, and in some cases daily and diurnal, variations. Likewise, flow rate, hydraulic residence time, and chemical doses in the treatment plant also change periodically. This can result in appreciable temporal differences in DBP concentrations in water entering the distribution system.

THMs and HAAs continue to form in distribution systems in which free chlorine is employed as a secondary disinfectant, due to continuing reactions between free chlorine and organic material in the bulk water, and with organic material and biofilms on the pipe walls. In most cases, as long as there is an appreciable chlorine residual present, THM and HAA concentrations increase with increasing water age (hydraulic residence time), resulting in significant spatial differences in DBP levels. This observation was reported by several authors such as Williams et al. (1994 and 1995), LeBel et al. (1995), Chen and Weisel (1998), McClellan et al. (1999), Singer (1999b), Uber et al. (2000), and Baribeau et al. (2001). One source of this variability is the dynamic nature of water distribution systems, i.e., as demands change in the water system and storage tanks fill and empty, water ages in the distribution system and thus affects chlorine residuals and DBP concentrations (McClellan et al. 1999, Uber et al. 2000, Singer et al. 2001). At some locations, tap water comes directly from the treatment plant for part of the day, while at other times, it is primarily aged water drawn from storage tanks. Another source of spatial variability is associated with DBP reactions. It has been well documented (e.g., Williams et al. 1994 and 1995; Baribeau et al. 2000) that HAAs, primarily the dihalogenated HAAs (X_2AAs), can be biodegraded in the absence of chlorine residual. Therefore, HAA levels can be reduced at distribution system locations with high water ages or at times when the tap water is being supplied by "old" water from storage tanks. Biofilms and pipe walls are expected to play a major role in this phenomenon. Because of this, locations with maximum THM levels (maximum water ages) are likely to be different than locations with maximum HAA concentrations. Also, within the HAA class, different HAA species are expected to peak at different locations because of their different chemical properties. Furthermore, preliminary evaluation of the ICR database indicates that the locations of these maximum concentration points vary from one distribution system to another (close to the point-of-entry in some cases, or at "maximum" residence time in others; Obolensky et al. 2002). This is the conceptual basis for the design of the initial distribution system evaluation (IDSE), which is part of the Stage 2 Disinfectants/Disinfection By-Products Rule (DBPR).

The lack of data on the fate of THMs and HAAs in full-scale distribution systems is also true for other DBPs. One of them, *N*-nitrosodimethylamine (NDMA), has been used in the production of 1,1-dimethylhydrazine for liquid rocket fuel and has a variety of other industrial uses such as in polymers and copolymers, batteries, solvents, and lubricants. It was found in a variety of foods, beverages, and products, including drinking water (NTP 2000). The presence of NDMA in drinking water is a serious concern to the water industry because it was identified as a "probable human carcinogen" by the U.S. Environmental Protection Agency (USEPA) (Class B2, USEPA 1997a), and the National Toxicology Program lists it as "reasonably anticipated to be a human carcinogen" (NTP 2000). Studies (e.g., Jobb et al. 1994a and 1994b; Graham et al. 1996; Najm and Trussell 2001; Sedlak and Mitch 2001; California Department of Health Services (DHS) 2002a; Mitch et al. 2003) suggest that NDMA's presence in water might be related to disinfection processes, as well as the use of cationic polymers and ion exchange resins to treat water. Also, technologies for NDMA removal or control of its formation are limited. Furthermore, little information is available about the fate of NDMA in distribution systems, although preliminary results suggest that they may increase with water age.

PROJECT GOALS AND OBJECTIVES

The goal of this project was to understand the formation and decay of DBPs in distribution systems focusing on the four THMs and the nine HAAs, as well as individual THM and HAA species. Because of its critical nature, NDMA was also included. The project objectives were as follows:

- Evaluate the critical factors that affect THM and HAA behavior in distribution systems.
- Determine the fate and behavior of NDMA in distribution systems.
- Evaluate the effect of pipe material and diameter on the fate of DBPs in distribution systems.
- Examine the effect of storage reservoirs/tanks and booster chlorination stations on THM, HAA, and NDMA concentrations.
- Evaluate the changes in DBP concentrations and speciation when a system seasonally switches from chloramines to free chlorine to limit potential nitrification episodes.

- Identify the compounds that result from the breakdown of degradable DBPs.
- Identify bacteria responsible for the degradation of HAA and other DBPs, and determine the degree and mechanisms of degradation.
- Determine a means of identifying the distribution system locations and times of maximum THM and HAA concentrations in order to address the Stage 2 DBPR requirements.

With the Stage 2 DBPR anticipated to regulate the sum of four THMs and five HAAs at specific distribution system locations, information is critically needed to determine and understand the formation and decay of these by-products, as well as others, within full-scale distribution systems.

PROJECT TASKS AND REPORT ORGANIZATION

To accomplish these objectives, this project was separated into four tasks. Task 1 consisted of a literature review focusing on the factors that impact the fate and behavior of DBPs in distribution systems. It is presented in Chapter 2 of this report.

Task 2 focused on the examination of full-scale distribution systems. A preliminary assessment of numerous systems was first conducted to select five systems that would best meet the goal and objectives of this project. Sampling locations were carefully selected in each distribution system. Details are provided in Chapter 3, Materials and Methods. The selected systems were then monitored for THM4, HAA9, NDMA, and several water quality parameters over a one-year period. A chloraminated system that periodically switches to free chlorine to limit the occurrence of nitrification was monitored before, during, and after the switch to free chlorine. The key results are presented in Chapter 4, Evaluation of Full-scale Distribution Systems. Additional information is presented in the appendices, including results of simulated distribution system tests (batch incubation in bottles), the effects of pipe material and diameter on the fate of DBPs, the effects of a storage reservoir and a booster chlorination station in one of the systems, correlation analyses between DBP concentration, water age and various water quality parameters, and results of inter-laboratory comparison experiments for THM4 and HAA9.

In Task 3, a desktop evaluation was conducted to develop guidelines to help water utilities determine the locations of maximum THM and HAA concentrations in their distribution systems. Such a tool is needed by water suppliers to help them meet the requirements of the Stage 2 DBPR, and particularly the IDSE. This work is presented in Chapter 5, Guidelines to Select Distribution System Sampling Locations for DBP Monitoring. Conclusions, recommendations, and suggestions for future work are presented in the last chapter of this report.

An additional task was undertaken during this project and included bench-scale work to examine the microbial degradation of HAAs and other DBPs more closely. This part of the project is presented in Appendix A, Microbial Degradation of HAAs and Other DBPs.

CHAPTER 2
LITERATURE REVIEW

It is well known that disinfection of drinking water provides protection against microbial diseases. However, disinfection also generates undesirable DBPs that may cause a long-term risk to public health. These DBPs include THMs, HAAs, chloral hydrate, haloacetonitriles (HANs), haloketones (HKs), chloropicrin, NDMA, and several others. THMs and HAAs have been identified as the two largest classes of DBPs detected on a weight basis in chlorinated finished water (Robert et al. 2002, Nieminski et al. 1993, Krasner et al. 1989, Koch et al. 1991, Carlson and Hardy 1998).

Considerable research has addressed DBP formation and control, including DBP precursor removal. Fewer studies have been performed on changes in DBP concentrations and speciation in distribution systems, particularly in full-scale systems. This literature review begins with a brief review of occurrence, health effects, and regulatory implications of the DBPs addressed in this study. A summary of information available at this point with regards to DBP formation, stability, and changes in concentration and speciation in distribution systems is then presented, followed by the effect of key water quality parameters and distribution system characteristics on DBPs.

DEFINITION, OCCURRENCE, AND HEALTH EFFECTS OF DISINFECTION BY-PRODUCTS

THM4, HAAs, HANs, and other DBPs have been shown to adversely affect reproduction and development in animals (Klinefelter et al. 2001). Studies have demonstrated alterations in sperm morphology; motility and count; decreased levels of fertility; spontaneous resorptions; decreased fetal body weight; and visceral, cardiovascular, and craniofacial malformations (USEPA 2000b). The groups presented in Table 2.1 were formed from evaluations of several data sets and suggest three general categories of effects that represent the most sensitive end points common for these chemicals:

1. Cardiovascular effects (e.g., interventricular septal defects, defects between ascending aorta and right ventricle, and levocardia);
2. Effects on fetus/litter (e.g., decreased fetal body weight and crown-rump length, full litter resorption);
3. Male reproductive effects (e.g., sperm alterations, testicular effects).

Data on the actual toxicological mechanisms causing these effects are generally not available. As such, the groups in Table 2.1 were formed to reflect the primary effect observed in animal studies for each DBP.

Table 2.1
DBPs grouped into subclasses by common mode of action (MOA)

Genotoxic carcinogens: • Bromodichloromethane (glutathione transferase activation, adduct formation is a distinct possibility) • Bromoform (mechanism unknown, potentially oxidative damage, glutathione transferase activation) • Chlorodibromomethane (mechanism unknown)	Non-genotoxic carcinogens: • Chloroform (gross tissue damage and regeneration, "necrotic" foci) • Dichloroacetic acid (no observed necrotic foci) • Trichloroacetic acid (no observed necrotic foci)
Developmental toxicants – Primary effect is cardiovascular: • Monochloroacetic acid • Dichloroacetic acid • Trichloroacetic acid • Monobromoacetic acid	Developmental toxicants – Primary effect is for whole organism: • Dibromoacetic acid (delayed parturition) • Bromochloroacetic acid (reduced pup viability) • Bromodichloromethane (full litter resorption) • Bromoform (full litter resorption) • Bromoform (full litter resorption) • Chloroform (reduced pup weight)
Reproductive toxicants – Primary effect in testis and sperm: • Dichloroacetic acid (testicular effects) • Dibromoacetic acid (testicular effects) • Bromochloroacetic acid (sperm effects) • Bromodichloromethane (sperm effects)	

Source: Adapted from Klinefelter et al. 2001, and USEPA 2000b

Trihalomethanes

Trihalomethanes are volatile substances defined as halogenated methane compounds that form during chlorination of natural water. The four THMs (THM4) are chloroform ($CHCl_3$), bromodichloromethane ($CHBrCl_2$), dibromochloromethane ($CHBr_2Cl$), and bromoform ($CHBr_3$).

THMs can be formed from numerous organic compounds present in water, including some ketones (relatively slow THM precursors), aromatic compounds with specific groups (e.g., resorcinol, which rapidly forms THMs), and humic and fulvic substances (Croué 1987, Doré 1989, Singer 1990, Krasner 1999).

THMs constitute the majority of the DBPs formed in most chlorinated waters. In the AWWA Water Industry Database (WIDB), which includes THM4 data from 815 American utilities serving a total of 141 million people, THM4 concentration from each system ranged from 1.5 to 71 μg/L (10th to 90th percentile), with a median value of 34 μg/L (Bull et al. 2001). In a study that included 25 water utilities across the U.S. and 10 water utilities in California, Krasner et al. (1989) measured quarterly median THM4 concentrations of 34, 44, 40, and 30 μg/L for the spring, summer, fall, and winter quarters, respectively. Samples were collected at the effluent of the clearwells, after final disinfection, but prior to distribution. Nieminski et al.

(1993) surveyed 14 medium and large treatment plants and 21 small plants in Utah. In these plant effluents, the mean and median THM4 concentrations were 31.3 and 22.0 µg/L, respectively; whereas, the 25th and 75th percentiles were 17.0 and 46.2 µg/L. As part of the monitoring program for the ICR, six quarters of DBP data were also collected between July 1997, and December 1998 (USEPA 1996). All U.S. utilities serving 100,000 people or more participated in the ICR monitoring, which involved 296 utilities (500 treatment plants). After collecting a total of 30,291 THM data points from the distribution system, the overall mean, median, and 90th percentile values were 38, 33, and 78 µg/L, respectively (McGuire and Graziano 2002).

According to the ICR monitoring data (McGuire and Graziano 2002), THM occurrence in distribution systems showed geographic variability among the regions of the USEPA. Mean THM concentrations ranged from 15 µg/L in the Pacific Northwest and Alaska (Region 10) to 47 µg/L in the Mid-Atlantic States of Virginia to Pennsylvania (Region 3). In general, the highest THM values stretched from Region 6 (Texas, Louisiana, Arkansas, Oklahoma, and New Mexico) to the regions dominated by the East Coast states due to high source water precursor concentration, high temperatures, and continued reliance on free chlorine for primary and secondary disinfections. The Great Lakes states (Region 5) showed the next highest THM concentrations, followed by the Western and Midwestern states, which had moderate THM concentrations. The Pacific Northwest and Alaska (Region 10) reported the lowest THM concentrations in the country.

THMs, and more specifically chloroform, are recognized as compounds susceptible to induce cancer in humans (NCI 1976, Trussell and Umphres 1978, Tuthill and Moore 1980, USEPA 1990, Bull and Kopfler 1991, Morris et al. 1993). Toxicological and carcinogenic studies indicated that THMs could cause liver or kidney cancer in laboratory animals (NTP 1986; Matsuoka et al. 1996; Chu et al. 1982a, 1982b; Condie, Smallwood, and Laurie 1983). Several epidemiological studies showed a positive relationship between consumption of chlorinated water and bladder, lung, breast, rectum, colon, and brain cancer (Yang et al. 1998, King and Marrett 1996, Cantor et al. 1998, Doyle et al. 1997, Vena et al. 1993, McGeehin et al. 1994). On the other hand, Marcus et al. (1998) and Foster et al. (1997) found no relationship between breast cancer and chronic myeloid leukemia, and studies involving bladder cancer induction have provided much more consistent results than studies involving other sites (Bull et al. 2001). These findings suggest that it may be too early to conclude that a causal relationship exists between THMs and the occurrence of cancer at specific organ sites.

An epidemiological study conducted by the California DHS suggested an increased spontaneous abortion risk to pregnant women exposed to THMs (Waller et al. 1998). A second study conducted by the California DHS showed that women who drank cold tap water had nearly a five-fold increased risk of miscarriage, compared to those who drank bottled water with very low THM concentrations (Swan et al. 1998). Bove et al. (1995) concluded that THMs at levels exceeding 80 µg/L may triple serious birth defects such as neural tube defects. However, several inconsistencies still exist in the studies involving reproductive and developmental effects of THMs such as fetal growth and spontaneous abortion (Reif et al. 1996, 2000). For example, Klotz and Pyrch (1999) found that the correlations between THM concentrations greater than 80 µg/L and neural tube defects were not statistically significant. Dodds et al. (1999) found no association between THM4 and fetal weight or gestational age. There was also no evidence that congenital abnormalities can be associated with THM4 concentrations. More recently, Savitz et al. (2005) observed that women who drank 5 or more glasses per day of tap water with

>75 μg/L THMs had the same risk of pregnancy loss as women who drank tap water with THM concentrations <75 μg/L, indicating no association between exposure and pregnancy loss. These results suggest that more data are required to ascertain the epidemiological implications of THMs.

Wilkes et al. (2002) developed a comprehensive exposure model to estimate population-based exposures and absorbed doses of 13 DBPs, incorporating parameters for chemical volatilization, human activity patterns, water use behaviors, ingestion characteristics, building characteristics, physiological measurements, and chemical concentrations in the water supply. Estimates were made for a three-person family based on data from women and men of reproductive age (ages 15 to 45) and children (age 6). The Total Exposure Model (TEM) uptake results provided information for comparing and contrasting uptakes as a function of the chemical, the population group and behavior, and the route of exposure. The study indicated that the inhalation route clearly dominates the absorbed dose estimates for THMs considering that these compounds are the most volatile of the DBP studied by Wilkes et al. (2002). The contribution of the ingestion and dermal routes are similar, and given the uncertainty of the parameters, it is unclear which route provides the most exposure.

Haloacetic Acids

HAAs are nonvolatile products formed by chlorination of organic materials. They have also been used as solvents, pesticides, chemical intermediates, and in the production of flame retardants. There are nine bromine- and chlorine-containing HAAs (HAA9) as presented in Table 2.2, and only five of them are currently regulated (HAA5). These HAA5 are ClAA, BrAA, Cl_2AA, Br_2AA, and Cl_3AA. BrClAA is considered an emerging contaminant. The sum of BrClAA and HAA5 is referred as HAA6. The remaining HAA species, including $BrCl_2AA$, Br_2ClAA, and Br_3AA, are known as HAA3.

Precursors of HAAs include humic substances, especially the organic hydrophobic fraction, as well as malonic acids (Ben Amor 1988, Croué 1987, Doré 1989, Cowman and Singer 1994). Some aromatic compounds (e.g., resorcinol) may form HAAs under high chlorine-to-precursor ratio. In addition, Cowman and Singer (1996) observed that HAA speciation for any classes of HAAs (XAAs, X_2AAs, or X_3AAs) is dependent on the bromide-to-chlorine ratio. However, the distribution among the XAA, X_2AA, and X_3AA classes appears to be independent of the bromide concentration over the range of concentrations tested (0 to 25 μM, 0 to 2 mg/L).

Table 2.2
Haloacetic acid species

Classes	Abbreviations	Compounds
Monohalogenated acetic acids (XAAs)	ClAA	Monochloroacetic acid
	BrAA	Monobromoacetic acid
Dihalogenated acetic acids (X_2AAs)	Cl_2AA	Dichloroacetic acid
	Br_2AA	Dibromoacetic acid
	BrClAA	Bromochloroacetic acid
Trihalogenated acetic acids (X_3AAs)	Cl_3AA	Trichloroacetic acid
	$BrCl_2AA$	Bromodichloroacetic acid
	Br_2ClAA	Dibromochloroacetic acid
	Br_3AA	Tribromoacetic acid

HAAs (mainly Cl_2AA and Cl_3AA) are often considered the second major group of DBPs after THMs; although, they are sometimes found at similar or even greater concentrations. When examining the ICR data, Obolensky et al. (2002) observed that HAA5 represented 66 to 81 percent of the THM4. Singer et al. (2002) studied the distribution and speciation of THMs and HAAs in 10 utilities from various geographic regions, with different water quality, treatment scenarios, and distribution system characteristics. Results showed that HAA concentrations exceeded THM concentrations only when chlorination was practiced after coagulation at pH values less than 6.5 to 7.0 or for chlorination of raw waters with high levels of hydrophobic natural organic matter (NOM). For all other cases, THM concentrations exceeded HAA concentrations.

In a survey of 35 U.S. utilities, Krasner et al. (1989) measured quarterly median HAA5 concentrations of 18, 20, 21, and 13 μg/L in spring, summer, fall, and winter, respectively. At the plant effluents of 35 water utilities in Utah, Nieminski et al. (1993) measured mean and median HAA5 concentrations of 17.3 and 12.8 μg/L, respectively, with 25th and 75th percentiles of 7.1 and 22.6 μg/L. X_2AAs and X_3AAs are commonly found at the highest concentrations; whereas, XAAs are generally measured at much lower concentrations.

In the ICR monitoring program, the median and mean HAA5 concentrations were 18 and 23 μg/L, respectively (Obolensky et al. 2002). Corresponding median and mean HAA6 concentrations were 21 and 26 μg/L, indicating that, at the central tendency of the data, inclusion of BrClAA increased HAA5 concentration by 15 percent. Although 90 percent of HAA5 concentrations measured as part of the ICR program were below 50 μg/L, the range of concentrations was wide with maximum concentrations as high as 280 μg/L. Nevertheless, the upper 10 percent of HAA5 results were attributable to only 4 percent of the associated treatment plants, suggesting that, for the large systems represented, elevated HAA5 and HAA6 concentrations in the U.S. are limited to a small fraction of utilities. The occurrence of HAA5 in the USEPA geographic regions was also studied as part of the ICR monitoring program (Obolensky et al. 2002). HAA5 concentrations were generally higher in the USEPA regions east of the Mississippi River compared with the western regions.

The toxicological properties of HAAs differ depending on the extent of halogen substitution, and the presence of bromine or chlorine. For example, XAAs are more acutely toxic than X_2AAs and X_3AAs; although, X_2AAs have been shown to produce a variety of chronic illnesses as well (Table 2.3) (Bull et al. 2001). Some of the differences observed within the X_2AAs and X_3AAs can be attributed to differences in metabolism or clearance from the body. When Cl_2AA is absorbed into the human body, the enzyme glutathione S-transferase zeta 1 (GSTZ1) is apparently responsible for its metabolism. DCAA undergoes hydrolysis and is converted to glyoxylic acid, which then inactivates further metabolism (Wempe et al. 1999). The X_3AAs are apparently metabolized at a much slower rate via a reductive mechanism to give rise to X_2AAs (Bull 2001).

Table 2.3
Toxicological effects of various HAAs

Haloacetic acid	Toxicological effect
Cl_2AA	Liver cancer
	Developmental effects
	Degeneration of prostrate gland
	Cysts in gall bladder
	Ocular lesions
	Brain lesions
	Aspermatogenesis
	Effects on nervous system
BrClAA	Liver cancer
	Effects on male reproduction
Br_2AA	Severe effects on sperm counts, motility, morphology
	Liver cancer
Cl_3AA	Cardiac malformation
	Liver cancer
	Developmental effects
$BrCl_2AA$	Liver cancer

Source: Adapted from Bull et al. 2001

Research has also shown that administration of Cl_2AA or Br_2AA induces the formation of hepatic tumors in laboratory animals. (Herren-Freund et al. 1987; Pereira 1996; De Angelo and McMillan 1989, De Angelo et al. 1996; Stauber and Bull 1997; Zavaleta et al. 1999). The same tumor-promoting activity was found in mice with Cl_3AA (Bull et al. 1990, Herren-Freund et al. 1987, De Angelo et al. 1996; Zavaleta et al. 1999). However, quantitative cancer risks for these compounds have not been estimated, and these results have not been successfully replicated in humans or nonhuman primates (Cattley et al. 1998). Thai et al. (2003) performed a mutation study and used a dosage of 2 g/L Cl_2AA to produce liver tumors in mice. Fertility was decreased in male rats receiving Br_2AA at doses of 10 mg/kg. It should be noted that this dose is several orders of magnitude higher than ambient exposure levels (Zavaleta et al. 1999). In addition, exposure to ClAA, Cl_2AA, Cl_3AA, and BrAA appear to produce more embryonic developmental effects, such as heart defects, in animals exposed *in utero*. These rather limited studies on various DBPs and their risks only enhance the need for continued research on the long-term toxicological effects on humans. It has been suggested that it may take 5 to 10 years after the initial exposure to the environmental hazard before one can detect the development of cancer (Sim 2002).

Using TEM results for exposure through ingestion, inhalation, and dermal exposure, Wilkes et al. (2002) indicated that the inhalation route has the least contribution to the absorbed dose of HAAs considering that these compounds are much less volatile. Given the large uncertainty in the dermal parameters, it is unclear whether ingestion or dermal exposure is the greatest contributor to the total absorbed dose.

N-nitrosodimethylamine

NDMA is a semi-volatile organic chemical that is soluble in water. It has been used as a commercial chemical for several decades in the production of 1,1-dimethylhydrazine for liquid rocket fuel and has a variety of other industrial uses: as a nematocide, a plasticizer for rubber, and an antioxidant. It is also used in polymers and copolymers, batteries, solvents, and lubricants. NDMA was found in a variety of foods, beverages, drugs, and in tobacco smoke. It has been detected as an air pollutant, in treated industrial wastewater, treated sewage in proximity to a 1,1-dimethylhydrazine manufacturing facility, deionized water, high-nitrate well water, and chlorinated drinking water (NTP 2000).

Studies (e.g., Jobb et al. 1994a, 1994b; Graham et al. 1996; Sedlak and Mitch 2001) suggested that NDMA's presence in drinking water might be related to disinfection processes. NDMA appears to be formed by several different reactions, depending on the disinfectant present, the water matrix, and chemicals present. Mitch and Sedlak (2002) and Choi and Valentine (2001) suggested that NDMA can be formed during the chloramination of dimethylamine (DMA) and other secondary amines. NDMA concentrations increased with increasing monochloramine concentration (Choi and Valentine 2001, Najm and Trussell 2001). According to Mitch and Sedlak (2002), the possible pathway involves the slow formation of 1,1-dimethyhydrazine (UDMH) by reaction between monochloramine and DMA, followed by its oxidation to NDMA and other products. Another possible pathway that may lead to NDMA formation is the reaction between sodium hypochlorite and DMA. Choi and Valentine (2001) also proposed a mechanism to explain NDMA formation in chlorinated water containing DMA and ammonia.

A report from the California DHS (2002a), suggested that chloramination, cationic polymers, ion exchange resins, and detention times appear to be the main factors responsible for NDMA formation. No correlations could be found between NDMA formation and chlorine residual (free or total), pH, temperature, alkalinity, total organic carbon (TOC), nitrate, nitrite, ammonia, or total Kjeldahl nitrogen. Chloramines provide nitrogen species that may trigger the formation of NDMA. Some cationic polymers may release precursors of NDMA in the water. In some instances NDMA was formed slowly, so long detention times in the distribution system may increase NDMA levels. The Ontario Ministry of the Environment, Canada, also indicated that NDMA may be formed as a result of the reaction of chlorine and process chemicals such as amine-based polymers. NDMA formation may cease once the use of those chemicals is terminated.

In November 1999, the California DHS conducted a survey of 19 California water utilities, primarily in the San Francisco and Los Angeles areas. Although two rounds of measurements were conducted, Table 2.4 presents only the data collected during the first, more complete round. Results showed NDMA concentrations varying between <1 and 63.7 ng/L at the plant effluent (California DHS 2002a). When NDMA was detected, the median concentration was 2.35 ng/L. The data obtained indicated that certain disinfection processes may lead to the formation of NDMA at levels greater than California DHS' action level of 0.01 μg/L. Additional data collected during the same survey suggested that NDMA concentrations were particularly high at the effluent of an ion exchange treatment facility.

Table 2.4
NDMA concentrations measured in California during the first round of sampling

Disinfection	Sample location	Range of concentrations (ng/L)	Average ± standard deviation of concentration* (ng/L)
None	Plant influent	<1 – 9.4	1.41 ± 1.54
Free chlorine	Plant effluent	<1 – 10.4	2.87 ± 2.66
Free chlorine	Distribution	<1 – 6.8	1.72 ± 1.66
Chloramine	Plant effluent	<1 – 63.7	6.72 ± 14.5
Chloramine	Distribution	<1 – 28.3	4.32 ± 6.43

Source: Adapted from California DHS 2002a
*For the calculation of these concentrations, concentrations of <1 ng/L were considered 1 ng/L

The Ontario Ministry of the Environment (MOE) began analyzing for NDMA as part of its Drinking Water Surveillance Program (DWSP) in 1989. During the first two years of sampling, concentrations exceeding 100 ng/L were sometimes observed (Jobb et al. 1993). Subsequently, NDMA occurrences were intermittent, generally related to colder weather or rain events, and while most occurrences were at trace levels, concentrations of up to 27 ng/L were observed (reported by Andrews and Taguchi 2000).

Research showed that NDMA may cause adverse health effects. In laboratory animal studies, rats, mice, hamsters, rabbits, and guinea pigs developed hepatocellular carcinomas and various tumors (primary liver, kidney, and lung tumors) when administered NDMA orally (reported by Najm et al. 2001). Because chemicals that cause cancer in laboratory animals may also increase the risk of cancer in people who are exposed over long periods, the National Toxicology Program (NTP) lists it as "reasonably anticipated to be a human carcinogen" (NTP 2000). NDMA is now considered potentially carcinogen by various regulatory agencies. Risk assessments from the California Office of Environmental Health Hazard Assessment (OEHHA) and USEPA identified lifetime *de minimis* (i.e., 10^{-6}) risk levels of cancer from NDMA exposures as 0.002 and 0.0007 μg/L, respectively (CCR 2000, USEPA 1997). The 10^{-6} risk level corresponds to up to one excess case of cancer per million people drinking 2 liters of water per day for a 70-year lifetime.

REGULATORY BACKGROUND

The first DBP regulation, the Total THM Rule, was promulgated by the USEPA in 1979. This rule set a maximum contaminant level (MCL) of 100 μg/L for THM4. This regulation pertained to systems serving a population of greater than 10,000 that use at least some surface water and add a disinfectant during any part of the treatment process.

The second rule that addressed halogenated DBPs was the ICR, which was finalized in May 1996 (USEPA 1996). The purpose of the ICR was to collect occurrence and treatment information to help evaluate the need for possible changes to the current SWTR and existing microbial treatment practices and to help evaluate the need for future regulation for disinfectants and DBPs. It required that large water systems that serve 100,000 people or more gather information on disease-causing microorganisms, DBPs (including THMs, HAAs, HANs, HKs, and others) and their precursors, as well as treatment plant design and operation to control these contaminants (USEPA 1996, McGuire et al. 2002).

The 1996 Amendments of the Safe Drinking Water Act (SDWA) require USEPA to develop rules to provide public protection against microbial pathogens while ensuring minimal health risks from DBPs. The Stage 1 DBPR and the Interim Enhanced Surface Water Treatment Rule (IESWTR) promulgated in December 1998, are the first set of these rules. The Stage 1 DBPR established the following MCLs:

- THM4: 80 µg/L
- HAA5: 60 µg/L (sum of ClAA, Cl_2AA, Cl_3AA, BrAA, and Br_2AA)
- Bromate ion (BrO_3^-): 10 µg/L
- Chlorite ion (ClO_2^-): 1 mg/L.

Compliance with the THM4 and HAA5 MCLs is based on the running annual average (RAA) of quarterly samples (the number of samples is based on system size) taken in the distribution system.

The Stage 2 Microbial/DBP Federal Advisory Committee Act (FACA) committee re-considered the strengths and limitations of the Microbial/DBP rules, as well as the related technical and policy issues. In September 2000, the stakeholders signed an Agreement in Principle that set the stage for the development of the Microbial/DBP Rules (USEPA 2000c). A proposed Stage 2 DBPR was presented on August 18, 2003, and a final rule is expected late 2005 or early 2006. All community water systems and non-transient, non-community water systems that add a primary or secondary disinfectant other than UV or distribute water that was treated with a disinfectant other than UV, must comply with the Stage 2 DBPR. Further information is presented in Chapter 6 Guidelines to Select Distribution System Sampling Locations for Disinfection By-Product Monitoring.

The Total THM Rule applied to systems serving a population greater than 10,000 and using at least some surface water. However, the Microbial/DBP Rules apply to all systems including those serving fewer than 10,000 people. In addition, the Ground Water Rule will establish disinfection requirements against microbial contamination for ground water systems. A Rule was proposed on May 10, 2000, and the Final Rule is expected in 2005. Thus, disinfectants and DBPs will soon be monitored by all water suppliers, and not only by large systems using surface water.

N-nitrosodimethylamine

NDMA historically has not been considered a common drinking water contaminant, so it does not have any state or federal MCL yet in the U.S. In the absence of drinking water standards, state health agencies may use a drinking water "action level" for the protection of public health. Specific action should be taken when NDMA concentrations exceed this level. As an example, the California DHS announced an action level of 0.002 µg/L in April 1998. However, analytical capabilities did not enable detection at that concentration, so any detectable quality of NDMA exceeded the action level. From 1999 to 2001, coincidental with more sensitive analytical detection methods becoming available, the California DHS established a temporary action level of 0.02 µg/L in response to studies on NDMA production in drinking water treatment. California DHS later revised the action level to 0.01 µg/L (California DHS 2002b).

The Ontario MOE has long been monitoring for NDMA in water treatment plants across this province of Canada (MOE 1998, Andrews and Taguchi 2000). Based on a $5x10^{-6}$ risk factor and analytical method limitations, the MOE set a maximum acceptable concentration (MAC) of 9 ng/L (MOE 2000).

CHLORINATION AND CHLORAMINATION

Chlorine Reactions

All reactions between chlorine and the various constituents of water are greatly influenced by temperature, pH, turbidity, NOM, chlorine dosage, and contact time. Chlorine reacts with organic compounds through three important routes: by oxidation of reduced functions, by addition onto unsaturated carbons, or by electrophilic substitution on nucleophilic sites. Thus, some DBPs represent halogen substitution by-products (THMs, HAAs, HANs, HKs, and chloral hydrate); whereas, others are oxidation by-products (aldehydes, aldo- and keto-acids, carboxylic acids, and bromate; Chowdhury and Amy 1999, Krasner 1999). Oxidation reactions result in a decrease in DBP precursors; whereas, substitution reactions result in the formation of organo-halogenated compounds by fixing one or more chlorine atom to the precursor molecule. The reactions result in the structural modification of the attacked molecules, leading to the formation of more oxygenated (and often more biodegradable) compounds. Organo-chlorinated molecules are also formed, specifically as a result of substitution reactions (De Laat, Merlet, and Doré 1982; Boyce et Hornig 1983; Johnson and Jensen 1986; Doré 1989; Bryant, Fulton, and Budd 1992). Johnson and Jensen (1986) observed that at relatively low chlorine doses (such as those applied in drinking water treatment plants), substitution reactions dominate; whereas, at high chlorine doses oxidation and cleavage reactions dominate. This observation suggests that in addition to forming greater concentrations of DBPs, changes in DBP speciation are also expected with increasing chlorine dosage.

Because aqueous chlorine species are electrophiles, they tend to react with electron-rich sites of organic structures. Thus chlorination reactions take place primarily on electron-rich sites such as carbonyl functions, reduced nitrogenated functions, and aromatic structures (Harrington et al. 1996).

It should be noted that the formation of organo-halogenated compounds only contributes to a small percentage of the chlorine demand. The remaining chlorine demand results from oxidation reactions, which may form organic acids and carbon dioxide in small concentrations (Merlet 1986, Doré 1989, Jadas-Hécart 1989).

Several mechanisms have been proposed to explain the formation of $CHCl_3$, Cl_2AA and Cl_3AA (Doré 1989, Merlet 1986, Rook 1980, Reckhow and Singer 1985, Adin et al. 1991). Some of the mechanisms proposed involve the formation of di- and tri-halogenated ketones as reaction intermediates. In general, THMs are the end-products of the numerous reactions that may occur between chlorine and organic materials (Johnson and Jensen 1986).

Chloramine Reactions

Chloramines are formed by combining appropriate quantities of chlorine and ammonia. Chloramine chemistry can be summarized by the following equations:

$$NH_3 + HOCl \leftrightarrow NH_2Cl + H_2O \quad (2.1)$$

$$NH_2Cl + HOCl \leftrightarrow NHCl_2 + H_2O \quad (2.2)$$

$$NHCl_2 + HOCl \leftrightarrow NHCl_3 + H_2O \quad (2.3)$$

The dominant chloramine species is a function of the chlorine-to-ammonia ratio and pH. Under typical drinking water conditions, monochloramine is the dominant chloramine species. The hydrolysis reactions (reverse of those presented above) are also important because they generate hypochlorous acid (HOCl), which may be involved in the formation of chlorine-substituted organics. Hypobromous acid (HOBr), hypobromite ion (OBr^-), bromamines, and thus bromine-substituted organics may also be formed during chloramine hydrolysis in the presence of bromide (Speitel 1999).

In general, chloramines form halogenated by-products to a considerably lower extent than free chlorine. Also, small concentrations of free chlorine and possible free bromine are always expected with chloramination. Therefore, formation of DBPs similar to those observed in chlorination is expected, but at much lower concentrations (Speitel 1999). In comparison to chlorine, chloramines are weaker oxidants and have a greater tendency to participate in chlorine substitution reactions, particularly with organic nitrogen compounds, rather than oxidation reactions. The identified DBPs mainly consist of the following small chlorinated and brominated compounds (Symons et al. 1998):

- Halogen-substituted amines, amino acids, and peptides
- Cyanogen chloride and bromide
- Halogenated nitriles, ethers, alcohols, and alkanoic acids
- Dihalomethanes.

Zhang et al (2000) observed THM4 and HAA9 concentrations less than 10 and 35 μg/L, respectively, in a model water with a combined chlorine dose of 5.0 mg/L Cl_2. THM4 and HAA9 concentrations were approximately 250 and 120 μg/L, respectively, in the same model water for a chlorine dose of 4.5 mg/L Cl_2. Cl_2AA constituted more than 90 percent of the HAA9 concentration by weight (Zhang et al. 2000).

Chloramination is more effective at controlling the formation of THMs and X_3AAs than the formation of X_2AAs (Cowman and Singer 1996, Symons et al. 1998, Speitel 1999). In general, DBP formation from chloramination decreases as the pH increases and the chlorine-to-ammonia ratio decreases (Symons et al. 1998; Speitel 1999).

When analyzing DBP data collected during the ICR from five locations in non-blended distribution systems (finished water, two locations with average water ages, one location with relatively high water age, and one location with well-characterized water age), Obolensky and Frey (2002) observed that the spatial variations in THM4 concentrations within individual systems were generally greater than the variations in HAA5 concentrations, regardless of source water (surface water or groundwater) or distribution system residual disinfectant type (free chlorine or chloramines) (Figure 2.1). The median THM4 concentration for utilities using surface water was 24.5 μg/L, compared with 14.1 μg/L for HAA5 (for systems using free chlorine as distribution system residual disinfectant). THM4 and HAA5 concentration ranges

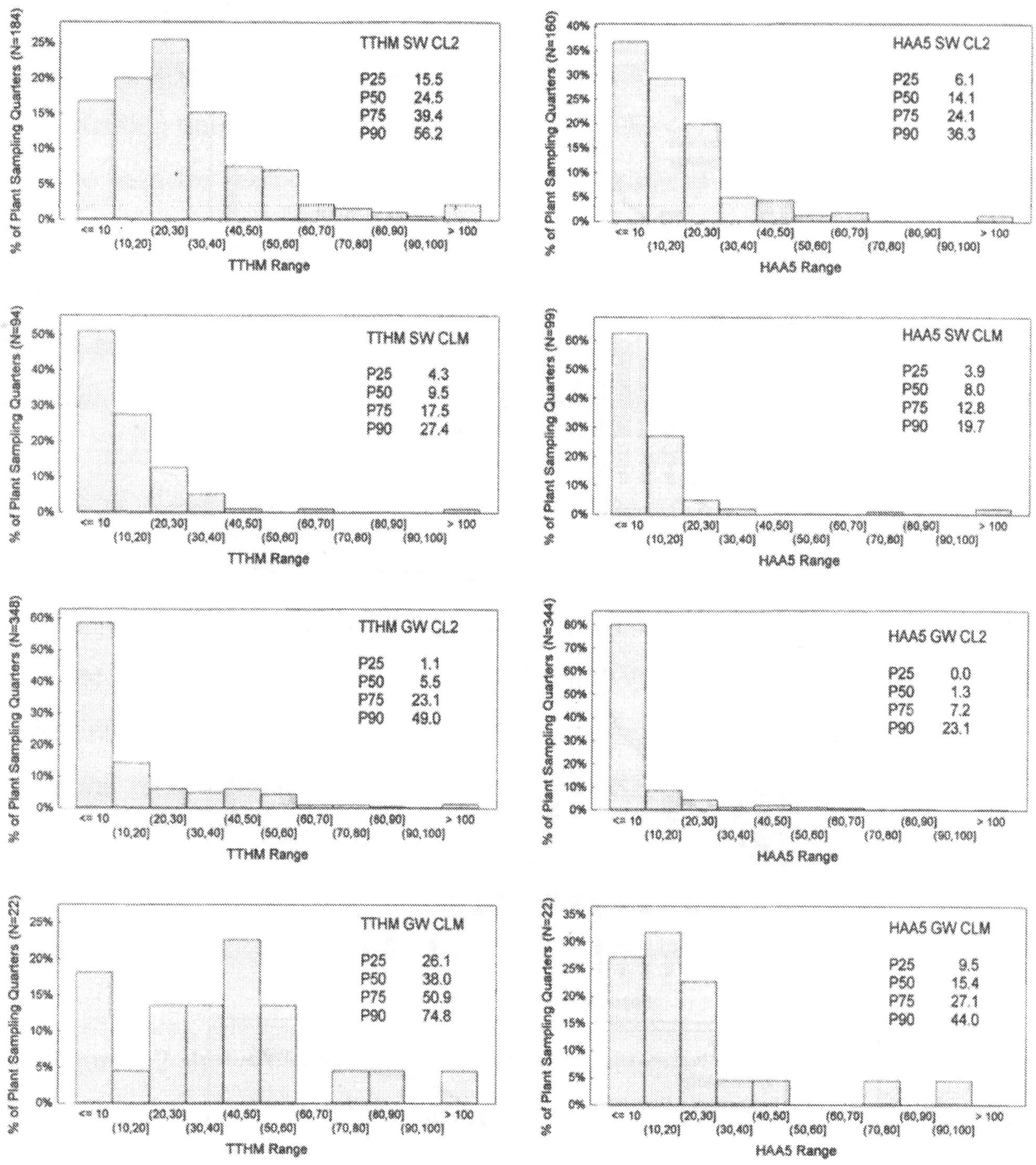

Source: Obolensky and Frey 2002

Figure 2.1 Distributions of ICR plant-quarter THM4 and HAA5 concentration ranges (difference between the highest and lowest observed concentrations in each plant-quarter). Surface water (SW), groundwater (GW), chlorine as secondary disinfectant (CL2), chloramines as secondary disinfectant (CLM), 25th percentile (P25), 50th percentile (P50 = median), 75th percentile (P75), and 90th percentile (P90)

were narrower and more comparable in surface water systems using chloramines than in systems using free chlorine. The median THM4 concentration calculated in surface water systems using chloramine was 9.5 µg/L, less than half the median concentration measured in the free chlorinated systems (24.5 µg/L), and close to the median HAA5 concentration calculated in the same systems (8.0 µg/L). The opposite trend was observed with respect to the type of secondary disinfectant in groundwater treatment plants, where THM4 and HAA5 concentration ranges were much wider in the systems using chloramines as distribution system residual disinfectant than in the free chlorinated systems (Obolensky and Frey 2002). The median THM4 concentration in the groundwater systems using free chlorine was 5.5 µg/L, compared with 38.0 µg/L in the groundwater systems using choramines. According to McGuire and Hotaling (2002) these results probably do not reflect the effect of the type of secondary disinfectant used at groundwater treatment plants but rather stem from differences in source water quality and type of treatment between the free chlorinated and the chloraminated treatment plants. The small number of systems using groundwater and chloramines may have also biased the results obtained.

Competition between bromide ion, ammonia, and chlorine determines the resulting chloramine concentration, bromamine, and DBP concentrations formed, as well as the DBP speciation. The speciation of the three bromamine species is a function of pH and the bromide-to-ammonia ratio (Johnson and Sun 1975). As the bromide-to-ammonia ratio increases, a greater degree of bromination occurs. At high pH, (9 to 10), monobromamine is the dominant species; whereas, at lower pH (<6) di- and tri- bromamines dominate. Bromamines form rapidly, relative to chloramines, and dissipate rapidly in the environment (Speitel 1999). Water chlorination may also produce bromochloramine in the presence of ammonia and bromide (Valentine 1986). In reactions with organic chemicals, bromochloramine should form by-products similar to those formed by free chlorine.

FORMATION AND STABILITY OF DISINFECTION BY-PRODUCTS

DBP formation depends on a variety of water quality and disinfection factors including:

- Concentration and chemical properties of the precursors
- Water temperature
- pH
- Bromide ion concentration
- Disinfectant type, dose, and residual
- Contact time

Accordingly, the concentration of DBPs at the POE to the distribution system depends on the chemical characteristics of the raw water, the treatment processes at the plant, chemical types and doses employed at the plant, and residence time in the various treatment units. The effect of these factors is further detailed in the paragraphs below.

Effect of Natural Organic Matter on Disinfection By-Products

NOM, a mixture of humic and nonhumic organic substances with high and low molecular weights, organic acids, and varying degrees of hydrophobicity and aromaticity, contributes to the DBP precursor levels and DBP speciation in drinking water. NOM can be classified as hydrophilic, transphilic, and hydrophobic depending on its affinity for water. The hydrophobic fraction consists of high molecular weight compounds and can be further classified as humic (less soluble), or fulvic (more soluble) substances. Humic substances have higher specific UV absorbance values (SUVA, which is the UV absorbance at 254 nm [cm^{-1}] times 100 divided by the dissolved organic carbon [DOC] concentration [mg/L]) and higher DBP formation potentials than the nonhumic fraction (Owen et al. 1993, Krasner et al. 1996). Thus, the UV-to-DOC ratio, which represents the aromatic content of the NOM, provides an indication of the possible reactivity of the organic matter to form DBPs. When isolating and chlorinating 10 aquatic humic and fulvic acids (pH 7, 20°C, 3-day contact time), Reckhow et al. (1990) observed that UV absorbance was greater for the humic acids, reflecting their higher aromatic content and greater molecular size. Humic acids also yielded greater DBP concentrations, such as total organic halides (TOX), THMs, and HAAs.

In a study conducted on the Horseshoe-Bartlett water supply reservoir system located on the Verde River, Arizona (Nguyen et al. 2002), the impact of hydrologic variability on reservoir annual DOC budgets was determined, and the DOC entering and leaving the reservoir system was characterized. The highest DOC concentrations were associated with spring snowmelt and late summer monsoon rains. During these high-flow periods, DOC concentrations ranged from 3 to 7 mg/L. The majority of DOC loading in the reservoir occurred in the spring, with short-term pulses of DOC loading observed during late summer monsoons. During this study, DOC loading from these monsoons was small in comparison to spring loading. Almost 90 percent of HAA5 precursors were loaded to the reservoir system during spring runoff (Figure 2.2). HAA5 precursors at downstream sites showed the delayed impact of surface runoff. HAA5 was slowly released from Bartlett Reservoir throughout the year. This slow release is evident in Figure 2.2, which shows the monthly mass loading of HAA5. A mass balance of THM precursors was conducted in parallel with the HAA5 mass balance, and similar conclusions were observed although slightly smaller losses were observed (20 to 32 percent THM loss in the two-reservoir system).

When analyzing the ICR data, Obolensky et al. (2002) observed that water treatment plants using a conventional process and unfiltered surface water treatment plants showed the highest median (25 μg/L for both types of plants) and 90th percentile (57 and 60 μg/L, respectively) HAA5 concentrations. Disinfected groundwater treatment plants had the lowest median (below the MRL) and 90th percentile (13 μg/L) HAA5 concentrations. Lower average source water TOC concentrations (1.9 mg/L versus 3.2 mg/L) and higher average source water bromide concentrations (101 μg/L versus 60 μg/L) at groundwater treatment plants helped account for the generally lower HAA5 concentrations at these plants compared with surface water treatment plants (Obolensky et al. 2002).

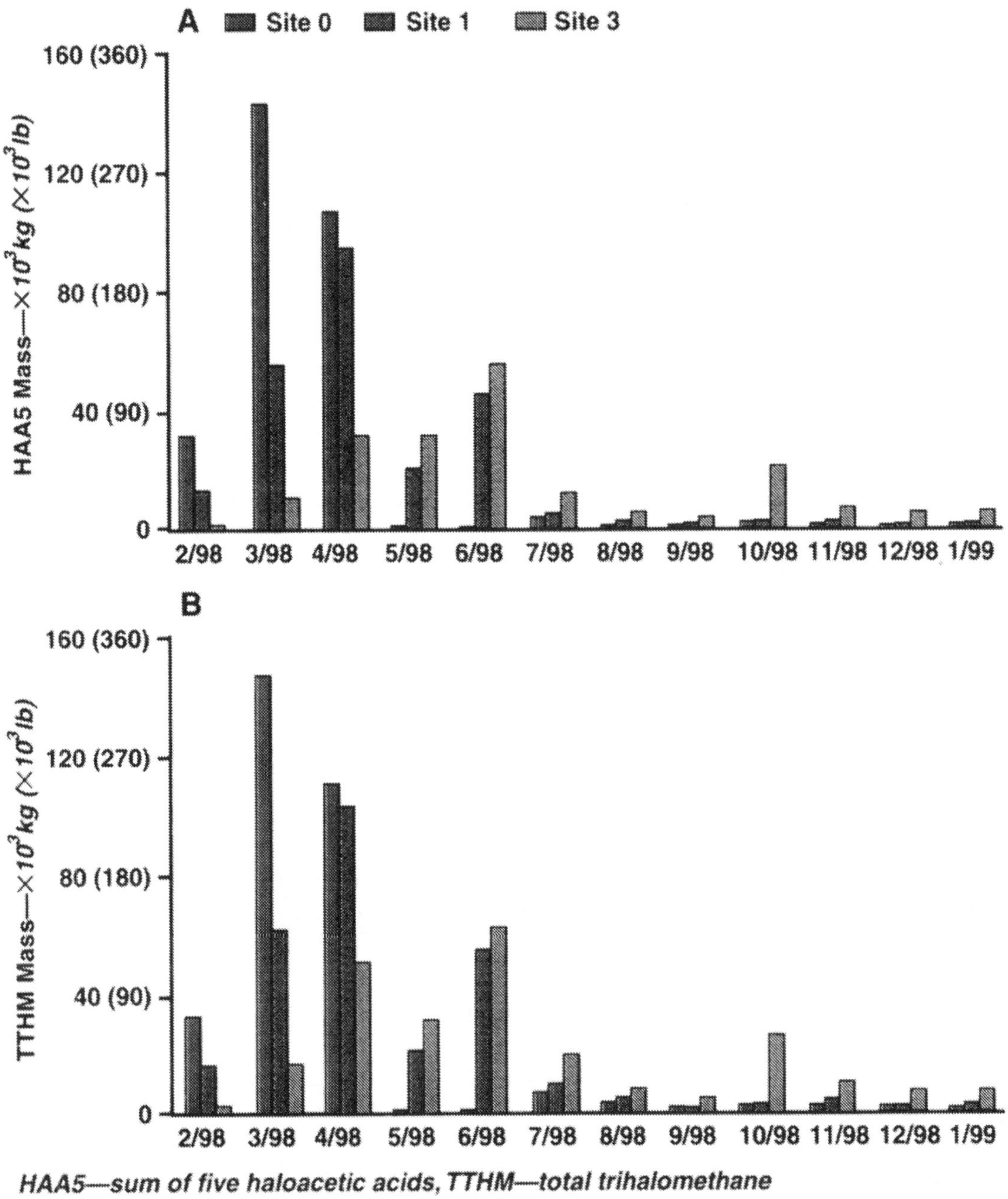

Source: Nguyen et al. 2002. DOC and DBP Precursors in Western U.S. Watersheds and Reservoirs. *Jour. AWWA,* 94(5):98-112. Published with permission from AWWA.

Figure 2.2 HAA5 (A) and THM4 (B) masses above Horseshoe Reservoir (Site 0), between Horseshoe and Bartlett Reservoirs (Site 1), and below Bartlett Reservoir (Site 3) from February 1998 to January 1999

Studies on DBP formation must include consideration for the nature of the organic precursors, which may vary in composition with the season (Singer 1999a). Zumstein and Buffle (1989) studied the seasonal evolution of NOM in a river and lake system. Increases in the DOC concentration and UV absorbance were observed after heavy rainfalls, and were attributed to increased leaching from soil organic matter during high river discharges. Bromide concentration can also vary seasonally (wet weather versus dry weather) (Singer 1999a).

Mathieu et al. (1992) observed significant seasonal variations in the quality and quantity of the organic material qualifiers (DOC, biodegradable dissolved organic carbon [BDOC], UV absorbance). These authors explained this observation by the fact that water treatment plants have difficulties correcting fluctuations in the organic matter concentrations in raw water, leading to significant variability in the concentration of these constituents in finished water. In fact, Meyer (1992) measured THM removals of less than 10 percent in the winter and up to 65 percent in the summer because of seasonal variation in the treatment process efficiency.

According to Huck et al. (1987); Lykins et al. (1991); and Alarcon-Herrera, Biswas, and Bewtra (1992), the significant increase in NOM observed in the spring may lead to different DBP precursors in raw waters. During this period, turbidity and chlorine demand increase, and semi-volatile and nonvolatile halogenated by-products (such as THMs and HAAs) are formed in greater concentration. This seasonal variation in organic matter also determines the speciation of the DBPs formed by influencing the bromide-to-ammonia ratio (Lykins et al. 1991).

Meyer (1992) and Ibrahim et al. (1994) measured much higher TOC and DOC concentrations in treated water in summer than in winter, leading to high concentrations in THMs, HKs, and HANs. Meyer (1992) noted that chlorinated by-products composed the majority of the DBPs in the summer; whereas, brominated by-products were dominant in the winter. This author suggests the following explanations: (1) chlorine dosages are lower in the winter, and the maximum formation of $CHBr_3$ is reached at the lower chlorine doses; (2) competitive reactions occur between chlorine and THM precursors or bromide ion (precursors are less abundant in the winter, and chlorine reacts preferably with bromide); or (3) bromide concentration increased significantly in the winter (also observed by Symons et al. 1987, Rebhun et al. 1988, Krasner et al. 1989, and Amy, Tan, and Davis 1991).

Effect of Bromide Ion on Disinfection By-Products

Bromide is a very important ion affecting chlorination chemistry. The two main reactions involved are the following:

$$HOCl + Br \rightarrow HOBr + Cl \qquad (2.4)$$

$$HOCl + HOBr + NOM \rightarrow DBPs \qquad (2.5)$$

Symons et al. (1993) and Nokes, Fenton, and Randall (1999) observed that the ratio of bromide to free available chlorine controls bromine substitution. Amy, Tan, and Davis (1991) and Cowman and Singer (1996) found that HOCl is a more effective oxidant; whereas, HOBr is a more efficient halogen substitution agent. Also, HOBr attacks more sites in the precursor than HOCl, and it reacts with them faster. A similar conclusion was reached by Trussell and Umphres (1978). Thus, it is expected that the presence of bromide ion increases the overall mass concentration of DBPs formed (Symons et al. 1998, Speitel 1999).

When conducting THM formation potential tests, Amy, Tan, and Davis (1991) observed that in general, less than 5 to 10 percent of the HOCl is incorporated in the THMs; whereas, as much as 50 percent or more of the bromide is incorporated in the THMs. The proportion of chlorinated vs. brominated by-products depends on the nature and concentration of precursors, bromide concentration, water temperature, pH, and contact time. In addition, the percentage of brominated DBPs increases as the bromide-to-TOC ratio increase, either as a result of increasing bromide ion concentration or decreasing TOC level (Amy, Tan, and Davis 1991; Cowman and Singer 1996). Several researchers observed that increasing bromide ion concentration generally decreases the concentration of $CHCl_3$, Cl_2AA, and Cl_3AA formed while increasing THM4 concentrations (Rook 1974; Oliver 1980; De Laat, Merlet, and Doré 1982; Croué 1987; Krasner et al. 1989, 1993; Jacangelo et al. 1989; Doré 1989; Koch et al. 1991; Siddiqui and Amy 1993). Symons et al. (1998) also confirmed that as the bromide concentration increases, the speciation within the individual DBP classes shifts toward the bromine substituted compounds.

Cowman and Singer (1996) chlorinated aquatic humic extracts under various bromide concentrations and measured the resulting HAA9 speciation. Results indicated that at low bromide concentrations (<3 μM or 240 μg/L), speciation is generally in the following order: $Cl_3AA > Cl_2AA > BrCl_2AA > BrClAA > Br_2ClAA > Br_2AA > Br_3AA$. At the higher bromide concentrations (>20 μM or 1.60 mM), speciation is in a nearly opposite order: $Br_3AA > Br_2AA > Br_2ClAA > BrCl_2AA > BrClAA > Cl_2AA, Cl_3AA$. The X_3AA species constituted the greatest mole fraction of the HAA9 concentration (61 to 67 percent), the X_2AAs accounted for 30 to 36 percent of the HAA9, and the XAAs constituted 3 to 5 percent of the HAA9. This distribution appeared to be independent of the bromide concentration over the range of concentrations studied (Cowman and Singer 1996).

McLain et al. (2002) analyzed DBP data collected in distribution systems during the ICR and compared them with bromide data collected at the treatment plant influent. In general, a shift in speciation to the more bromine-substituted DBPs among THMs, HAAs, and HANs, was observed as average influent bromide concentrations increased from <20 to >100 μg/L. For all DBP classes, however, the range of speciation was quite large, often including a minimum of 0 percent and a maximum of 100 percent for each of the three DBP categories. The proportion of bromine-substituted species followed the order: HANs > THMs > HAAs. The average percentage of bromine-substituted THMs was 18 percent for the lowest bromide concentrations (<20 μg/L) and 60 percent for the highest bromide concentrations (>100 μg/L). For HAA6, the respective percentages were 7 to 12 percent for low and high bromide concentrations, respectively, and 47 to 50 percent for HAA9 (McLain et al. 2002).

McLain et al. (2002) reported that the average percentage of bromine-substituted THMs increased as influent bromide concentrations increased (Figure 2.3). Even with bromide concentrations below the MRL (<20 μg/L), a small percentage (1 percent) of $CHBr_3$ and 17 percent of mixed halogen-substituted THMs ($CHBrCl_2$ and $CHBr_2Cl$) were produced. At plants with the highest concentrations of bromide (>100 μg/L), more than half of the THMs were bromine-substituted. In addition, plants that produced low concentrations of THMs had the largest percentage of bromine-substituted THMs ($CHBrCl_2$, $CHBr_2Cl$ and $CHBr_3$). Bromine incorporation decreased in plants with low THM concentrations compared with plants with high THM concentrations (from <10 to >60 μg/L), presumably because of the decrease in the ratios of bromide-to-TOC. The trend of increasing bromine incorporation as bromide concentrations increased was also apparent (McLain et al. 2002).

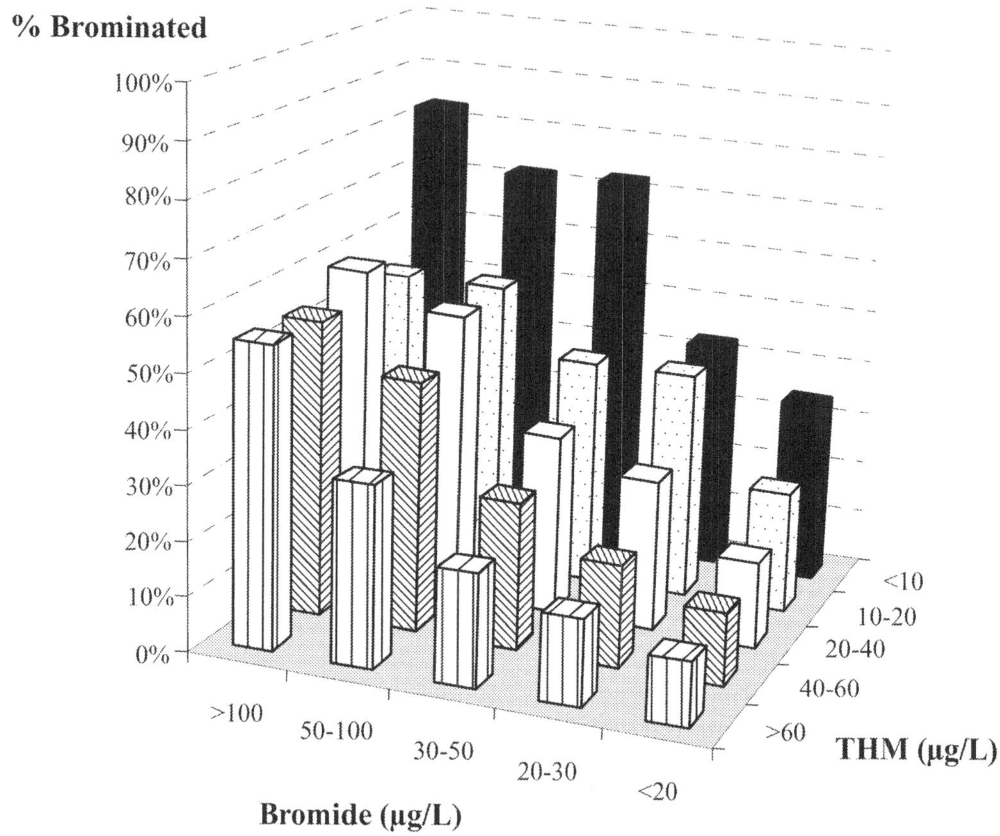

Source: McLain et al. 2002

Figure 2.3 Averaged ICR THM4 data showing the relationship between influent bromide level, average THM4 concentration and total percent of brominated THM4 species

Although the overall percentage of bromine-substituted HAA6 was lower than that of THM4, a similar pattern was observed when the data were analyzed as a function of DBP concentrations (Figure 2.4) (McLain et al. 2002). On average, plants that produced very low concentrations of HAA6 (<5 μg/L) had the highest levels of bromine incorporation. For these plants, the concentrations of bromine-substituted HAAs were almost twice those produced by plants in the next highest HAA6 concentration.

McLain et al. (2002) also used the ICR data to examine the effect of bromide level on the average concentration of each HAA9 species. For low bromide level (<30 μg/L), the concentration of each HAA9 species increased in the following order: Cl_3AA > Cl_2AA >> $BrCl_2AA$ > BrClAA > ClAA > Br_2ClAA > Br_2AA > BrAA = Br_3AA. For bromide concentration in the midrange (30 to 100 μg/L), the most significant change was that Cl_2AA production exceeded Cl_3AA production. For the highest bromide concentrations (>100 μg/L), the order was as follows: Cl_2AA > BrClAA > Cl_3AA = $BrCl_2AA$ > Br_2AA > Br_2ClAA > ClAA > Br_3AA > BrAA. This analysis of ICR HAA9 data indicated that Br_2ClAA and especially $BrCl_2AA$ are significant components of HAA formation and speciation (McLain et al. 2002).

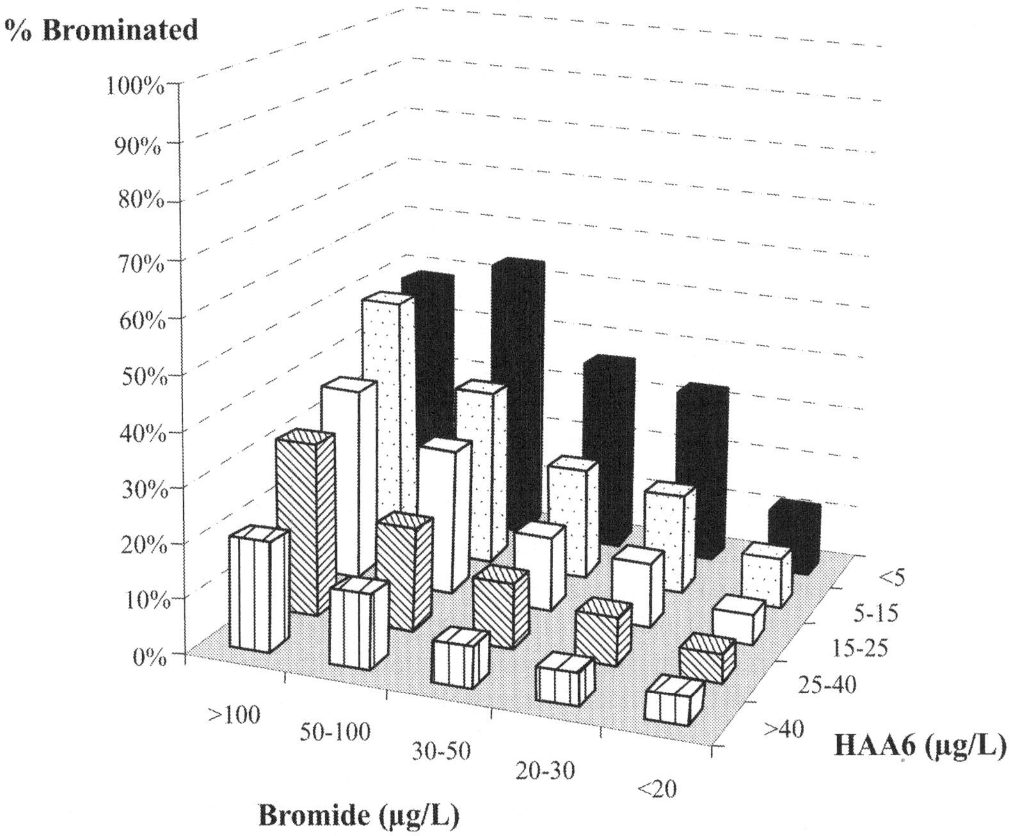

Source: McLain et al. 2002

Figure 2.4 Averaged ICR HAA6 data showing the relationship between influent bromide level, average HAA6 concentration (ClAA, Cl2AA, Cl3AA, BrAA, Br2AA, BrClAA), and total percent of brominated HAA6 species

Effect of Water Temperature on Disinfection By-Products

As discussed in the following paragraphs, various studies analyzed the impact of temperature on the concentration of DBP precursors, DBP intermediates, and DBPs. Considering that reaction kinetics are slower at lower temperature, it is possible to measure higher concentrations of reaction intermediates and lower levels of terminal products such as THMs in the winter. In general, higher temperatures were associated with greater increases in THM4 concentrations with increasing water age in free chlorinated distribution systems; whereas, changes in other DBPs were relatively insensitive to temperature except in the highest temperature ranges (above 25°C). In chloraminated distribution systems, higher temperatures were generally associated with more pronounced DBP losses with increasing water age. In addition, chlorine-substituted THMs and HAAs showed greater increases in formation with increasing temperature than bromine-substituted species.

Because seasonal variations change the concentration of DBP precursors (i.e., NOM and bromide), Krasner et al. (1989) sorted the data collected from 35 U.S. utilities in temperature ranges (samples were collected at the effluent of the clearwells after final disinfection but prior to distribution). The median THM concentration of the high-temperature group (24 to 31°C) was significantly higher than the concentration of the other temperature groups. Most of the DBPs, except for HKs, followed the same trend as THMs. HKs were measured in higher concentration in the winter than in the summer, which can be explained by the fact that these DBPs are intermediates of reaction and THM precursors. Considering that reaction kinetics are slower at lower temperature, it is possible to measure higher concentrations of reaction intermediates and lower levels of terminal products such as THMs in the winter (Krasner et al. 1989, Koch et al. 1991).

Temperature was also found to have a significant impact on the relative dominance of HAAs and THMs (Singer et al. 2002). Analysis of the ICR data indicated that the HAA species tended to increase moderately with temperature with the exception of Cl_3AA (Obolensky et al. 2002). Mean HAA5 and HAA9 concentrations of samples collected at temperatures above 25°C were 27 and 34 μg/L, respectively, compared with 21 and 25 μg/L for those collected at temperatures below 15°C. Analysis of the ICR data also indicated that for surface water treatment plants using only free chlorine, the HAA9/THM4 ratio increased with decreasing temperature. The ratios were found to be lowest in the winter (January to March) and highest in the summer (July to September). When analyzing the ICR data as a function of distribution system residual disinfectant, Obolensky and Frey (2002) showed that higher temperatures were associated with greater increases in THM4 concentrations with increasing water age in free chlorinated distribution systems; whereas, changes in Cl_3AA, Cl_2AA, and HAN species were relatively insensitive to temperature except in the highest temperature ranges (above 25°C). In chloraminated distribution systems, higher temperatures were generally associated with more pronounced DBP losses with increasing water age. This was evident for some DBPs, including Cl_2AA.

Laboratory studies involving water samples collected at the effluent of three different water treatment plants were conducted by Hooper et al. (1994) and Summers et al. (1996). Results showed that an increase in sample incubation temperature from 6 to 36°C increased the concentrations of THM4, HAA6, and dissolved organic halides (DOX) formed at pH 8.0 with a chlorine residual of 1.0 mg/L measured after 24 hours of incubation (Figures 2.5 and 2.6). This trend was more pronounced for THMs (especially $CHCl_3$) and HAA6 with increases of more than 100 percent, than for DOX (increases of 10 to 75 percent). Also, the slope of the curve representing the increase in THM levels increased with temperature; whereas, the HAA and DOX curves were linear. At 20°C, a change of incubation temperature of 1°C (between 5 and 35°C) led to concentration variations of 3, 2, and 1 percent for THM4, HAAs, and DOX, respectively (Hooper et al. 1994). Summers et al. (1996) observed that the more chlorine-substituted THMs and HAAs showed greater increases in formation with increasing temperature than the more bromine-substituted species.

Grulois, Matejka, and Parinet (1989) confirmed these trends by showing THM concentrations of 10 μg/L at 15°C, and 30 μg/L at 45°C. Other authors also observed increases in THM concentrations of 35 to 80 percent for each 10°C increase in temperature (Stevens et al. 1976, Kavanaugh et al. 1980, Oliver 1980).

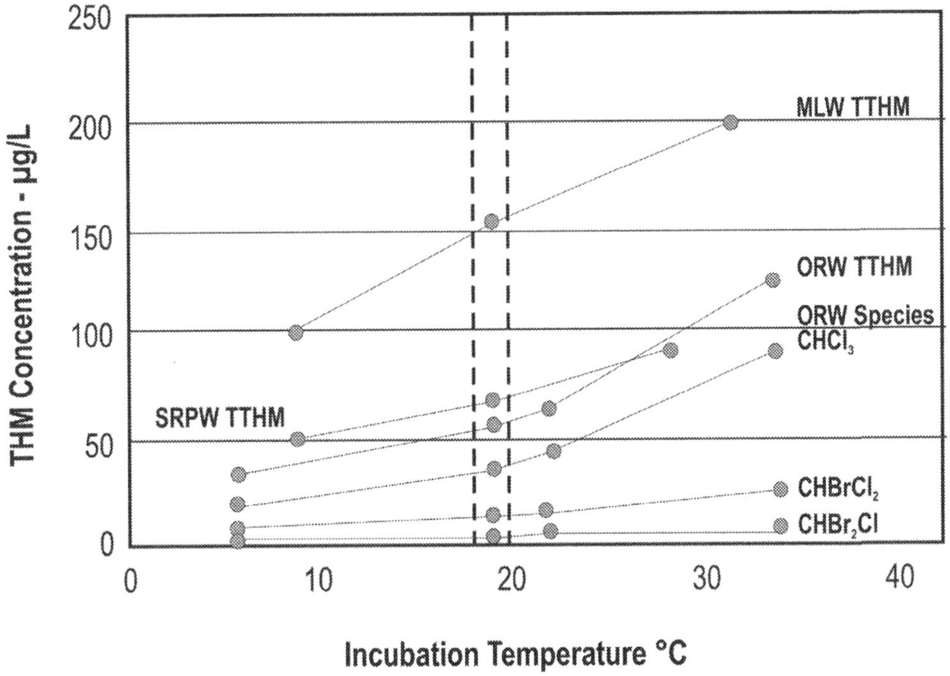

Source: Summers et al. 1996. Assessing DBP Yield: Uniform Formation Conditions. *Jour. AWWA*, 88(6):80-93. Published with permission from AWWA.

Figure 2.5 Effect of incubation temperature on THM4 formation

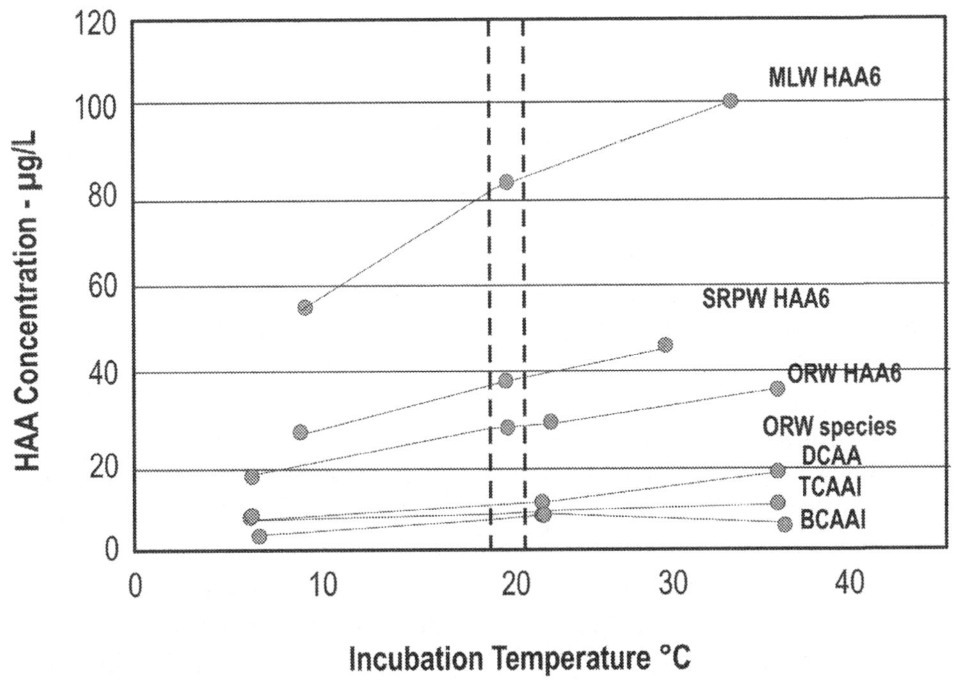

Source: Summers et al. 1996. Assessing DBP Yield: Uniform Formation Conditions. *Jour. AWWA*, 88(6):80-93. Published with permission from AWWA.

Figure 2.6 Effect of incubation temperature on HAA6 formation

Effect of pH on Disinfection By-Products

Water pH affects DBP formation. Several authors observed increases in THM concentrations from 20 to 240 percent associated with increasing pH from 7 to 10, depending on the organic materials present and chlorination conditions (Rook 1976; Oliver 1978; Oliver and Lawrence 1979; Boyce and Hornig 1983; Johnson and Jensen 1986; Hoigné and Bader 1988; Stevens et al. 1989; Koch et al. 1991; Bontrager, Chowdhury, and Hayes 1992). Chlorination of several compounds (acetone, butanone, fulvic acids, etc.) and natural waters lead to these results, although Croué (1987) mentioned that the pH effect on $CHCl_3$ formation may vary depending on the type of precursors present. In the case of HAAs, studies showed that their concentration tends to decrease with the increasing pH, while some HAA species such as X_2AA and Cl_3AA were relatively insensitive to changes in pH (Singer et al. 2002, Obolensky and Frey 2002, Hooper et al. 1994, Croué 1987, Cowman and Singer 1996, Stevens et al. 1989). The effect of pH on Cl_3AA and Cl_2AA tends to vary depending on experimental parameters.

Laboratory studies involving water samples collected at the effluent of three water treatment plants were conducted by Hooper et al. (1994) and Summers et al. (1996). Results indicated increasing THM4 concentrations (mainly $CHCl_3$) associated with increasing pH from 6.5 to 8.7 (at 20°C with a chlorine residual of 1.0 mg/L after 24 hours of incubation). According to these authors, a 2-unit pH increase led to an increase in THM4 concentration of 50 to 60 percent. When chlorinating fulvic acids, Reckhow and Singer (1985) and Croué (1987) observed that $CHCl_3$ forms rapidly between pH 2 and 8, then stabilized at higher pH values (up to pH 11). Stevens et al. (1989) also measured increasing THM4 concentrations following chlorination of raw and filtered water samples from a pilot plant (incubation of 4, 48, and 144 hours), from pH 5 to 7 to 9.4.

According to Singer et al. (2002), HAA9 concentrations tend to decrease in treated waters with increasing pH; whereas, THM4 concentrations tend to increase with increasing pH. However, among the HAA species, the formation of X_3AA decreases with increasing pH; whereas, the formation of the X_2AA was relatively independent of pH. Analysis of the first 12 months of ICR data indicated that the HAA9/THM4 ratio increased with decreasing chlorination pH. Chlorinated surface water systems with finished water pH of less than 7 showed higher HAA9/THM4 ratios than systems with pH greater than 8.

When analyzing the ICR data collected at five locations in non-blended distribution systems (finished water, two locations with average water ages, one location with relatively high water age, and one location with well-characterized water age), Obolensky and Frey (2002) observed that in the distribution systems fed with free chlorine as the final disinfectant, THM4 concentrations increased moderately with sample pH. The median value of the change in THM4 concentrations measured between the treatment plant effluent and the distribution system samples for the lowest pH range (<7.3) was 6.4 μg/L, compared with 9.2 μg/L at pH 7.3 to 7.8, 10 μg/L at pH 7.8 to 8.3, and 12.3 μg/L at pH >8.3. Median change in HAA5 concentrations did not vary with pH; although, the 90th percentile of the change in HAA5 concentrations for samples above pH 8.3 (11.3 μg/L) was markedly lower than the 90th percentile of the change in HAA5 concentrations (18.5 to 20.9 μg/L) for samples in the lower pH ranges. Changes in Cl_2AA and Cl_3AA concentrations in the distribution systems were relatively insensitive to pH in free chlorinated systems except at the highest pH range (pH >8.3), where increases in the concentration of these DBP species were somewhat smaller than those measured at lower pH ranges. In chloraminated systems, high pH values appeared to suppress both Cl_2AA and Cl_3AA

losses with increasing water age (Obolensky and Frey 2002). In chloraminated systems, although higher pH values appeared to promote continued Cl_2AA formation, they did not appear to promote continued Cl_3AA formation.

In the case of HAAs, laboratory studies conducted by Hooper et al. (1994) and Summers et al. (1996) suggested that HAA6 may not be influenced by chlorination pH (between pH 6.5 and 8.7); although, a slight decrease in concentration with increasing pH (mainly Cl_3AA) was measured above 8.2. A pH increase of 0.2 unit led to a 3-percent decrease in HAA concentrations at 20°C with a chlorine residual of 1.0 mg/L after 24 hours of incubation (Hooper et al. 1994).

Several authors observed that water pH affects HAA speciation. When chlorinating fulvic acids, Reckhow and Singer (1985) and Croué (1987) observed maximum formation of Cl_3AA at pH ~6, whereas its concentration was not detected at high pH (11 to 12). Following the chlorination of raw and filtered water samples collected from a pilot plant (incubation of 4, 48, and 144 hours), Stevens et al. (1989) indicated that Cl_3AA concentrations were similar at pH 5 and 7, and decreased at pH 9.4. pH did not have a significant effect on Cl_2AA (Stevens et al. 1989). According to Croué (1987), Cl_2AA formation following chlorination of fulvic acids was much less dependent on pH, and only a slight decrease in Cl_2AA concentration was observed with increasing pH from 2 to 12. Cl_3AA and Cl_2AA formation decreased with increasing pH from 4 to 10; although, the decrease in Cl_2AA concentration was much less pronounced than that of Cl_3AA (Miller and Uden 1983; Koch et al. 1991; Andrews, Huck, and Coutts 1992).

While chlorinating aquatic humic extracts under bromide concentrations varying from 0 to 25 μM (0 to 2 mg/L) and chlorine doses of 2 mg Cl_2 per mg TOC, Cowman and Singer (1996) observed that the formation of several HAA species, most notably Br_2AA, BrClAA, $BrCl_2AA$ and Br_2ClAA, was enhanced by lowering the pH from 8 to 6. In general, the brominated X_2AA and X_3AA species appeared to be more affected by pH than their chlorinated counterparts (Cl_2AA and Cl_3AA), which increased only slightly at the lower chlorination pH. The opposite was observed for the XAA species, for which ClAA formation was decreased by lowering the pH to 6, whereas BrAA formation appeared to be little affected (Cowman and Singer 1996).

Effect of Disinfectant Concentration on Disinfection By-Products

Several researchers observed that the rate of formation, extent and distribution of DBPs are affected by the chlorine dose and residual. Johnson and Jensen (1986) observed that at relatively low chlorine doses (such as those applied in drinking water treatment plants), substitution reactions dominate, whereas at high chlorine doses, oxidation and cleavage reactions dominate. This observation suggests that in addition to forming greater concentrations of DBPs, changes in DBP speciation are also expected with increasing chlorine dosage. For example, higher doses and residuals favor the formation of HAAs over THMs. In addition, higher chlorine doses result in a greater proportion of X_3AAs as compared to XAAs or X_2AAs (Krasner 1999).

When analyzing the ICR data collected at five locations in non-blended distribution systems (finished water, two locations with average water ages, one location with relatively high water age and one location with well-characterized water age), Obolensky and Frey (2002) observed that the change in DBP concentrations were strongly associated with finished water total chlorine residual in the systems using free chlorine as distribution system residual disinfectant. Concentrations of THM4, HAA5, Cl_3AA and Cl_2AA increased with finished water total chlorine residual in the free chlorinated systems. In the chloraminated distribution systems, finished water total chlorine residual did not appear to be associated with changes in THM concentrations (Obolensky et al. 2002); however, higher residuals appeared to suppress Cl_2AA decay and, to a lesser extent, Cl_3AA decay in these systems. Higher residuals also appeared to promote increased concentrations of Cl_2AA in chloraminated systems, though to a much smaller extent than observed in the free chlorinated systems.

Stability of Disinfection By-Products

In order to evaluate the changes in DBP concentrations and speciation in distribution systems, it is important to first examine their stability in water under controlled, constant conditions. Pereira et al. (2001) collected chloraminated water from the POE of a distribution system and held it at 22°C for over 13 days. Results showed that THM and HAA concentrations remained constant, with coefficients of variability ranging between 3 and 6 percent.

Koch et al. (1988) conducted two laboratory studies to examine the effect of sample preservation techniques on unchlorinated filter effluent water collected from a water treatment plant (Study I), and to determine the use of the best preservation technique with chlorinated filter effluent water (Study II). In the first study, most DBPs remained stable for at least 7 days at 4°C in the presence of NH_4Cl, which was proved to be the best preservation agent tested. In Study II, chlorinated filter effluent water samples were incubated at 4°C for up to 48 hours. Results showed that $CHCl_3$ and $CHBrCl_2$ concentrations were stable up to 48 hours in the presence of NH_4Cl. $CHBr_2Cl$, $CHBr_3$, and other DBP concentrations were stable under all conditions (chlorinated and dechlorinated).

Koch et al. (1988) measured the concentration of DBPs formed under the simulated distribution system (SDS) test. Chlorinated filter effluent water samples were collected from a water treatment plant and incubated over a 5-day period at 25°C. SDS test results showed that $CHCl_3$ formed the most rapidly during the first day of incubation and reached a maximum of about 50 μg/L after 5 days. $CHBrCl_2$ formed up to a maximum of 29 μg/L after 5 days. $CHBr_2Cl$ concentrations increased slightly for the first day and then remained constant at approximately 14 μg/L up to 5 days. $CHBr_3$ did not form at all and remained constant at a very low level of 1 μg/L.

CHANGE IN WATER QUALITY IN THE DISTRIBUTION SYSTEM

Water quality in distribution systems depends on the quality of the water at the POE as well as on the system itself. Studies have indicated that the influence of the water is most pronounced close to the treatment plant; whereas, the effect of the pipes is more important at the end of the distribution system (Lavoie 1991, Desjardins et al. 1991). Several factors influence water quality in the distribution system:

- Treatment processes used at the plant
- Type of disinfectant(s) and doses used at the treatment plant and in the distribution system at booster chlor(am)ination stations

- Physical and chemical characteristics of the water: temperature, pH, dissolved oxygen concentrations, etc.
- Type and quantity of NOM and nutrients present in the water
- Concentration and type of microorganisms
- Presence and density of biofilm inside the distribution system
- Pipe materials and presence of corrosion in pipes
- Presence of sediments introduced in pipes during their installation or repair
- Hydraulic conditions: water velocity, pressure, and residence time in clearwells, pipes, reservoirs, and tanks
- Size, age, and configuration of the network, including pipe diameter
- Distribution system operation and maintenance.

Change in Disinfectant Concentrations in the Distribution System

Treatment processes, disinfecting conditions at the treatment plant, and varying distribution system characteristics influence the residual disinfectant stability. Water quality also affects chlorine speciation and demand, and therefore DBP formation and fate in distribution systems. As an example, Andrews, Huck, and Coutts (1992) observed that at pH 6, the chlorine demand of the organic matter hydrophilic fraction is higher than at pH 8. With increasing pH values, Noack and Doerr (1978) observed that the 7-hour chlorine demand of fulvic acids increased slightly from pH 6 to 9.5. Reckhow (1984) showed that the 72-hour chlorine demand of fulvic acids decreased significantly from pH 7 to pH 12. Reckhow and Singer (1985) measured rapid increases in chlorine consumption of humic substances from pH 3 to 8, following by a slight decrease at pH $\geq$9. At a water treatment plant effluent, Jadas-Hécart (1989) measured a decrease in chlorine demand kinetic constant with increasing pH, which was associated with a slight variation of long-term chlorine consumption. This author believes that chlorine speciation is the main phenomena responsible for the differences of kinetic constants calculated.

Decreases in chlorine residual (free and total) in pilot- and full-scale distribution systems have been observed in many studies (Haas et al. 1983; Olivieri et al. 1986; Lavoie 1991; Desjardins et al. 1991; Coallier et al. 1991; Barthe et al. 1991; Colbourne et al. 1991; Mathieu et al. 1992; Capellier, Picoche, and Deguin 1992; Meyer et al. 1993; Baribeau et al. 2001). Olivieri et al. (1986) observed free chlorine concentrations dropping from 1.3 to 0.5 mg/L in less than 4 hours in a pilot distribution system. Several authors observed that free and total chlorine concentrations were depleted more slowly in cold water than in warm water (Lavoie 1991; Desjardins et al. 1991; Colbourne et al. 1991; Chen and Weisel 1998; Baribeau et al. 2001). In a study conducted by Baribeau et al. (2001), free chlorine demand was monitored as water flowed in full-scale distribution systems. At similar water age, a greater demand was measured in small-diameter pipes than in the main pipes, and similarly a greater demand was measured in the main pipes than in batch incubations in bottles, particularly in warm water (Figure 2.7). The rate of chlorine decay was calculated at $0.29\text{-}1.00 \times 10^{-5}$ mg/L/m^2/h in the main pipes, and $120\text{-}190 \times 10^{-5}$ mg/L/m^2/h in the small-diameter pipes.

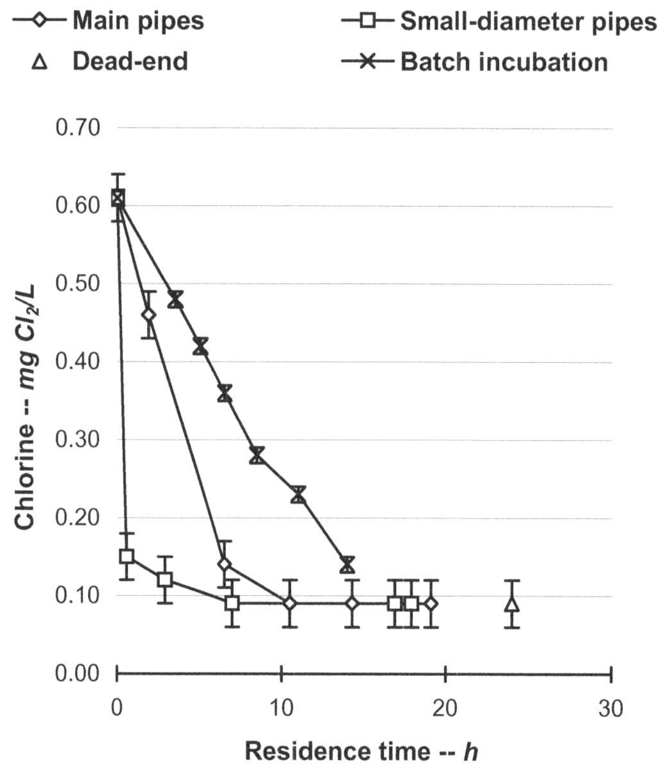

Source: Baribeau et al. 2001. Changes in Chlorine and DOX Concentrations in Distribution Systems. *Jour. AWWA*, 93(12):102-114. Published with permission from AWWA.

Figure 2.7 Free chlorine residual as a function of water age in a full-scale distribution system

In a distribution system, chlorine decays due to reactions with compounds contained in the bulk water, and/or reactions at the pipe walls. As such, chlorine consumption has been associated with the presence of sediments introduced in the transmission lines during their installation or repair, the presence of biofilm inside the pipes, oxidation of pipe materials, corrosion by-products present on some materials, and increase in temperature.

Powell et al. (2000) conducted over 200 investigations of bulk chlorine decay against time in waters collected from 32 locations in the U.K. Results showed that temperature, the initial chlorine concentration, and the organic content (TOC and UV absorbance at 254 nm) of the water were the parameters that influence the most chlorine decay. Vasconcelos et al. (1996) reported that the following water quality parameters increase chlorine decay: high TOC and temperature, low chlorine dose, presence of ferrous iron from corrosion, and presence of biofilm. Wable et al. (1991) observed that biofilm and tubercles could result in significant chlorine decay. Trussell (1992) reported that corrosion could play an important role in controlling chlorine residual stability in metallic conduits.

According to Kiéné, Lu, and Lévi (1998), chlorine decay in synthetic pipes is mainly associated with the presence of deposits (which explained 56 percent of the decay), followed by the effect bulk water (26 percent), the fixed biomass (16 percent), and the pipe material itself. In cast iron pipes, corrosion of the material appeared to be the main factor affecting chlorine decay (explaining 57 percent of the decay), followed by the presence of deposits (25 percent), the effect of bulk water (11 percent), and the fixed biomass (7 percent). The TOC level and temperature are the main parameters affecting chlorine decay in the bulk phase. The influence of the biofilm depends on the BDOC content of the water and the pipe diameter (Kiéné, Lu, and Lévi 1998).

LeChevallier et al. (1988 and 1990) noted that mass transport between the bulk flow and pipe wall can significantly affect chlorine consumption in distribution systems. Rigorous mass transfer models under steady and nonsteady states have been developed to describe bulk flow reaction, radial diffusion, and pipe wall reaction with chlorine (Biswas, Lu, and Clark 1993; Rossman et al. 1994).

Changes in Disinfection By-Product Concentrations and Speciation in the Distribution System

Continuous formation of DBPs in the distribution system depends upon (Singer 2001):

- Type of disinfectant used (free chlorine or combined chlorine).
- Disinfectant residual: The higher the residual, the faster is the extent of DBP formation.
- pH: An elevated pH tends to favor THM formation; whereas, X_3AA formation is favored at lower pH (this effect is discussed in greater detail in the following sections).
- Temperature: the rate of chemical reactions, including DBP formation, increases with increasing temperature (this effect is discussed in greater detail in the following sections).
- Contact time from the POE: Because the formation of most DBPs continues in the presence of a disinfectant residual, the water age in the distribution system impacts DBP concentrations and speciation at the consumer tap. The water age to any point in the system is a complex function of water demand patterns in the system, pipe diameter, system size, network design, storage volume in the system, and water production at the plant.
- DBP species: The formation kinetics of some DBPs (e.g., HAAs) tends to be faster than the kinetics of formation of others (e.g., THMs). As a result, more HAA formation tends to take place in the treatment plant and less in the distribution system. In addition, some DBPs such as HKs, tend to decay in the distribution system due to hydrolysis and continuing reaction with chlorine.

In the distribution system, overall change in DBP concentration and speciation represents the summation of DBP formation and degradation. In the presence of free chlorine and depending on the residual present and the quality of water, an increase in DBP concentration can be observed when DBP formation exceeds DBP decay (e.g., in the summer when the chlorine doses are higher). This is due to continuing reactions between available chlorine and organic material in the bulk water and biofilms on the pipe walls. When the chlorine residual is lower, degradation of DBPs already formed predominates. DBP decay may be attributed to several factors, with the relative importance of each varying according to the conditions prevailing in the distribution system. These factors include (Baribeau et al. 2001):

- DBP chemical degradation
- Water pH
- Interaction with corrosion by-products
- Adsorption in the biofilm matrix, biodegradation, and/or bioaccumulation by microorganisms.

Recent publications indicate that DBP concentrations at various points in the distribution system can vary significantly from a spatial standpoint (Williams et al. 1994 and 1995; LeBel et al. 1995; Chen and Weisel 1998; McClellan et al. 1999; Singer 1999b; Uber et al. 2000; Baribeau et al. 2001). One source of this variability is the dynamic nature of water distribution systems, e.g., as demands change in the water system and storage tanks fill and empty, water age at the tap can vary greatly (McClellan et al. 1999, Uber et al. 2000, Singer et al. 2001). At some locations, tap water comes directly from the treatment plant for part of the day, while at other times it is primarily older water drawn from storage tanks. Such fluctuations lead to significant variations in chlorine residual and DBP concentration.

Singer (2001) explained that significant temporal variability in DBP concentrations can also be expected at any point in the distribution system. These variations can be seasonal, daily, and even hourly. Seasonal raw water quality change is one cause of this variability. For example, organic carbon concentration, the nature of the organic material, bromide ion concentration, pH, and temperature can all exhibit significant seasonal, and in some cases daily and diurnal variations. Likewise, flow rates, residence times, and chemical doses at the treatment plant also change periodically. This can result in appreciable temporal differences in DBP concentrations of water entering the distribution system.

Pereira et al. (2001) collected samples every 4 hours over a five-day period at the POE to a chloraminated distribution system. Results showed that THM and HAA concentrations and speciation changed dramatically over the course of a day, with coefficients of variation of 35 percent for THMs and 29 percent for HAAs. A similar test was conducted with water collected at the POE to the distribution system, as well as at two locations within the system at retention times of approximately 1 and 5 days. During this experiment, the coefficient of variation for THM concentrations at the POE was 24 percent, whereas it was only 9 percent in the distribution system (Figure 2.8). Additional experiments indicated a coefficient of variation for THM concentrations of 14 percent at the POE and 9 percent in the distribution system (one-day water age). HAAs showed a similar trend with a coefficient of variation of 12 percent at the POE and 7 percent after approximately one day in the distribution system. These results showed that temporal variations in THM and HAA concentrations appeared to be dampened in the distribution system for utilities using chloramines for final disinfection.

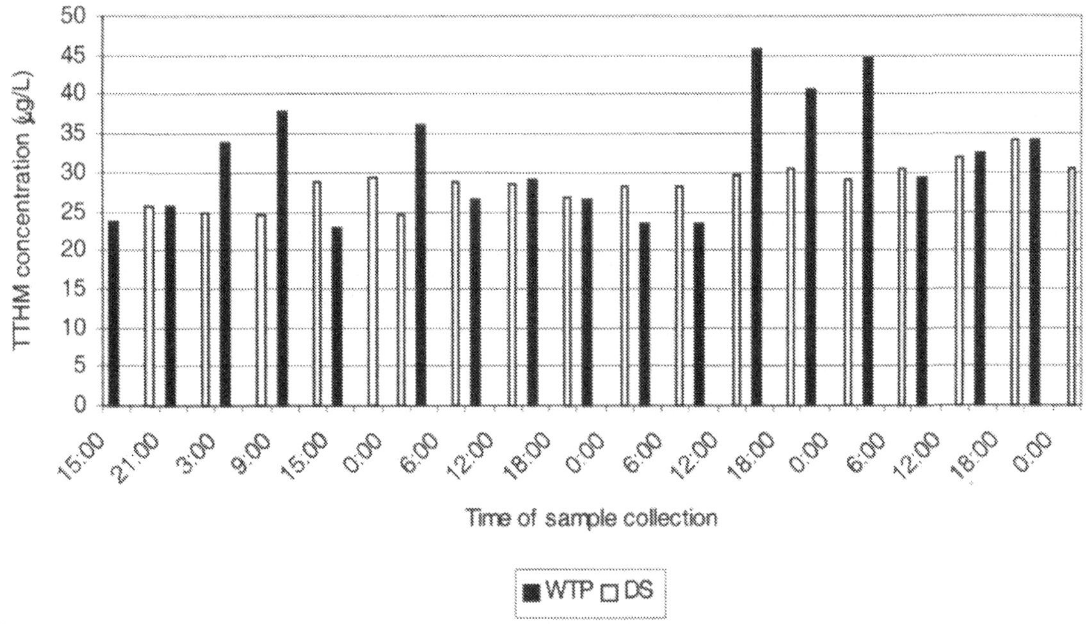

Source: Pereira et al. 2001. Development of a Sampling Approach to Assess Patterns of Exposure to Disinfection By-Products in a Chloraminated Drinking Water System. *In proc. of AWWA Water Quality Technology Conference*. Nashville, TN, November 11-15 (poster). Published with permission from AWWA.

Figure 2.8 Comparison of THM4 concentrations leaving the treatment plant and in the distribution system

Obolensky (2004) analyzed three sources of distribution system THM4 and HAA5 data: (1) utility distribution system DBP study involving a total of 23 distribution system sites in three systems examined over a 18-month period, (2) ICR data involving five sampling sites per treatment plant during an 18-month period, three plants were examined, and (3) compliance data involving four locations per plant, three plants were examined. Results indicated that changes in DBP concentrations in the distribution system may be related to timing of sample collection rather than simply the sampling location itself. Different sampling events in the same quarter showed very different results at the same sites. Moreover the short-term fluctuation in finished water HAA5 concentration was more important than the short-term fluctuation in THM4 concentration. Obolensky (2004) suggested that changes in bromide concentration, TOC level, and/or temperature patterns may be responsible for these trends.

Trihalomethanes

Several authors observed increasing THM concentrations with increasing distance from the treatment plant or water age in the distribution system (LeBel et al. 1996, Harms and Loogenga 1977, Brett and Calverly 1979, Levine et al. 1987, Koch et al. 1991, Williams et al. 1994, Chen and Weisel 1998). In a simulated pipe loop system, Vasconcelos et al. (1996) observed an increase in THM concentrations and a significant drop in THM formation potential related to DOC consumption, along with a significant decrease in chlorine residual.

In a pilot distribution system with a three-day residence time, Meyer et al. (1993) observed an increase in THM4 concentrations with increasing water age in the presence of free chlorine. At the POE to this pilot, the average chlorine residual was 1.6 mg/L, temperature was 16.5°C, and pH was 7.5. In the absence of a measurable chlorine residual, THM concentrations were stable or increased more slowly. When a pilot distribution system was compared with laboratory batch incubations (SDS tests), Meyer et al. (1993) observed lower THM concentrations in the batch incubations than in the distribution system. According to Meyer (1992), the higher THM concentrations measured in the pilot distribution system were attributed to the higher organic material content present in the pipes (mainly due to the presence of biofilm) and water agitation in the pipes; both the batch incubations and pilot distribution system received similar chlorine dosages. This author noted that the reaction between chlorine and biofilm is a function of the organic material content of the water. Chlorine reacts preferably with THM precursors, but it does not react or reacts very slowly with the biofilm when DOC and THM precursor concentrations of the bulk liquid are high. In the winter when the organic material content is lower, chlorine reacts preferably with the biofilm. Meyer (1992) noted that the percentage of each individual THM in relation to THM4 does not vary or varies only slightly with increasing water age. This author explained that in their particular case, pH variations were not responsible for the change in DBP concentrations observed in the distribution system. The pH showed similar variations in both testing devices (pilot distribution system and batch incubations); whereas, the changes in DBP concentrations were different.

Using the ICR data collected from five locations in non-blended distribution systems (finished water, two locations with average water ages, one location with relatively high water age, and one location with well-characterized water age), Obolensky and Frey (2002) observed that the greatest increases in DBP concentrations relative to those measured in the finished water of treatment plants were seen for $CHCl_3$ in the systems using free chlorine as distribution system residual disinfectant. $CHBrCl_2$ concentrations also exhibited significant increases in these distribution systems. Distribution system increases in $CHCl_3$ concentrations were substantially smaller in the chloraminated systems than in the free chlorinated systems. Obolensky and Frey (2002) also observed that the variations in THM4 concentrations increased with water age in the free chlorinated systems that participated in the ICR, but they were relatively insensitive to water age in the chloraminated systems. Increases in DBP concentrations with water age were greatest for $CHCl_3$, and decreased in the following order: $CHCl_3 > CHBrCl_2 > CHBr_2Cl > CHBr_3$.

Singer et al. (1993) observed decreases in THM concentrations with increasing water age in aquifer storage and recovery (ASR) projects. The authors suggested that THM removal was associated with reducing conditions under which halogenated aliphatic compounds such as $CHBr_3$ may be transformed to less halogenated forms under anoxic conditions when nitrate was present as an electron acceptor. $CHCl_3$, on the other hand, was reported to be resistant to dehalogenation under denitrifying conditions and required the more severe reducing conditions of methanogenesis before dehalogenation occurred. It was speculated that the electron donor for this particular reaction was most likely organic carbon. The more highly brominated THM species tend to be removed first, suggesting a biological mechanism for reductive dehalogenation. In most cases, removal of the THM species did not appear to occur until anoxic conditions developed and followed the onset of denitrification.

Haloacetic acids

When measuring HAA concentrations at various locations in full-scale distribution systems, Chen and Weisel (1998), LeBel et al. (1995), Williams et al. (1994, 1995) reported lower concentrations at the end of the distribution system where low levels of chlorine residual and high heterotrophic plate counts (HPCs) were measured compared to locations closer to the water treatment plant. Cl_2AA was more affected than Cl_3AA, and results suggested that biodegradation was responsible for the decrease in HAA concentration.

Jammes et al. (1994) observed that maximum concentrations of HAAs were measured in various French distribution systems (chlorine doses at the plant effluent ranged from 0.8 to 2.2 mg/L), but not immediately at the treatment plant effluent, indicating a slow formation kinetic. Following an increase in HAA concentration, a 50 percent decrease was measured between the beginning and the end of the distribution system, and was attributed to biodegradation, chemical instability, or adsorption.

Singer et al. (2002) and Koch et al. (1991) noted that in the presence of chlorine residual, HAAs continued to form with increasing water age; however, in the absence of chlorine residual, X_2AAs were biodegraded under warm water conditions, but X_3AAs were not biodegraded under similar conditions (Figure 2.9).

HAAs are not as stable as THMs, and some of the HAA species are known to degrade when chlorine residual is low or absent. Using the ICR data collected from five locations in non-blended distribution systems (finished water, two locations with average water ages, one location with relatively high water age, and one location with well-characterized water age), Obolensky and Frey (2002) observed that Cl_3AA and Cl_2AA exhibited significant increases in concentrations in the free chlorinated distribution systems relative to the concentrations measured in the finished water of treatment plants. Several distribution systems also showed decreases in DBP concentrations, including Cl_2AA, BrClAA, and Cl_3AA. Cl_2AA showed the greatest decrease in concentration in the free chlorinated systems. Increases in Cl_3AA concentrations were substantially smaller in the chloraminated systems than in the free chlorinated systems; whereas, such differences with respect to the type of disinfectant was less evident for Cl_2AA. Obolensky and Frey (2002) also observed that the variations in HAA5 concentrations increased with distribution system water age in the free chlorinated systems that participated in the ICR, but were relatively insensitive to water age in the chloraminated systems. Also, the losses of HAA5 were more pronounced at distribution system locations with maximum water age, regardless of the residual disinfectant used.

Hozalski et al. (2001) measured the degradation rates of all X_3AAs, Br_3AA, Cl_3AA, Br_2ClAA, and $BrCl_2AA$ in distilled and deionized waters in the presence of zero-valent iron. pH was allowed to fluctuate and rose from 3.6 to 6.2 over the course of the experiments. Incubations were conducted at room temperature for up to 94 hours. Bromine was preferentially removed over chlorine, and Br_3AA was the only compound completely dehalogenated to acetic acid. Given that drinking water systems primarily consist of unlined cast iron pipes, this may account for some of the loss of HAA concentrations observed in full-scale distribution systems.

Zhang and Minear (2002) studied the decomposition of X_3AAs ($BrCl_2AA$, Br_2ClAA, and Br_3AA) and the formation of their corresponding THMs ($BrCl_2CH$, Br_2ClCH, and Br_3CH). The X_3AAs were found to decompose via a decarboxylation pathway, and THMs were demonstrated to be the only decomposition products. The effect of pH in the range of 6 to 9 and the effect of drinking water matrix on the decomposition of X_3AAs in water were found to be insignificant. Incubations were conducted at 4, 36, and 50°C over an 84-day period.

Singer et al. (1993) studied DBP fate in ASR projects. Results showed that in contrast to THMs, HAAs may be degraded aerobically, which may explain their rapid disappearance at each of the ASR sites, even before the onset of anoxic denitrificating conditions.

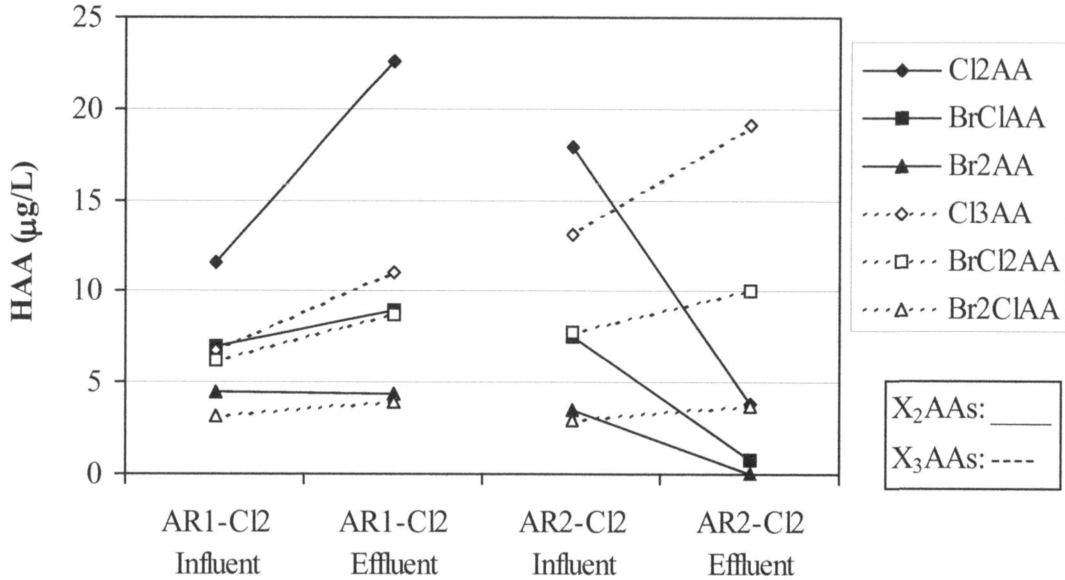

Source: Singer et al. 2002

Figure 2.9 Example of HAA results obtained in warm water in the chlorinated train (ClAA ≤2 µg/L, BrAA ≤3 µg/L, Br_3AA not detected) (temperature of 19.2°C, 66.6°F). AR1: Upstream annular reactor in presence of chlorine residual. AR2: Downstream annular reactor in absence of chlorine residual

In the spring of 1993, Newport News Waterworks, Virginia, began a study during which samples from nine representative distribution system sites and three maximum residence time locations (MRTLs) were analyzed for HAAs and THMs (Williams et al. 1994). The MRTLs exhibited high THM levels but the HAA levels were unexpectedly low. In the summer and fall of 1993, the HAA levels at these MRTLs were basically nondetectable. This unexpected decrease in HAA levels continued through the summers of 1994 and 1995 (Williams et al. 1994 and 1995). Similar decreases in HAA concentrations were also observed in a large number of utilities, such as Norfolk/Virginia Beach, Portsmouth, Philadelphia, City of Laval, and Ontario (LeBel et al. 1995). In the case of the Virginia systems, an analysis of the water quality parameters generated two similar features of the MRTLs: low levels of free chlorine and high HPC (Williams et al. 1994), which lead to the concept of possible bacterial degradation of the HAAs within the distribution system. This prompted a closer examination of the biofilm's bacterial constituents.

Initial studies indicated that the bacteria in the biofilm were capable of using Cl_2AA as an energy source for growth. Two different bacteria were isolated from water samples taken from the MRTL, a white bacterium and a distinctly yellow bacterium. The white bacterium was found to be Gram negative, citrate positive, oxidase positive, urease positive, and was found to be non-fermentative. The yellow bacterium tested Gram negative, oxidase positive, and was also found to be non-fermentative (Williams et al. 1994). Subsequent studies found that crude enzyme fractions isolated from both the yellow and white bacterium were effective in degrading HAAs (Williams et al. 1998). It was later shown that the white bacterium was capable of degrading 93 percent of six of the HAAs in under 12 hours, and it was able to completely degrade all of the HAAs, except Cl_3AA, within 24 hours. The yellow bacterium could successfully degrade the HAAs but at a somewhat slower rate (48 hours). In the preliminary studies with both bacteria, the degradation of the HAAs was accompanied by a detectable decrease in pH and an increase in inorganic chloride levels. Further information on this topic is presented in Appendix A, which addresses the microbial degradation of HAAs and other DBPs.

N-nitrosodimethylamine

Information about the fate of NDMA in full-scale distribution systems is very limited. Studies suggested that NDMA concentrations may increase in distribution systems with increasing water age. Andrews and Taguchi (2000) reported that even though UV irradiation is effective at reducing NDMA concentrations to below detection limits, NDMA may continue to form in distribution systems supplied with UV-irradiated water (10~32 ng/L). The elevated levels of nitrite present in the particular system that was described might have caused the observed increases in NDMA levels. Soroushian et al. (2001) and Liang et al. (2000) observed that the addition of hydrogen peroxide slightly improved NDMA destruction by UV irradiation and substantially limited NDMA reformation in the presence of chlorine.

Wilczak et al. (2002) noted that NDMA concentrations can be reduced in the distribution system by lowering the cationic polymer dose at the treatment plant, adding a small amount of flocculant aid at the flocculator effluent instead of the rapid mix and limiting water age in the distribution system. Other studies have also found amine-based polyelectrolytes used in drinking water treatment behave as NDMA precursors (Child et al. 1996, Najm and Trussell 2001, Loveland et al. 2001, Kohut and Andrews 2002).

A survey of 24 water utilities in North America found that NDMA formation occurred at treatment plants that used chlorine or chloramines for disinfection (Barrett et al. 2003, Valentine et al. 2004) and that distribution system samples contained higher levels of NDMA than plant effluent samples (except where effluent water was blended with other water prior to distribution). NDMA concentrations increased with water age for those utilities that had detectable NDMA; although, most samples remained below the Ontario drinking water standard of 9 ng/L and the California action level of 10 ng/L. Results showed that 50 percent of the data were below 2 ng/L for chloraminated drinking water distribution system samples and below 1 ng/L for chlorinated drinking water distribution systems. There were insufficient data to determine the effect of distribution pipe material, if any; however, the use of chloramines in the distribution system appeared to cause slightly higher NDMA levels than the use of free chlorine.

Influence of Seasons and Water Temperature on Disinfection By-Products in the Distribution System

As discussed earlier, the concentration and composition of NOM and the chlorine dose stability and demand vary with seasons and water temperature, which in turn affects DBP formation and fate. Several researchers studied the fate of DBPs in the distribution system during warm and cold seasons and observed higher concentrations of THMs, HANs, HKs, chloropicrin, and HAAs in warm waters (Chen and Weisel 1998, Singer 1999a, Stevens et al. 1976, Huck et al. 1987, Krasner et al. 1989).

In a full-scale distribution system fed with chloraminated water and a free chlorine residual of 0.5 mg/L Cl_2 (ammonia was added to convert all but approximately 0.5 mg/L of the free chlorine to chloramine), Chen and Weisel (1998) measured increases in THM concentrations with increasing water age but decreases in HAA concentrations. These changes occurred faster (closer to the treatment plant) in warm water (>10°C) than in colder water (≤10°C).

When studying the fate of DOX in two distribution systems fed by treatment plants using free chlorine as the final disinfectant, Baribeau et al. (2001) observed that in the presence of free chlorine, DOX concentrations generally increased in the distribution system with increasing water age in warm and cold water. In lukewarm water (8-14°C), levels remained stable. In the absence of measurable chlorine residuals, DOX concentrations decreased in warm and lukewarm water; in cold water, the decrease in DOX levels was much less pronounced. The lower DOX decay in colder water was also confirmed during bench-scale experiments (Baribeau et al. 2001). To explain this trend, the authors suggested that two phenomena occur simultaneously in the distribution system: DBP formation in the presence of free chlorine and DBP decay. In the presence of chlorine and depending on the chlorine dose applied, the organic matter present, the water temperature, and other factors, an increase in DOX concentration is observed when DBP formation exceeds DBP decay (example in the summer when the chlorine doses are high). In the absence of a chlorine residual, the decrease in DOX already formed predominates.

When examining the influence of water temperature in distribution systems, it is important to remember the buffering effect of the ground around the pipes. Several authors observed increases in water temperature with increasing water age in distribution systems (Rizet et al. 1984; Barthe et al. 1991; Coallier et al. 1991; Desjardins et al. 1991; Lavoie 1991; Capellier, Picoche, and Deguin 1992). The opposite can be observed in the summer: a decrease in temperature can be measured with increasing water age.

Influence of pH on Disinfection By-Products in the Distribution System

Several authors have observed various trends of pH variations in distribution systems. In a pilot distribution system made of cement-lined cast iron pipes with a three-day residence time, Meyer (1992) measured increases in pH from 7.5 to 8.1. At the POE to this pilot, the average chlorine residual was 1.6 mg/L, and temperature was 16.5°C. Lavoie (1991) did not measure any significant pH variation with increasing water age in a distribution system consisting mainly of cement or cast iron pipes, but Haas et al. (1983) observed decreases in pH with increasing water age in two New Hampshire distribution systems. Capellier, Picoche, and Deguin (1992) noticed higher pH values in the distribution system than at the plant effluent with almost no differences between the various sampling locations within the network.

Snoeyink (1994) mentioned that pH may fluctuate by more than 2 units depending on distribution system conditions. He attributed a pH increase to the following phenomena: Kuck's mechanism (oxidation of iron into Fe^{2+} and hydroxide ion, OH^-), dissolution of pipe cement (leading to Ca^{2+} and OH^-), denitrification (transformation of nitrate into nitrogen gas and carbon dioxide), and/or cathodic corrosion. A pH decrease can be associated with an increase in carbon dioxide concentration, nitrification (transformation of ammonia into nitrite and/or nitrate along with hydrogen ion), and/or anodic corrosion.

When studying water quality in two distribution systems, Baribeau et al. (2001) conducted bench-scale experiments during which treatment plant effluent water (containing DBPs) was incubated at two pH levels (7.2 and 9.8) and two temperatures (4 and 20°C), in absence of residual disinfectant. Following a 24-hour incubation period, DOX were measured. Results showed much greater DOX decay at pH 9.8 (24 percent at 4°C and 43 percent at 20°C), than at pH 7.2 (4.4 percent at 4°C and 15 percent at 20°C). In the same conditions (absence of chlorine residual) in two distribution systems, however, DOX levels did not follow this trend. DOX concentrations usually decreased in absence of chlorine residual in the distribution system; whereas, the pH did not follow any specific trend. Only small pH variations of 0.2 to 0.6 pH units were observed between the entrance and the end of the network (minimum and maximum values of 7.3 and 9.0, respectively). Baribeau et al. (2001) concluded that pH by itself does not explain the change in DOX concentrations observed in the distribution system; a conclusion also reached by Meyer (1992).

Influence of Pipe Material on Disinfection By-Products

Disinfectant residual may be consumed by pipe material or corrosion by-products. Because the standard redox potential of Cl^-/HOCl is higher than that of Fe^{2+}/Fe^{3+}, HOCl may be transformed into chloride in the presence of Fe^{2+}, which is then oxidized into Fe^{3+} (Kiéné 1994). This author observed that iron reacts preferably with chlorine instead of oxygen, leading to a significant chlorine demand.

Rossman et al. (2001) compared the rate of THM4 and HAA6 production in a simulated pipe environment (an unlined ductile iron pipe with a contact time of 24 hours) to that observed for the same water held in glass bottles. The pipe simulator was operated at room temperature and at two pH values, 7.3 and 8.0. Results indicated that although the chlorine decay rate measured in the pipe was an order of magnitude higher than in the bottle (Figure 2.10a), there was no decrease in the concentration of HAA produced (the rate of HAA6 production in the pipe was similar to that measured in the bottle; Figure 2.10b) and THM levels actually increased by an average of 15 percent (Figure 2.10c). Separate tests confirmed that this increase in THM concentrations was due to a reservoir of organic precursor materials associated with deposits on the pipe walls. As for DBP speciation, the distribution of THM species in the pipe was similar as that observed in the bottle after 24 hours of contact. At earlier times; however, there was a shift toward a higher fraction of $CHCl_3$ produced in the pipe as compared to the bottle. Although HAA6 concentrations were similar, there was a tendency for the pipe to produce more Cl_2AA and less Cl_3AA than the bottle as contact time progressed.

a) Chlorine decay curve

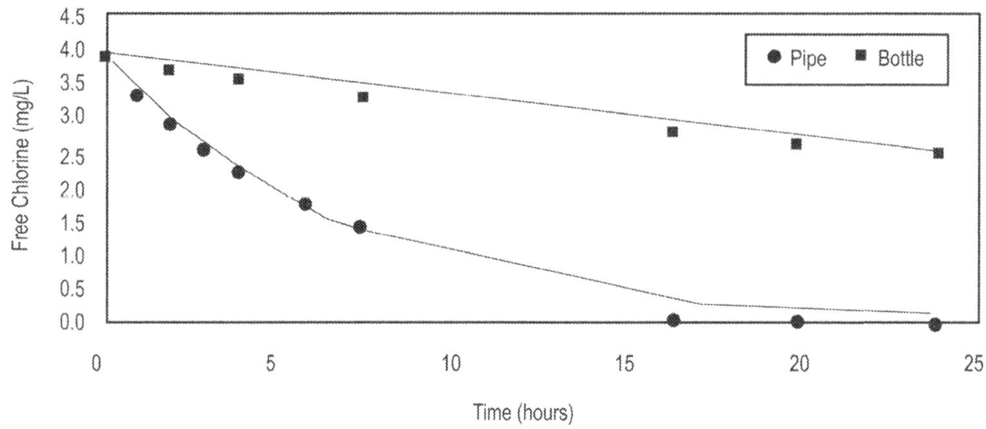

b) THM4 production curve

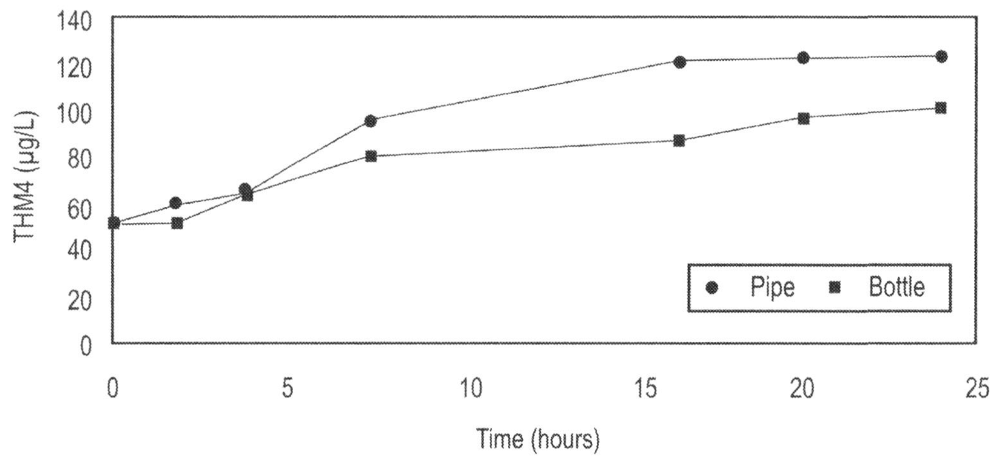

c) HAA6 production curve

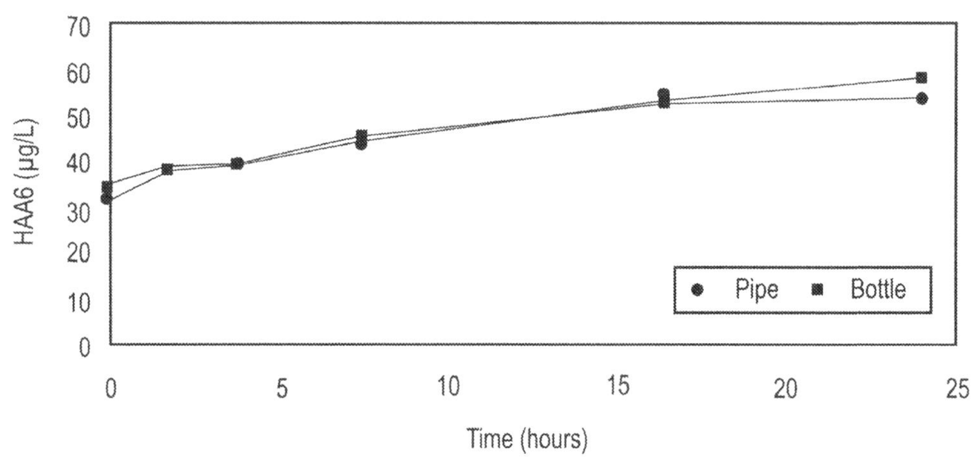

Source: Reprinted from *Water Research,* 35(14):3483-3489, Rossman, L.A., R.A. Brown, P.C. Singer, and J.R. Nuckols, DBP Formation Kinetics in a Simulated Distribution System, 2001, with permission from Elsevier.

Figure 2.10 Chlorine decay curve, THM4 production curve and HAA6 production curve

Baribeau et al. (2001) presented results of free chlorine decay and DOX concentrations in two distribution systems. Samples were collected in large-diameter (main) pipes (concrete pipes) and small-diameter pipes (unlined gray iron with signs of corrosion), and the results were compared with laboratory batch incubations. The rate of free chlorine decay was estimated by dividing the slope of the linear sections of the free chlorine curves by the total surface area of pipe and by the water age. Results indicated that in warm water, chlorine demand of the biofilm was much lower than that of the corrosion by-products, 0.16×10^{-5} and 96×10^{-5} $mg \bullet L^{-1} \bullet m^{-2} \bullet h^{-1}$, respectively. DOX samples collected during the same study indicated higher concentrations in the batch incubations than in the distribution system, but no significant difference between the DOX levels measured in the main pipes and the small-diameter pipes. The similarity in DOX concentrations measured in the main and small-diameter pipes (made of non-corrosive and corrosive materials, respectively) suggests that corrosion does not influence DOX. This suggestion was confirmed by bench-scale experiments during which effluent water collected at one of the treatment plants was incubated with corrosion by-products, in the absence of biofilm, at 4 and 20°C, and pH 7.2 and 9.8. Overall, corrosion by-products did not influence the DOX concentration over a 24-hour incubation period (Baribeau et al. 2001).

Influence of Pipe Diameter on Disinfection By-Products

To date, research aiming at evaluating the effect of distribution system pipe diameter has mainly focused on microbial growth rather than DBP fate. Vasconcelos et al. (1997) indicated that the chlorine reaction rate at the pipe wall is inversely related to the pipe diameter, and can be limited by the rate of chlorine transfer to the wall. This observation suggests that DBP fate may differ between small- and large-diameter pipes; although, Baribeau et al. (2001) did not observe significant differences between DOX concentrations measured in small- and large-diameter pipes of two free chlorinated distribution systems.

Influence of Hydraulic Conditions on Disinfection By-Products in the Distribution System

Water flow inside distribution system pipes affects water quality. High flows increase nutrient availability at the pipe walls, disinfectant transport, and biofilm detachment. On the other hand, water stagnation in pipes (i.e., in dead-ends) increases disinfectant consumption and increase biomass growth (LeChevallier 1990). In addition, distribution system configuration (cross connections, etc.) influences significantly the change in water quality inside the pipes (Olivieri et al. 1986, Smith et al. 1989a). Considering this, it is possible that hydraulic conditions affect DBP fate in distribution systems; however, information on these effects is not available.

CORRELATIONS BETWEEN DISINFECTION BY-PRODUCTS AND WATER QUALITY CHARACTERISTICS

Continuous effort has been devoted to find surrogates for DBP precursors in raw water and facilitate monitoring of their concentration during treatment and distribution. The most common surrogates are TOC, UV absorbance, and the specific UV absorbance (SUVA), which is defined by the UV absorbance at 254 nm (cm^{-1}) times 100 divided by the DOC concentration (mg/L).

According to Ben Amor (1988), chlorination by-products formation potential is better correlated with UV absorbance than with any other parameter. Similarly, laboratory studies conducted by Najm et al. (1994) showed better correlations between THM4 and UV absorbance (r of 0.93) than with TOC concentration (r of 0.67). Similar trends were observed for HAA5 with higher correlation with UV absorbance (r of 0.81) as compared to TOC concentration (r of 0.64). Najm et al. (1994) explained these observations by the fact that UV absorbance is a better representation of organic matter reactivity than other parameters including TOC. For these experiments, 16-hour SDS tests were conducted on four low-bromide waters incubated at room temperature and pH 8.0 in the presence of 0.5 to 1.0 mg/L free chlorine residual (measured at the end of the incubation period).

Singer et al. (1981), Ben Amor (1988), and Najm et al. (1994) observed that TOC and color can still be well correlated to DBP formation. In the raw water of 12 U.S. treatment plants, Singer and Chang (1989) observed good correlations between TOC concentration and THM formation potential (r of 0.83). Correlation coefficients increased to 0.89 between UV absorbance and THM formation potential. When analyzing results obtained at the plant effluent and in distribution systems, Singer and Chang (1989) obtained lower correlation coefficients because of varying chlorination conditions at the different treatment plants studied.

Amy, Chadik, and Chowdhury (1987) reported that THM4 are better correlated with the product of TOC and UV absorbance (r^2 of 0.93 after 2 hours of contact time, r^2 of 0.95 after 96 hours) than with any other parameters in natural waters under different conditions of temperature (10, 20, and 30°C), pH (ambient ± 1.5) and chlorine dosage (chlorine-to-non-purgeable organic carbon mass ratios of 0.5, 1.0, 3.0, and 5.0). Similarly, Edzwald, and Van Benschoten (1990) observed that SUVA is a good surrogate for water's humic content, and Krasner et al. (1996) showed that the higher the SUVA, the higher the THM formation potential.

Nguyen et al. (2002) examined DBP formation potential (DBPfp) using raw water from a reservoir located on the Verde River, Arizona. Correlations were determined between THMfp and DOC, THMfp and UV absorbance at 254 nm, HAA5fp and DOC, and HAA5fp and UV absorbance (Figure 2.11). Reasonable correlations were found between DOC and THMfp and between DOC and HAA5fp (r^2 of 0.79 and r^2 of 0.82, respectively). As DOC concentrations increased, DBPfp increased. UV absorbance was better correlated with HAA5fp (r^2 of 0.91) than with THMfp (r^2 of 0.69). This is especially significant because SUVA varied by a factor of two in some samples and indicated that UV absorbance was a better predictor of HAA precursors than was DOC. THMfp levels were consistently higher than HAA5fp for the Verde River water. In general, high HAA5 yields, expressed as HAA5/DOC, were observed during the initial surface runoffs (116 μg/mg during spring runoff and 102 μg/mg for summer monsoon rains). THM yields were also higher during spring runoff and summer monsoon rains (150 and 193 μg/mg respectively).

Meyer et al. (1993) observed that the THM concentration variations measured in a three-day pilot distribution system were correlated to chlorine demand measured in the system, which explained 88 percent of the increase in THM4 concentrations measured (the consumption of 1 mg/L Cl_2 led to 19.8 μg/L THM4). DOC was also an important parameter that explained 46 percent of the THM4 concentrations formed (each mg/L C of DOC formed 8.5 μg/L THM4). At the POE to this pilot, the average chlorine residual was 1.6 mg/L, temperature was 16.5°C, and pH was 7.5.

When analyzing the ICR data, Obolensky et al. (2002) calculated mean HAA5 concentrations of 4.5, 22.5, 32.8, and 39.2 μg/L for successively higher TOC concentration ranges of <1, 1 to 3, 3 to 5, and >5 mg/L, respectively (Figure 2.12). The concentration of the fully chlorinated HAA6 species (Cl_2AA and Cl_3AA) increased most strongly with TOC concentration. Similar patterns were observed for HAA9.

Several researchers have attempted to model DBP formation and fate in water. Because of the complex nature of DBP precursors and their reactions with disinfectants, models for quantification of DBPs have mainly been developed using empirical approaches. Empirical models involve statistical analysis of data derived from controlled laboratory or field experiments to develop functional relationships. One of the drawbacks of these empirical models is that they do not perform very well under conditions other than those from which they were developed. Most of the time, they do not consider the effect of the distribution system and thus cannot predict DBP formation in this environment. For this reason, efforts have been undertaken to develop mechanistic models, which are expected to be applicable under a much broader range of conditions, for predicting DBP concentrations.

More recently, some researchers have investigated the use of artificial neural networks (ANN) for the water industry. According to Zhang and Stanley (1997), ANN modeling presents the following advantages: (1) no mathematical algorithm is required because the network learns from the sample data and generates a black-box-type relationship; (2) ANN modeling is fast and flexible; (3) ANN can handle nonlinear relationships well because of its inherent non-linear data structure and computational process; (4) ANN tend to be fault tolerant because their data structure is loosely organized and there is no boundary limit on the input parameters; and (5) the water industry already has extensive databases that can be directly used. Disadvantages to ANN modeling have also been identified as follows: (1) many researchers consider the developed models to be of a black-box type as ANNs do not yield explicit mathematical formulae (Harvey and Harvey 1998); (2) little is known about the applicability of the models to data that lies outside the domain on which the models were trained; (3) each modeler may incorporate different modeling techniques as no set protocol exists for developing ANN models; and (4) the ANN technique is data intensive and is best suited to problems where large data sets already exist (Zhang and Stanley 1997). Current research efforts aim at eliminating or reducing the effects of these disadvantages. Results obtained appear promising and additional applications are being considered.

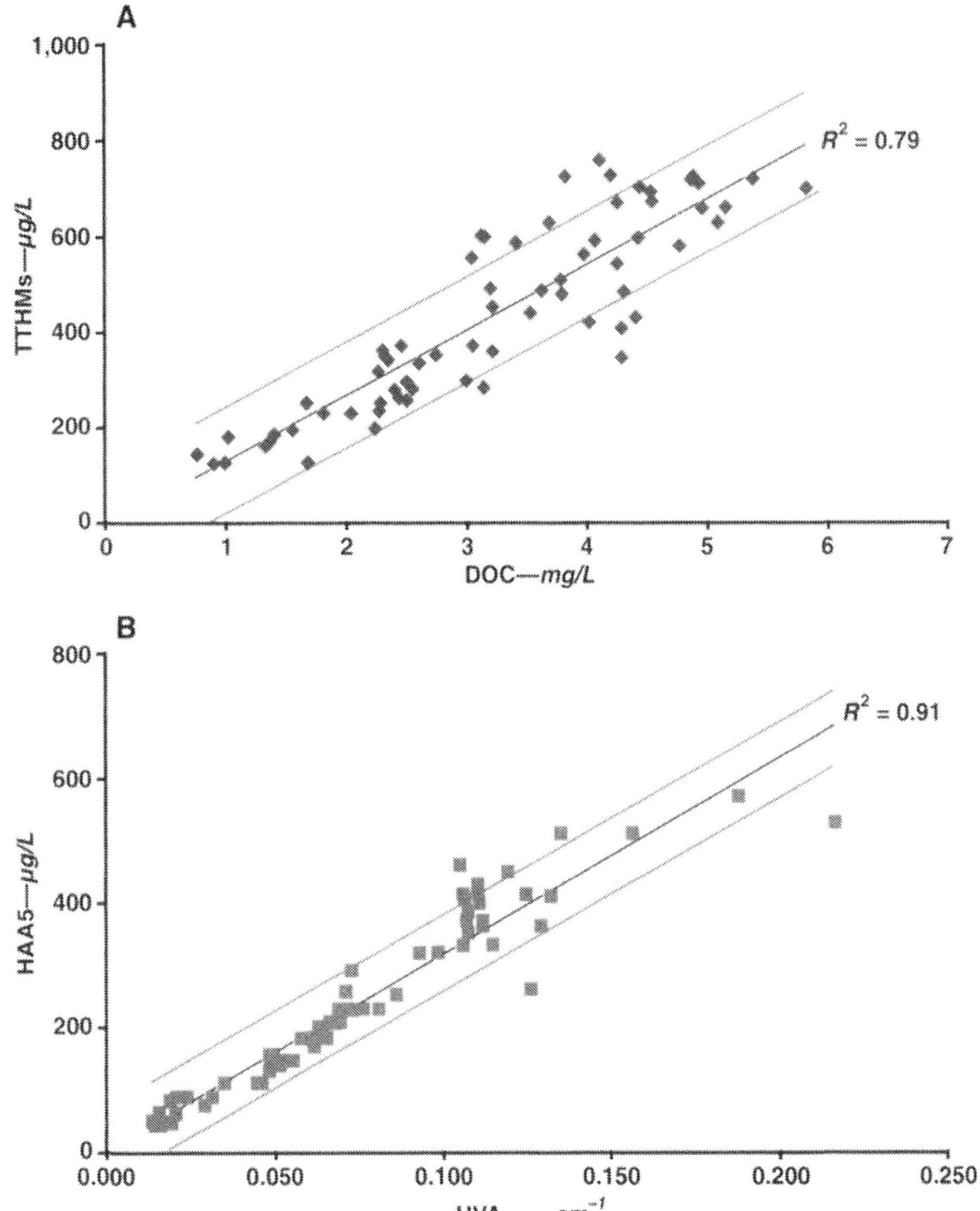

The upper and lower lines (lighter) represent the 95% confidence limit. TTHMs—total trihalomethanes, DOC—dissolved organic carbon, HAA5—sum of five haloacetic acids, UVA_{254}—ultraviolet absorbance at 254 nm

Source: Nguyen et al. 2002. DOC and DBP Precursors in Western U.S. Watersheds and Reservoirs. *Jour. AWWA,* 94(5):98-112. Published with permission from AWWA.

Figure 2.11 Correlations between THMs and DOC (Graph A), and HAA5 and UV absorbance at 254 nm (Graph B) using simple linear regression

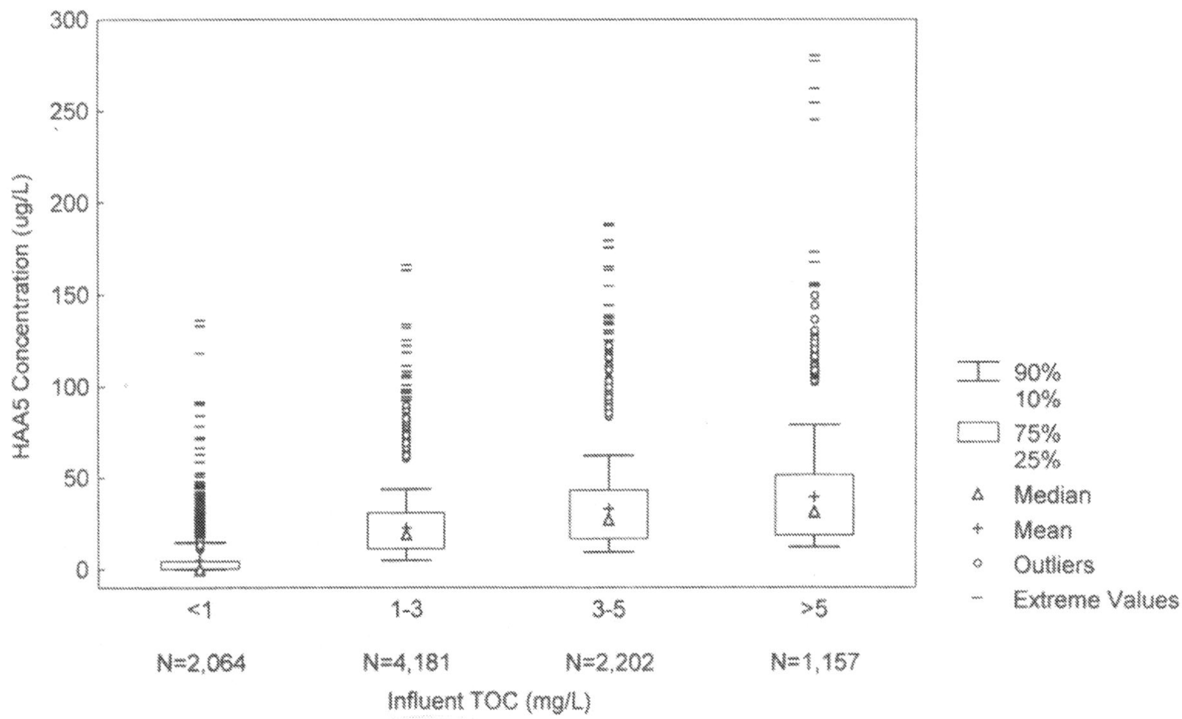

Source: Obolensky et al. (2002)

Figure 2.12 Distribution of HAA5 concentrations as a function of influent TOC level (all finished water and distribution system samples for six quarters)

CONCLUSIONS

The following conclusions can be drawn from this literature review:

- All of the DBPs examined in this study (i.e., THM4, HAA9, and NDMA) have shown adverse health effects in animals, including carcinogenic, reproductive, and developmental effects. THMs and HAAs constitute the majority of the DBPs formed in most chlorinated waters. The occurrence of NDMA in drinking water is still being examined. Studies to date suggest that its occurrence is intermittent and while most occurrences were at trace levels, high concentrations were also observed.
- Several rules have addressed halogenated DBPs including the Total THM Rule, the ICR, and the Stage 1 DBPR. The upcoming Stage 2 DBPR and its IDSE are also expected to significantly impact the drinking water industry. There are no federal or state MCL for NDMA; although, some state health agencies have adopted action levels to protect public health.
- DBP formation depends on a variety of water quality and disinfection factors including concentration and chemical properties of the precursors; water temperature; pH; bromide concentration; disinfectant type, dose and residual; and contact time.

The fate of these factors in distribution systems also influences DBP concentrations and speciation.

- Humic substances are characterized by high SUVA values and have higher DBP formation potentials than the non-humic fractions.
- Bromide tends to increase the overall mass concentration of DBPs formed and shift DBP speciation towards the bromine-substituted DBPs.
- Reaction kinetics are slower at lower temperature. Thus, relatively higher concentrations of reaction intermediates and lower levels of terminal products, such as THMs, are likely to occur in colder water.
- THM concentrations increase with increasing chlorination pH. HAA concentrations tend to decrease with increasing pH; although, some HAA species are relatively insensitive to changes in pH while others are influenced by chlorination conditions.
- In general, chloramines form halogenated by-products to a considerably lower extent than free chlorine. Greater concentrations of halogenated DBPs are observed with increasing chlorine dosages. The distribution of DBPs and speciation within each DBP class are also affected by the chlorine dose.
- DBPs continue to form in the presence of a disinfectant residual. As such, greater contact times lead to higher concentrations of terminal DBPs such as THMs, but lower concentrations of reaction intermediates such as HANs and HKs.

- In the distribution system, DBP formation and degradation occur, leading to changes in spatial and temporal DBP concentrations and speciation.
  - THM concentrations generally increase with increasing water age or distance from the treatment plant.
  - HAA concentrations also increase with increasing water age in distribution systems; although, decreases have been observed at maximum water age locations. Decreases in HAA concentrations were mainly associated with biodegradation of the X_2AAs.
  - Information about the fate of NDMA in distribution systems is very limited; although, it suggests that NDMA concentrations may increase with increasing water age.
- Several hypotheses have been proposed to explain DBP degradation in distribution systems, including chemical degradation (instability of the compounds themselves), DBP hydrolysis at specific pH values, interaction with corrosion by-products, adsorption in the biofilm matrix, biodegradation, and bioaccumulation by microorganisms.
- Continuous effort has been devoted to find surrogates for DBP precursors. The most common surrogates include TOC, UV absorbance, and SUVA.

CHAPTER 3
MATERIALS AND METHODS

This chapter presents the materials and methodologies used to assess the spatial and temporal variability of DBPs in five full-scale distribution systems.

SELECTION OF WATER UTILITIES

It is understood that treatment processes affect subsequent water quality and chemical and biological reactions observed in the distribution system; however, in order to limit the number of factors investigated and focus this project on the distribution system, utilities were selected based on specific finished water characteristics; not on the treatment processes used. Nonetheless, it happened that all of the utilities selected use a conventional process, including coagulation, flocculation, sedimentation, filtration, and final disinfection. One of them also uses lime softening (Table 3.1).

Four selection conditions were mandatory and common to all systems selected:

- All utilities possessed a distribution system hydraulic model available for the determination of residence time (water age) throughout the network.
- Systems were simple and well characterized in order to maximize system control and limit the number of factors affecting the changes in water quality (i.e., limited number of pressure zones, limited number of inter-connections).
- Water sources of different quality could not be blended in the distribution system (e.g., no untreated groundwater mixed with treated surface water) to reduce system complexity and to better separate and identify fundamental causes of DBP fate and transport in distribution systems.
- Moderate to high level of assimilable or total organic carbon (AOC or TOC), such that DBP levels were moderate to high (THM4 >40 μg/L and HAA5 >20 μg/L).

The following criteria or inter-system variables were also considered:

- Secondary disinfectant type, free chlorine, or monochloramine.
- Presence of brominated DBPs in some sites.
- Variation in geographical locations to include different seasonal patterns and temperatures.
- Presence of storage reservoirs and booster chlorination stations in one system at least.
- Inclusion of a chloraminated site that switches to free chlorine periodically to limit nitrification in the distribution system.

Table 3.1
Water utilities selected for the full-scale distribution system monitoring

Utility	Source water	Pre-oxidant	Treatment processes	Final disinfectant	Plant capacity (mgd, m^3/d)	Population served in 2002 in area sampled	Finished water reservoirs	No. of pressure zones
Utility A	60% SW, 40% GW	Free chlorine	Lime-soda softening, CFS, recarbonation, filtration, chlorine	Free chlorine	9.12 mgd (34,523 m^3/d) capacity, 5.9 mgd (22,334 m^3/d) average delivery	16,730 customers (67,000 persons)	2 elevated, 1 ground level	4
Utility B	100% SW	Chlorine dioxide	CFS, chlorine, filtration, chlorine, lime	Free chlorine	90 mgd (340,687 m^3/d) capacity, 32 mgd (121,133 m^3/d) average delivery	189,032 persons	14	4
Utility C	95% SW, <5% GW	None	CFS, chlorine, filtration, chlorine, ammonia	Chloramines	1.13 mgd (4,278 m^3/d) average delivery at POE	16,900 persons	2	1
Utility D	100% SW	Free chlorine	CFS, filtration, chlorine, ammonia	Chloramines	Not applicable*	6,109 service connections (31,733 persons†)	None	1
Utility E[1]	100% SW	None	2 WTPs: CFS (alum), chlorine, filtration, chlorine, ammonia at POE (no polymers)	Chloramines periodically switching to free chlorine	WTP No.1: 30 mgd (113,562 m^3/d); WTP No.2: 22 mgd (83,279 m^3/d)	190,000 persons	4	2

*Purchase 100 percent of water distributed from a wholesaler
†As of 2003
SW: Surface water
GW: Ground water
CFS: Coagulation, Flocculation, Sedimentation
POE: Point of entry
WTP: Water Treatment Plant

The participating utilities were selected using these criteria, and their water quality characteristics are presented in Table 3.2. For confidentiality purposes, the participating utilities are not named in this report, but are referred as follows:

- Utility A uses free chlorine as final disinfectant and includes a storage reservoir and a booster chlorination station.
- Utility B uses free chlorine as final disinfectant.
- Utilities C and D use chloramines as final disinfectant.
- Utility E uses chloramines as final disinfectant, with a one-month switch to free chlorine once per year to limit nitrification occurrence.

Below is a detailed description of each utility selected.

Utility A – Free Chlorine

Utility A is located in the mid-western portion of the U.S. The raw water source of supply for Utility A is obtained from the Little Scioto River, the Scioto River and 16 groundwater wells. These sources are all blended for treatment at a water treatment plant, and the plant effluent is the only POE to the distribution system (there is no blending of sources in the distribution system). In general, the water used is approximately 60 percent surface water and 40 percent groundwater; although, these relative percentages change as surface water quality and availability changes. It is estimated that over the course of the study however, the ratio of surface water to ground water used was consistent.

The Little Scioto River is a tributary of the Scioto River, and it has a watershed area of approximately 86 square miles at the intake. In turn, the Scioto River is a tributary of the Ohio River, and it has a watershed area of approximately 409 square miles at the intake. Both watersheds are broad and flat, and they are subject to high runoff and stream flow during precipitation events. The combined raw water supply from the Scioto and Little Scioto Rivers contains low to moderate amounts of turbidity, moderate alkalinity, high hardness, and a slightly basic pH. Runoff from upstream agricultural areas, however, can raise nitrate, pesticide, and coliform levels significantly.

Table 3.2
Water quality at the participating utilities

Utility	Final disinfectant	Temperature (°C)	pH at POE	TOC at POE (mg C/L)	AOC at POE‡ (mg C/L)	Bromide (mg/L)	THM4 at POE‡ (µg/L)	% $CHCl_3$ of THM4³	HAA5 at POE‡ (µg/L)	% Cl(1-3) of HAA5‡
Utility A	Free chlorine	4 - 25	8.7 - 9.6	1.2 - 2.9	0.20 - 0.74	<0.02 - 0.03	13 - 55	56 - 87	5 - 33	100
Utility B	Free chlorine	8 - 31	8.7 - 9.2	<0.1 - 5.5	0.07 - 0.86	<0.1 - 0.5	39 - 76	51 - 82	17 - 60	65 - 92
Utility C	Chloramines	4 - 27	7.0 – 7.8	2.5 - 4.0	0.21 - 0.42	<0.01 - 0.03	43 - 88	11 - 91	26 - 80	38 - 94
Utility D	Chloramines	16 - 24	7.7 - 8.2	1.6 - 2.6	0.11 - 0.25	0.20 - 0.55	74 - 94	7.0 - 15	19 - 27	42 - 80
Utility E	Chloramines periodically switching to free chlorine	5 – 35*	6.9 - 9.4†	1.6 - 2.6*	0.19 - 0.52¹	Not available	37 - 64† 38 - 74§	85 - 91† 85 - 92§	49 - 58† 55 - 90§	100† 100§

*Observed throughout the year
†Observed during the chloramination period
‡Data obtained from this study
§Observed during the free chlorination period
POE: Point of entry

The 16 groundwater wells range in size from 10- to 14-inch (25- to 35-cm) diameter with depths from 140 to 270 feet (43 to 82 m). The total rated pumping capacity is 8.7 mgd (32,933 m^3/d). Eleven of the wells were installed between the years 1913 and 1926, and the remaining five wells were installed between 1950 and 1956.

Treatment processes at the plant include lime-soda softening (to reduce the hardness of the ground water from 760-1,235 mg/L $CaCO_3$ to 135-160 mg/L), clarification, recarbonation, filtration, disinfection, and fluoridation. The capacity of the treatment facility is 9.12 mgd (34,523 m^3/d). In 2001, the average day system delivery was approximately 5.9 mgd (22,334 m^3/d), the maximum daily flow was 7.1 mgd (26,876 m^3/d), and the minimum daily flow was 4.5 mgd (17,034 m^3/d).

All water delivered by Utility A to the distribution system is processed at the treatment plant and is pumped from the plant to the Main Service pressure zone. The Main Service pressure zone covers the majority of Utility A's distribution system. In 1994, Utility A extended a pipeline approximately 5 miles (8 km) east from the Main Service pressure zone to serve a village of approximately 300 customers. The distribution system was extended further eastward in 1995 to serve approximately 50 customers in the east-end village. In 1997, a 12-inch (30-cm) main of approximately 1.5 miles (2.4 km) was extended south to serve approximately 100 customers. Fifty additional new customers were supplied along the pipeline route. More recently this main was extended approximately 6.5 miles (10 km) further south to serve approximately 600 to 700 customers. In general, pipe sizes range from 4 to 18 inch (10 to 46 cm) in diameter. Typical pipe materials include ductile iron, cast iron (lined and unlined), polyvinyl chloride (PVC), asbestos cement, and galvanized iron. The most prevalent size is 6-inch (15-cm) diameter, which comprises the major portion of the distribution grid. Three distribution storage tanks are located within the service area (Table 3.3).

Over the course of the study, there were no significant changes at the water treatment plant or in the distribution system that may have affected distribution system hydraulic conditions or water quality. The only change that occurred was the elimination of the chlorine addition at settled water at the plant in 2003.

Table 3.3
Distribution storage information for Utility A

Tank	Pressure zone	Type	Capacity (MG, m^3/d)	Pumping?	Pipe diameter to and from the storage
Reservoir No.1	Main	Elevated tank	0.30 MG, 1,136 m^3/d	None	16 inch (40 cm) ductile iron
Reservoir and Booster No.2	Main	Ground level reservoir	2.20 MG 8,238 m^3/d	Yes	8.5 inch (22 cm)
Reservoir No.3	Main	Elevated tank	0.10 MG 379 m^3/d	None	16 inch (40 cm) ductile iron; and 6 inch (15 cm) undesignated

Utility B – Free Chlorine

Utility B is located in the southeast portion of the U.S. Source water used for drinking water is 100 percent surface water. Utility B's water treatment process consists of the following: chlorine dioxide is injected in the raw water as it is pumped to the plant. Raw water is stored at the plant before being moved in the treatment processes. As the water moves in the flash mixer, alum is added for coagulation. Depending on raw water conditions, powdered activated carbon (PAC), lime, chlorine dioxide, chlorine, or polymer may also be added. Following coagulation, the water enters flocculation chambers and sedimentation basins. Settled water is chlorinated and then filtered on 30-inch (76 cm) beds. More free chlorine is added to reach a residual of 1.9 to 2.0 mg/L Cl_2 in fall and winter, and 2.0 to 2.2 mg/L in spring and summer. Fluoride and lime are also added to the filtered water. Finished water is stored in a clearwell prior to distribution.

Utility B has only one POE to the distribution system. The distribution system has 14 storage reservoirs, one booster pump station, and no booster chlorination station. The system has 3 pressure zones: one fed by gravity, and two pressured zones. The studied portion of the distribution system consisted of one of the pressured zones without booster pump station.

Over the course of the study, there were no significant changes at the water treatment plant or in the distribution system that may have affected distribution system hydraulic conditions or water quality.

Utility C – Chloramines

Utility C is located in the southeastern portion of the U.S. This system was selected as it has historically observed decreases in THM and HAA concentrations at remote locations of the distribution system. Utility C purchases 100 percent of its water from a wholesaler located 1 to 2 miles from the POE and does not boost disinfectant at the POE. The distribution system is looped and consists of approximately 46 miles of mains.

Water used for drinking water is primarily surface water from the Chicahomony River, occasionally supplemented by small amounts of reverse-osmosis treated groundwater (<5 percent). Blending occurs in the finished water clearwell. The water treatment plant uses a conventional process without pre-oxidation and consisting of coagulation, flocculation, sedimentation, free chlorine disinfection, filtration, and addition of chlorine, lime and ammonia. The contact time between additions of chlorine and ammonia is practically instantaneous, and the target chlorine-to-ammonia-N ratio is 4.5:1 to 5:1. The capacity of the treatment facility is 30 to 45 mgd (113,562 to 170,344 m^3/d).

Finished drinking water is supplied to the studied portion of the distribution system through one 14-inch (35-cm) main. There is one booster pump station that provides pressure to fill three storage reservoirs: two of the reservoirs have capacities of 0.20 MG (757 m^3/d), and the third reservoir has a capacity of 0.50 MG (1,893 m^3/d). There is no booster chlorination station within the studied portion of the distribution system and no periodic switch to free chlorine.

Over the course of the study, there were no significant changes at the water treatment plant that may have affected distribution system hydraulic conditions or water quality. In the distribution system, a pipe replacement program aiming at replacing old cast iron pipes with new PVC pipes may have affected sampling Site No. 15 (See Table 3.9 below) during the samplings conducted on January 7, February 25, and April 29, 2003.

Utility D – Chloramines

Utility D is located in the western portion of the U.S. This utility distributes treated surface water from a wholesaler, supplemented with their own well water. In the studied section of the distribution system, however, Utility D purchases 100 percent of its distributed water. In this portion of the distribution system, the water may come from either one of three possible water treatment plants; the water can also be a blend of the three plant effluents occasionally. These plants are fed by either one of two surface water sources or a blend of both sources. They use a conventional treatment consisting of coagulation using alum in a rapid mixer, flocculation, sedimentation, free chlorination, and filtration on anthracite and sand, followed by final free chlorination and addition of ammonia to form chloramines. The target chlorine-to-ammonia-N ratio is 5:1. These plants have a combined capacity of 1,670 mgd (6,321,638 m^3/d). They do not routinely switch to free chlorine. There were no significant changes at the water treatment plants nor in the distribution system during the study period.

Approximately 53 percent of Utility D's pipes are unlined cast iron and steel pipes installed during the 1930's, 40's, and 50's. The remainder of the pipes are made of ductile iron, PVC, and transite. The section of Utility D's distribution system that was chosen for sampling was a residential area fed by a 14-inch (35-cm) main. There are two possible POEs to this area; although, they both distribute water of similar quality. The studied portion of the distribution system contained no storage reservoir or booster chlorination station.

Over the course of the study, there were no significant changes at the water treatment plants or in the distribution system that may have affected distribution system hydraulic conditions or water quality.

Utility E – Chloramines Periodically Switching to Free Chlorine

Utility E is located on the east coast. Its population is served by two treatment facilities, which produce an annual average total flow of 32 mgd (121,133 m^3/d; plant capacities are 30 and 22 mgd, 113,563 and 83,279 m^3/d). Both treatment facilities use the same blended raw water from two lakes, which is treated by conventional surface water treatment processes of coagulation with ferric chloride (rapid mix), flocculation, sedimentation, and filtration. Free chlorine is added to the treated water after sedimentation and again after filtration, and ammonia is added at the POE to the distribution system to form chloramines. The target chlorine-to-ammonia-N ratio entering the distribution system is between 3.5:1 and 4:1 on a weight basis for both water treatment plants. Every year, Utility E switches from chloramines to free chlorine for a four-week period during the month of March. Zinc orthophosphate is added at the plant effluent for corrosion control in the distribution system with an average phosphate residual from 0.30 to 0.35 mg/L. Fluoride and a base are also added at the plant effluent.

The distribution system consists of two pressure zones with four elevated tanks located throughout the city. The downtown area contains lined and unlined cast iron pipes, while the surrounding area has newer ductile iron pipes.

Over the course of the study, there were no significant changes at the water treatment plant or in the distribution system that may have affected distribution system hydraulic conditions or water quality.

SELECTION OF SAMPLING LOCATIONS

Within each distribution system studied, the sampling locations were carefully selected to provide as complete an understanding as possible of the fate of DBPs considering the complexity of full-scale distribution systems. The sampling locations were based on the following intra-system variables:

- Various water ages from close to the treatment plant to dead-ends. The utilities' hydraulic models were used for this evaluation.
- Consideration of the pipe material and diameter (further information is presented in Appendix D).
- Preferably sampling locations within the same pressure zone.
- Areas with low or non-measurable disinfectant residuals (spatially and temporally).
- Critical areas of the network, such as locations with coliform or nitrification occurrences, or presence of DBP species that might not be present throughout the entire system.
- The sampling locations selected in Utility A distribution system specifically addressed the effect of a booster chlorination station and a storage reservoir (further information is presented in Appendix D).
- Areas outside the zone of influence of potential major interference from external routine activities such as autoflushers.

The following is a summary of the procedure used to determine the water ages in each distribution system and select the sampling locations.

Utility A – Free Chlorine

WaterCAD 4.5 hydraulic model was used to model the distribution system and estimate water age. The model included most 6-inch (15-cm) diameter and larger mains in the system; some 4-inch (10-cm) mains were included as needed to provide connectivity. Pump curves for existing pumps were used in the model where available. Metered consumption data were used to distribute demands among the model node points. The model was last calibrated in 1997 using November 21, 1995, peak hour and minimum hour demands and the hydrant flow test field data. Tracer studies were, however, not conducted. For this research project, the following modifications were made to the model:

- Addition of a main south to the south-end village;
- Use of maximum, minimum, and average daily flows for the year 2001; and
- Modification for use as extended period simulations (EPS).

To obtain the range of water ages at various locations of the distribution system, the model was run for 14 days (336 hours), and the water ages were obtained at the minimum, average, and maximum flow hours on the 13th day (i.e., Hours 315, 327, and 320, respectively). The model was run for three different flow scenarios, corresponding to the minimum, average, and maximum daily flows of the year 2001. Water ages were then obtained for the following five scenarios, which were representative of the scenarios encountered during the samplings conducted for this study (Table 3.4):

- Minimum daily flow, minimum flow hour (4.5 mgd, 17,034 m^3/d, at 3:00 AM)
- Average daily flow (5.9 mgd, 22,334 m^3/d, at 3:00 PM) and:
  - Minimum flow hour
  - Average flow hour
  - Maximum flow hour
- Maximum daily flow, maximum flow hour (7.1 mgd, 26,876 m^3/d, at 8:00 AM).

The diurnal pattern used for all model simulations (minimum, average and maximum daily flow simulations) was based on the maximum daily flow scenario developed:

- Maximum hour: 8:00 – 9:00 AM (factor of 1.197 relative to the average flow hour)
- Average hour: 3:00 – 4:00 PM (factor of 0.993 relative to the average flow hour)
- Minimum hour: 3:00 – 4:00 PM (factor of 0.554 relative to the average flow hour).

Thirteen sampling sites were selected to represent a cross-section of residence times found in the distribution system, as described in Table 3.5. Figure 3.1 illustrates the distribution of estimated water age obtained at each sampling locations, with the exception of the sites located around the reservoirs that are not shown on this figure. The rationale for selection of these sites is explained below.

Table 3.4

Summary of plant production during the samplings conducted at Utility A

Sampling date	Plant production (mgd, m^3/d)	System delivery (mgd, m^3/d)
October 16, 2002	5.410 mgd (20,479 m^3/d)	5.320 mgd (20,138 m^3/d)
December 3, 2002	5.450 mgd (20,630 m^3/d)	Not available
January 7, 2003	5.665 mgd (21,444 m^3/d)	5.530 mgd (20,933 m^3/d)
February 25, 2003	6.005 mgd (22,731 m^3/d)	5.870 mgd (22,220 m^3/d)
June 10, 2003	5.895 mgd (22,315 m^3/d)	Not available
August 12, 2003	5.750 mgd (21,766 m^3/d)	Not available
September 23, 2003	7.100 mgd (26,876 m^3/d)	5.230 mgd (19,798 m^3/d)

Table 3.5
Sample site characteristics for Utility A

Site identification number	Estimated water age, based on average daily flow (hours)			Pipe diameter (in, cm)	Pipe material
	Minimum	Average	Maximum		
EP001	0.0	0.0	0.0	16 in, 40 cm	CI
RD002	5.9	8.4	7.3	10 in, 25 cm	Assumed CI
RD004	18.6	18.0	13.7	8 in, 20 cm	TR
RD005	18.2	8.2	10.9	16 in, 40 cm	DI
RD007	4.7	5.5	4.8	8 in, 20 cm	CI
RD013	5.3	5.2	4.8	8 in, 20 cm	TR
RD018	3.2	3.5	3.2	16 in, 40 cm	CI
RD025	313	323	317	8 in, 20 cm	DI

CI:Cast Iron
DI:Ductile Iron
TR:Transite

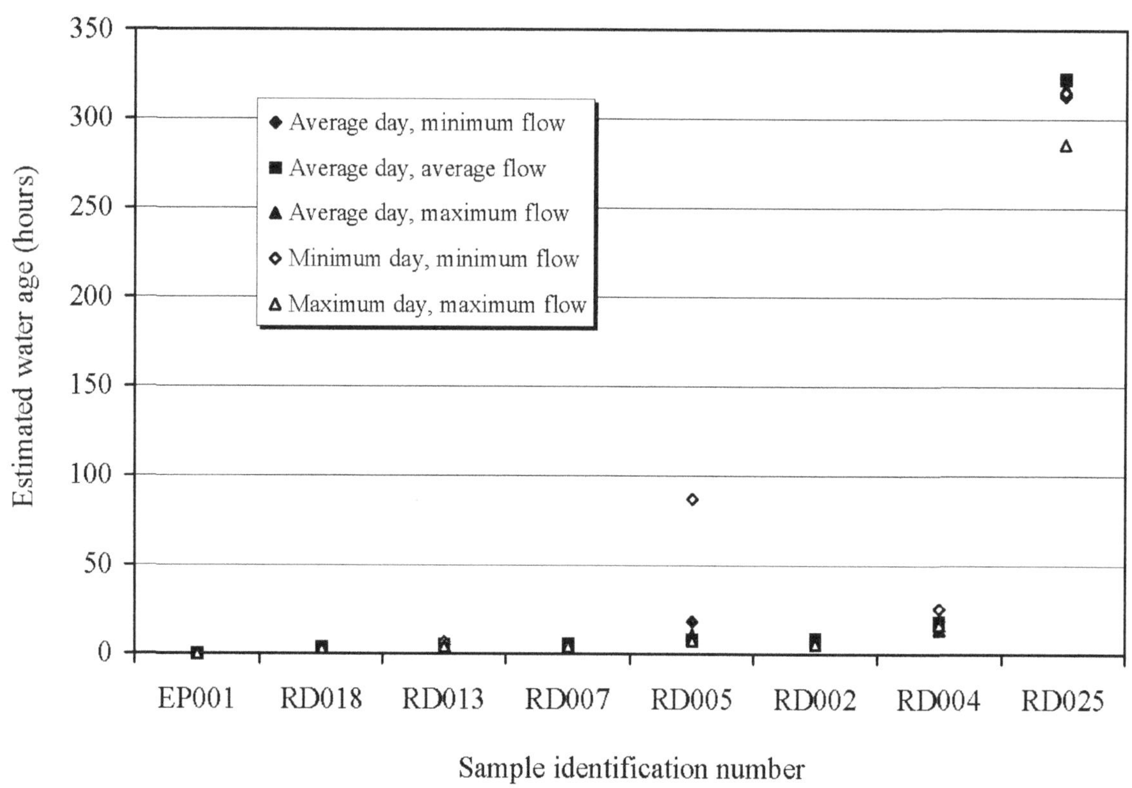

Figure 3.1 Distribution of estimated water age obtained at Utility A sampling locations (sampling sites located around Reservoirs No.2 and 3 are not shown)

The four regulatory compliance DBP sampling locations and the water treatment plant effluent were selected first because of the existence of historical DBP data at these sites. It was then possible to select sampling locations with very high water age (up to 323 hours) in the long transmission main extending eastward. To assess the effects of water age and also the effect of the booster chlorination station and storage reservoirs, additional sampling sites were selected around Reservoirs No. 2 and 3 where a DBP compliance sampling site already exists.

The remaining three sites were selected from the existing chlorine and/or coliform sampling locations based on similar estimated water ages, but differing pipe materials (cast iron, ductile iron, or transite) and diameters (from 8 to 16 inch, 20 to 40 cm). All samples were collected from regularly used sampling taps inside buildings or dedicated sampling taps.

Utility B – Free Chlorine

Utility B utilizes the HAESTAD (Waterbury, Conn.) hydraulic model for its distribution system and to estimate water age. The model was calibrated in 2002 using 3 years of flow data from the distribution system. Average water ages were obtained by running the model at average flow and demand conditions (32 mgd, 121,133 m^3/d). Tracer studies were not conducted at Utility B. Table 3.6 summarizes the plant production and water delivery to the studied portion of the distribution system. This table suggests that the plant production during the various samplings conducted was similar to that use to estimate the water age at the sampling locations, except in the winter when plant production was lower.

Fourteen sampling locations were selected in the northern pressure zone of the system. This zone is fed by a single main coming out of the water treatment plant effluent and does not have pumping stations or storage reservoirs. Average residence times of the sampling locations selected ranged from 0 to 47 hours. In addition to residence time, pipe material and pipe diameter were included in the criteria for selecting sampling locations. Ten sampling sites were located along a 30-inch (75-cm) diameter ductile iron transmission main and four sampling sites were located on 4-inch (10-cm) diameter PVC pipes. Table 3.7 presents water age results obtained from the model, as well as pipe material and diameter data for each sampling location. Figure 3.2 illustrates the distribution of estimated water age obtained at each sampling location. All samples were collected from fire hydrants.

Utility C – Chloramines

Utility C utilizes H2ONET to model their distribution system and obtain water age. The model network was generated from the map used for the unidirectional flushing program and was last updated in 2002. It was calibrated with field measurements of system pressures and tank hydraulic grade lines against values predicted by the model. The model was further verified using historical water quality data (disinfectant residual). Table 3.8 summarizes the plant production during the samplings conducted for this study.

Table 3.6
Summary of plant production during the samplings conducted at Utility B

Sampling date	Plant production (mgd, m^3/d)	Delivery in study section of the distribution system (MG, m^3)
October 15, 2002	27.4 mgd (103,720 m^3/d)	6.1 MG (23,091 m^3/d)
November 18, 2002	25.7 mgd (97,285 m^3/d)	5.5 MG (20,820 m^3/d)
January 7, 2003	24.9 mgd (94,257 m^3/d)	5.2 MG (19,684 m^3/d)
February 25, 2003	24.7 mgd (93,500 m^3/d)	5.6 MG (21,198 m^3/d)
April 29, 2003	28.9 mgd (109,398 m^3/d)	7.6 MG (28,769 m^3/d)
June 10, 2003	31.9 mgd (120,755 m^3/d)	7.8 MG (29,526 m^3/d)
August 12, 2003	28.5 mgd (107,884 m^3/d)	7.5 MG (28,391 m^3/d)
October 21, 2003	33.6 mgd (127,190 m^3/d)	7.3 MG (27,634 m^3/d)

Table 3.7
Sample site characteristics for Utility B

Site identification number	Estimated average water age (hours)	Pipe diameter (in, cm)	Pipe material
308	4.7	30 in, 75 cm	DI
975	15.4	30 in, 75 cm	DI
883	11.3	30 in, 75 cm	DI
979	19.3	30 in, 75 cm	DI
208	24.2	30 in, 75 cm	DI
232	28.8	30 in, 75 cm	DI
100	33.8	30 in, 75 cm	DI
116	38.7	30 in, 75 cm	DI
251	43.5	30 in, 75 cm	DI
227	47.1	30 in, 75 cm	DI
X-3	8.8	4 in, 10 cm	PVC
R-1	27.0	4 in, 10 cm	PVC
D-6	33.8	4 in, 10 cm	PVC
X-11	11.7	4 in, 10 cm	PVC

DI:Ductile iron
PVC:Polyvinyl chloride

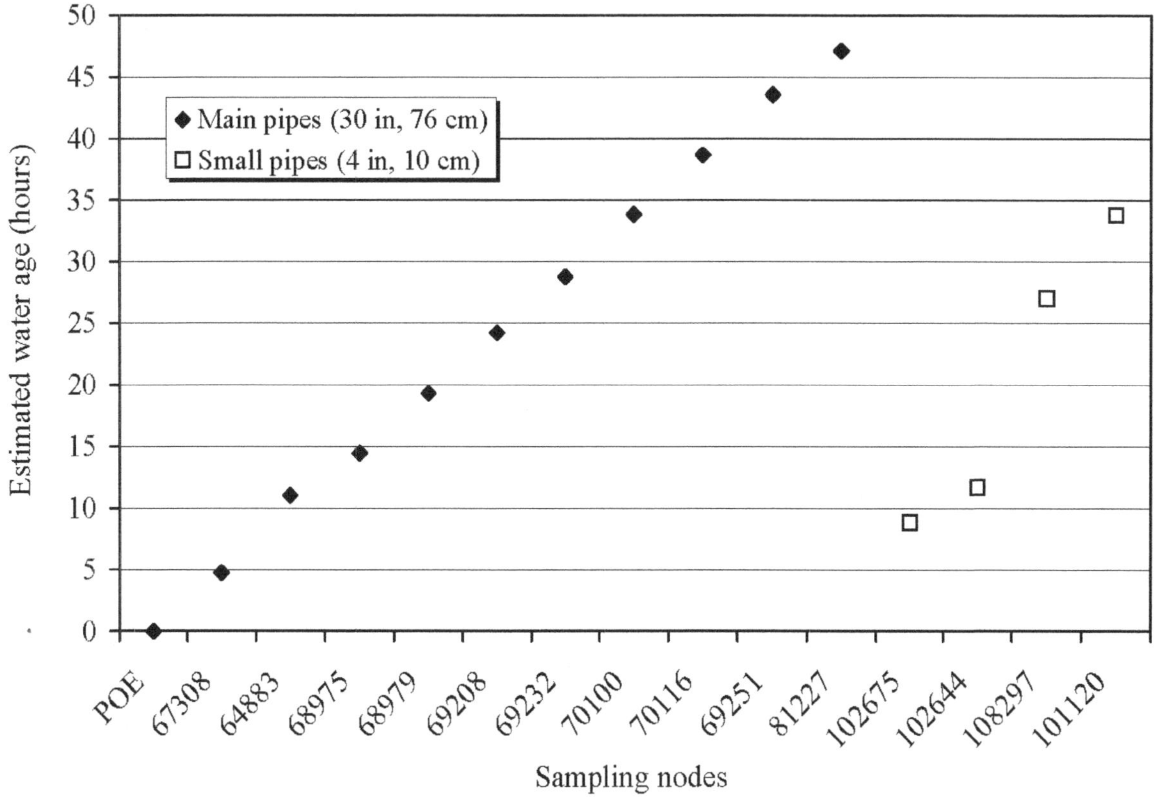

Figure 3.2 Distribution of estimated water age obtained at Utility B sampling locations

Table 3.8
Summary of plant production during the samplings conducted at Utility C

Sampling date	Plant production (mgd, m^3/d)	Minimum flow at the plant	Maximum flow at the plant
October 15, 2002	25.2 mgd (95,392 m^3/d)	22.9 mgd (86,686 m^3/d) at 4h00	31.1 mgd (117,726 m^3/d) at 1h00
November 18, 2002	28.1 mgd (106,370 m^3/d)	22.8 mgd (86,307 m^3/d) at 17h00	31.1 mgd (117,726 m^3/d) at 2h00
January 7, 2003	28.8 mgd (109,020 m^3/d)	26.1 mgd (98,799 m^3/d)	31.9 mgd (120,755 m^3/d)
February 25, 2003	21.9 mgd (82,901 m^3/d)	20.4 mgd (77,222 m^3/d) at 14h00	26.0 mgd (98,421 m^3/d) at 20h00
April 29, 2003	23.4 mgd (88,579 m^3/d)	19.8 mgd (74,951 m^3/d) at 10h00	26.6 mgd (100,692 m^3/d) at 23h00
June 11, 2003	26.3 mgd (99,556 m^3/d)	24.7 mgd (93,500 m^3/d) at 4h00	27.2 mgd (102,963 m^3/d) at 14h00
August 12, 2003	28.9 mgd (109,398 m^3/d)	26.9 mgd (101,828 m^3/d) at 22h00	32.5 mgd (123,026 m^3/d) at 11h00
October 21, 2003	27.2 mgd (102,963 m^3/d)	24.1 mgd (91,228 m^3/d) at 4h00	30.2 mgd (114,319 m^3/d) at 17h00

Thirteen sampling locations were selected for this study, with the following characteristics:

- Water treatment plant effluent
- POE to the distribution system section studied
- Locations with various water ages ranging from 0 to >120 hours
- Locations with similar water age and varying pipe diameters from 6 to 14 inch (15 to 35 cm)
- Locations known to experience some HAA degradation
- The pipe material was the same (unlined cast iron) for all sampling locations.

To calculate the water ages, H2ONET uses a numerical scheme called the Discrete Volume Element Method (DVEM). Within each hydraulic time step when flows are constant, DVEM computes a shorter water quality time step and divides each pipe into a number of completely mixed volume segments. Results include an estimate of the water age at each junction or node (at the head of each pipe segment) and an estimate of water flows leaving the node.

Water ages were estimated from EPS at the following conditions under a typical yearly average flow rate condition:

- Average day demand (1.16 mgd, 4,391 m^3/d)
- Minimum day demand
- Maximum (2.32 mgd, 8,782 m^3/d) day demand (simulations also included consideration for fire flow demands).

The characteristics of each sampling location and distribution of estimated water age are presented in Table 3.9 and Figure 3.3, respectively. All samples were collected from outside sampling taps, except for the location with an estimated water age of 24 hours. At this location, samples were collected inside a building from a sink tap with the aerator removed.

There are currently seven autoflushers located on the distribution system to avoid non-compliance episodes with the Total Coliform Rule. These sites provide an automatic daily release of water at specific locations in the distribution system for a total of 30 minutes to 6 hours at each site. Each autoflusher runs once or twice per day, either in the middle of the night (between 1 and 5 AM) or late in the evening (8 PM to midnight). Each autoflusher flow rate is 185 gpm (700 L/min). Although all sites selected were located as far as possible from the zones of influence of the autoflushers, the status of each autoflusher was documented during each sampling. Auto flushing occurred on October 15, 2002, June 11, August 12, and October 21, 2003.

Table 3.9
Sample site characteristics for Utility C

Site identification number	Estimated water age (hours)			Pipe diameter (in, cm)
	Minimum	Average	Maximum	
WTP effluent	0	0	0	Not applicable
1	1	6	9	14 in, 35 cm
2	4	24	41	8 in, 20 cm
3	3	32	45	12 in, 30 cm
4	38	43	64	8 in, 20 cm
7	19	39	51	8 in, 20 cm
8	10	39	52	8 in, 20 cm
9	7	32	49	6 in, 15 cm
10	2	21	32	8 in, 20 cm
12	3	42	57	10 in, 25 cm
13	2	11	16	8 in, 20 cm
20	3	8	11	6 in, 15 cm
15[1]	Not available	>120	Not available	8 in, 20 cm
Tap 1721*	Not available	24	Not available	10 in, 25 cm

*Estimated water ages based on the location of these sites and the map of water age contours
WTP:Water treatment plant

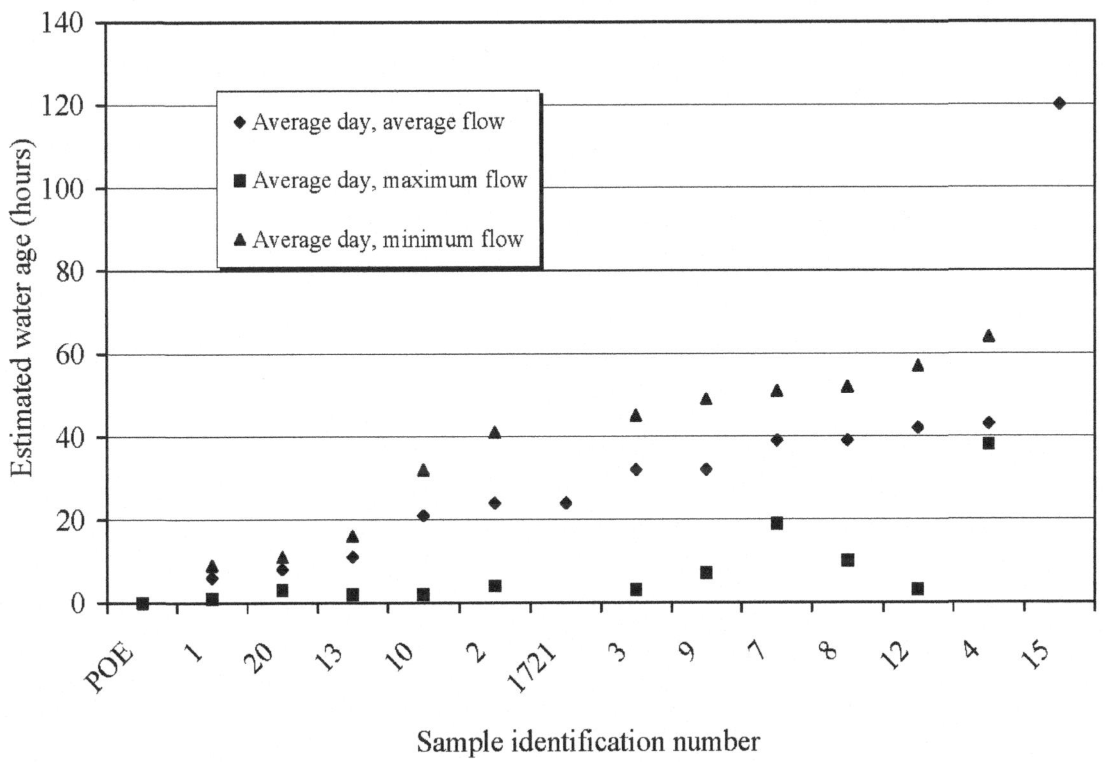

Figure 3.3 Distribution of estimated water age obtained at Utility C sampling locations

Utility D – Chloramines

One section of Utility D distribution system was selected for this project. The section is simple, with no influence from storage reservoirs nor booster chlorination stations. Utility D employs H2ONET as a hydraulic model. The model was last updated and calibrated in 1998, and tracer studies were not conducted. In the studied section of the distribution system, all pipes are included in the model. Water ages were obtained by running the model for 72 hours at average flow and demand, and the evaluation was then focused on the following conditions:

- Minimum hourly flows (Hour 51, corresponding to 3:00 AM)
- Average hourly flows (Hour 69, corresponding to 9:00 PM)
- Maximum hourly flows (Hour 67, corresponding to 7:00 PM).

Flow distributions and directions were also calculated throughout the system. Water ages at the minimum, average, and maximum hourly flow demands were estimated for all nodes. A number of sampling locations were chosen to represent a cross-section of water ages, pipe diameter, and pipe materials (cast iron, PVC, or asbestos cement) found in the studied portion of the distribution system. Diurnal distributions of water age were plotted over the 72-hour simulation period.

Over the course of the study, the studied portion of Utility D's distribution system had to supplement their typical POE with a second POE. Both POEs pulled water of similar quality from the same transmission main. The addition of a second POE changed the flow patterns in the distribution system, which changed the water age at each sampling location. All samplings conducted were grouped into four distinct patterns (Table 3.10), and the hydraulic model was re-run to produce four different hydraulic scenarios, each dependent on the relative flow at the POE into the portion of the distribution system studied. At each sampling location, a weighted water age was calculated by the model based on the proportion of water that came from each source. The average water ages calculated for each hydraulic scenario were used in the subsequent analysis of sampling data. Table 3.11 presents the average water ages for each sampling location under each of the four hydraulic scenarios. Figure 3.4 illustrates the distribution of estimated water age at each sampling location. For this figure, the four hydraulic scenarios were combined together.

Table 3.10
Hydraulic scenarios for the samplings conducted at Utility D

		Typical POE (production and flow in ft^3)				Second POE (production and flow in ft^3)			
Scenarios	Sampling date	Production	Day high	Day low	Day max	Production	Day high	Day low	Day max
No.1	November 18, 2002	54,700	3.8	0.0	20	0.0	0.0	0.0	0.0
No.1	January 7, 2003	72,700	4.2	0.0	20	0.0	0.0	0.0	0.0
No.2	April 8, 2003	106,700	4.4	0.0	20	229,700	2.3	1.5	5.0
No.3	June 10, 2003	0.0	0.0	0.0	0.0	130,100	2.4	1.0	5.0
No.4	July 22, 2003	129,000	2.0	0.6	20	148,700	3.4	1.2	5.0
No.4	September 23, 2003	111,400	1.9	0.2	20	97,800	2.2	0.6	5.0

Table 3.11
Average residence time for all hydraulic scenarios for Utility D

Site identification	Estimated average water age (hours)						Pipe diameter (in, cm)	Pipe material
	November 18, 2002	January 7, 2003	April 8, 2003	June 10, 2003	July 22, 2003	September 23, 2003		
No.1 - POE	0.0	0.0	0.0	0.0	0.0	0.0		
No.2	1.6	1.6	2.6	5.2	2.0	2.0	12 in, 30 cm	AC
No.3	2.4	2.4	2.7	6.3	2.5	2.5	8 in, 20 cm	CI
No.4	3.3	3.3	3.8	7.5	3.3	3.3	8 in, 20 cm	CI
No.5	5.7	5.7	3.7	8.3	4.0	4.0	8 in, 20 cm	CI
No.6	10.5	10.5	6.3	12.6	7.0	7.0	6 in, 15 cm	CI
No.7	18.4	18.4	13.1	15.2	10.9	10.9	6 in, 15 cm	CI
No.8-A	50.0	50.0					8 in, 20 cm	CI
No.8-B			7.9	15.4	8.8	8.8	6 in, 15 cm	CI
No.9-A	15.3	15.3					8 in, 20 cm	CI
No.9-B			13.5	21.1	12.7	12.7	6 in, 15 cm	CI
No.10-A	8.3	8.3					8 in, 20 cm	PVC
No.10-B			2.2	3.8	3.4	3.4	6 in, 15 cm	AC
No.11-A	37.2	37.2					6 in, 15 cm	AC
No.11-B			1.5				6 in, 15 cm	CI
No.11-C				19.6	10.0	10.0	8 in, 20 cm	CI
No.12-A	11.1	11.1					6 in, 15 cm	CI
No.12-B - POE					0	0		

AC:Asbestos cement
CI:Cast iron
PVC:Polyvinyl chloride

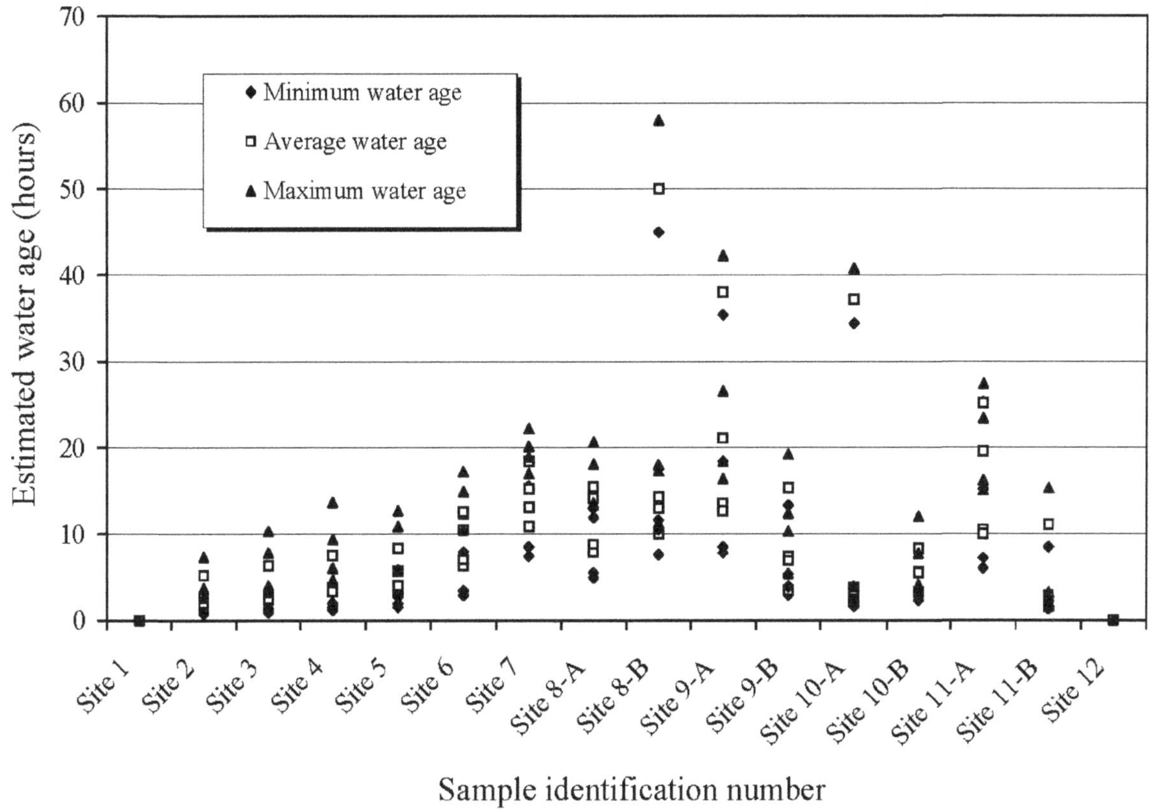

Figure 3.4 Distribution of estimated water age obtained at Utility D sampling locations. All four hydraulic scenarios were pooled together

Samples collected from the main POE were taken directly from the pipe, via a corporation stop. Samples collected from the alternate POE were taken at the meter box, via a 5 to 6-ft (1.5- to 1.8-m) service line.

Flushing is routinely performed in Utility D's distribution system. Flushing information was documented and possible effects on sampling results are summarized in Table 3.12.

Table 3.12
Flushing information during the samplings conducted in Utility D's distribution system

Sampling date	Flushing date	Sampling location affected	Duration of flushing (minutes)
January 7, 2003	January 6, 2003	Site No.9A	10
June 10, 2003	June 9, 2003	Site No.10B	20
July 22, 2003	July 21, 2003	Sites No.7, 9B and 11C	10

Utility E – Chloramines Periodically Switching to Free Chlorine

The hydraulic model used (EPANET Version 2.0) was calibrated using flow measurements in the distribution system during the summer of 2002. The pump curves, speed, and efficiency were input to the model for the pumps located at both treatment plants. The storage tanks in the distribution system had minimum and maximum volume levels and were modeled as completely mixed reactors.

For this study, the model was run for 168 hours (one week) at the average daily demand measured in the summer of 2000, with diurnal variations in demand. The range of water ages was determined at minimum, maximum, and average daily demand of the 4th and 5th day of this average week. Utility E's distribution system is fed by two water treatment plants, receiving a constant blend of the two sources over the course of a year. A summary of the treatment plant production on the days of sampling is presented in Table 3.13.

Fourteen sampling locations including the two POE locations were selected with average water age ranging from 6 to 74 hours, pipe diameters ranging from 2 to 16 inch (5 to 40 cm), and cast iron and ductile iron pipe materials. In addition, known problem areas within the distribution system were sampled, including both pressure zones. Sites 1 and 2 correspond to the POEs at the treatment plants, while the remaining sites are located throughout the distribution system.

Low, medium, and high categorical limits were established for the pipe diameter and water age ranges identified for each of the sampling locations. The pipe diameter stratification ranges were small (2 to 4 inch; 5 to 10 cm), medium (6 to 8 inch; 15-20 cm), and large (≥12 inch; 30 cm). Water age categories were established for low (≤12 hours), medium (32 to 45 hours), and high (≥64 hours) to ensure that no overlap in water age categories occurred due to changes in water production and demand. Each sampling location water age was entirely defined by the categorical limits established regardless of treated water flow and system-wide demand, and the minimum and maximum residence times of any sampling location did not extend into the adjacent categorical definition. Table 3.14 summarizes the characteristics of each sampling location in terms of water age, pipe diameter, and material. Figure 3.5 illustrates the distribution of water age at each sampling location. All sampling taps were either outside taps from residential homes or inside taps in commercial or public facilities. POE samples were collected from taps at the water treatment plants.

Because the distribution system was fed by two treatment plants, a modified tracer study was conducted to determine the source water for each sampling location. The Site 1 treatment plant used ferric chloride as a coagulant at the time the tracer study was conducted, while the Site 2 treatment plant used alum. This difference in coagulants results in significantly different chloride ion concentrations leaving each treatment plant; therefore, chloride was used as a tracer to determine the source water for each sampling location. Results showed that Site 2 appears to be the dominant source water for sample Sites 3, 4, and 7, while all of the remaining sampling locations receive predominantly Site 1 source water (Nichols 2004). The predominant source water for each sampling location determined by the tracer study was assumed to remain valid for the entire four-month sampling period.

Table 3.13
Summary of plant production during the samplings conducted at Utility E

	Sampling date	Plant production at POE No.1 (mgd, m^3/d)	Plant production at POE No.2 (mgd)
Before switching to free chlorine:	February 11, 2003	16.00 mgd (60,567 m^3/d)	7.00 mgd (26,498 m^3/d)
	February 28, 2003	17.25 mgd (65,298 m^3/d)	6.80 mgd (25,741 m^3/d)
During free chlorination period:	March 11, 2003	13.00 mgd (49,210 m^3/d)	6.85 mgd (25,930 m^3/d)
	March 28, 2003	22.5-27.0 mgd (85,172-102,206 m^3/d)	7.00 mgd (26,498 m^3/d)
After return to chloramines:	April 29, 2003	22.30 mgd (84,415 m^3/d)	14.50 mgd (54,888 m^3/d)
	May 20, 2003	19.15 mgd (72,491 m^3/d)	7.20 mgd (27,255 m^3/d)

Table 3.14
Characteristics of sample collection sites in Utility E

Sample identification	Estimated water age (hours)			Pipe diameter (in, cm)	Pipe material
	Maximum	Average	Minimum		
Site 1	0	0	0		
Site 2	0	0	0		
Site 3	76.6	73.7	69.5	12 in, 30 cm	DI
Site 4	31.9-32.8	33.3-38.2	32-36.7	12 in, 30 cm	DI
Site 5	38	42.7	40.5	4 in, 10 cm	DI
Site 6	44	41.7	45	6 in, 15 cm	CI
Site 7	10.2	8.9	8.6	2 in, 5 cm	CI
Site 8	6.3	7.1	6.4	4 in, 10 cm	DI
Site 9	7.9	6.8	9.5	8 in, 20 cm	DI
Site 10	5.2-5.5	5.6-6	5.4-5.8	16 in, 40 cm	DI
Site 11	69.35	80.9	67.6	12 in, 30 cm	DI
Site 12	74	91.8	69.7	8 in, 20 cm	DI
Site 13	11.7	11.4	11.8	8 in, 20 cm	CI
Site 14	62.5	76.5	63.6	4 in, 10 cm	DI

DI:Ductile iron
CI:Cast iron

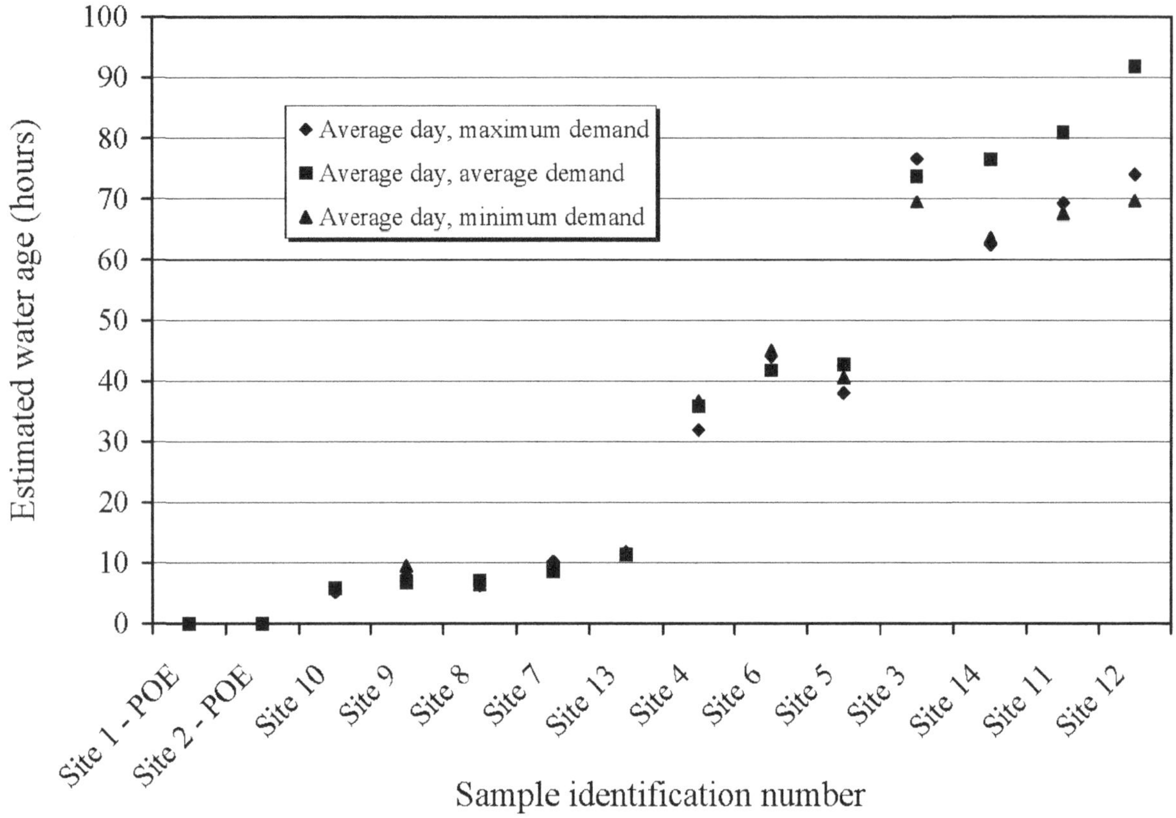

Figure 3.5 Distribution of estimated water age obtained at Utility E sampling locations

The distribution system is impacted by two storage tanks in the low-pressure zone and one storage tank in the high-pressure zone. An additional storage tank in the low-pressure zone was under construction when this study was conducted. All of the storage tanks were operated on a fill and draw basis, and the tanks were regularly turned over to ensure that low water ages were maintained. All four of the storage tanks were included in the model used to estimate water age at all sampling locations. Sampling of the water entering or leaving the storage tanks, however, was not conducted. Therefore, the water age at many of the sampling locations within the distribution system may have been affected by water being released from the storage tanks.

SAMPLING STRATEGY

To address the spatial and seasonal variability of DBPs in distribution systems, samplings were conducted regularly for one full year. Originally, one fall sampling, three winter samplings, one spring sampling, and three summer samplings were planned at each utility. However following data collection, some of the samplings needed to be re-categorized as presented in Table 3.15.

Utility E uses chloramines as final disinfectant, and switches to free chlorine every year on March 1 for a one-month period to decrease the occurrence of nitrification. The effect of such change on DBP formation and fate in the distribution system was evaluated. Two samplings were conducted just before the switch to free chlorine, and two samplings during the free chlorination period (Table 3.16). After the switch to free chlorine, a period of 11 days was planned prior to the first sample collection to allow the distribution system to recover from the transition. Two samplings were scheduled following the return to chloramines on April 1. A period of 4 weeks was planned prior to the first sampling following conversion to chloramines to allow distribution system water quality to recover from the transition.

Table 3.15
Monitoring program to address the seasonal variability of DBPs in distribution systems

Season	Date of sampling	Water temperature at the POE (°C, °F)	Utilities and final disinfectant
Fall	October 16, 2002	13.3°C (56°F)	Utility A – Free chlorine
	November 18, 2002	17°C (63°F)	Utility B – Free chlorine
	November 18, 2002	16.0°C (61°F)	Utility C – Chloramines
Winter	December 3, 2002	5.6°C (42°F)	Utility A – Free chlorine
	January 7, 2003	6.1°C (43°F)	
	February 25, 2003	4.0°C (39°F)	
	January 7, 2003	11°C (52°F)	Utility B – Free chlorine
	February 25, 2003	12°C (54°F)	
	January 7, 2003	10.5°C (51°F)	Utility C – Chloramines
	February 25, 2003	8.0°C (46°F)	
	January 7, 2003	15.6°C (60°F)	Utility D – Chloramines
	April 8, 2003	18.3°C (64°F)	
	June 10, 2003	17.8°C (64°F)	
Spring	June 10, 2003	15.0°C (59°F)	Utility A – Free chlorine
Summer	August 12, 2003	20.1°C (68°F)	Utility A – Free chlorine
	September 23, 2003	16.6°C (62°F)	
	October 15, 2002	26°C (79°F)	Utility B – Free chlorine
	April 29, 2003	22°C (72°F)	
	June 10, 2003	26°C (79°F)	
	August 12, 2003	27°C (81°F)	
	October 21, 2003	22°C (72°F)	
	October 15, 2002	22.0°C (72°F)	Utility C – Chloramines
	April 29, 2003	20.0°C (68°F)	
	June 11, 2003	25.0°C (77°F)	
	August 12, 2003	27.0°C (81°F)	
	October 21, 2003	19.0°C (66°F)	
	November 18, 2002	21.1°C (70°F)	Utility D – Chloramines
	July 22, 2003	21.1°C (70°F)	
	September 23, 2003	21.1°C (70°F)	

Table 3.16
Monitoring program to examine the effect of a switch in final disinfectant (Utility E)

Date of sampling	Water temperature at the POEs (°C, °F)	Final disinfectant
Prior to switch:		
February 11, 2003	6.4°C (44°F) at POE 1 9.0°C (48°F) at POE 2	Chloramines
February 28, 2003	7.0°C (45°F) at POE 1 8.0°C (46°F) at POE 2	Chloramines
Free chlorine:		
March 11, 2003	9.9°C (50°F) at POE 1 10.2°C (50°F) at POE 2	Free chlorine
March 31, 2003	13.3°C (56°F) at POEs 1 and 2	Free chlorine
After switch:		
April 29, 2003	15.6°C (60°F) at POE 1 15.5°C (60°F) at POE 2	Chloramines
May 20, 2003	17.7°C (64°F) at POE 1 17.5°C (64°F) at POE 2	Chloramines

NDMA was measured in a chloraminated system (Utility C) because this DBP is mainly found in chloraminated water. NDMA was monitored in Utility E distribution system before, during, and after the switch to free chlorine to address the effect of a change in disinfectant on this DBP. NDMA was also measured in a free chlorinated system (Utility A) as a control.

Sampling occurred only during "normal" operation. If unusual circumstances such as fire or main break (leading to changes in distribution system water quality or hydraulic characteristics) were noticed, the sampling was postponed until normal conditions were re-established.

To ensure consistency and reproducibility, a standard operating procedure (SOP) for sample collection and shipping was provided to each utility before each sampling. Ice chests prepared by the analytical laboratories and containing all sample containers, ice packs (blue ice), sufficient packing material to prevent bottle breakage, instruction sheet, and a chain-of-custody form, were sent to the appropriate utilities prior to the sampling date.

On the morning of sampling, all vials were placed in coolers with frozen ice packs. At a given sampling location, any faucet aerators were removed, and the faucet, hydrant, or other water spigot was turned on and allowed to run for at least 5 minutes if it was not running continuously. Samples were collected at outside taps when at a residence and inside faucets for businesses and treatment facilities. After 5 minutes, chlorine, temperature, and pH were measured. All sampling containers (including caps) that did not contain a preservative or dechlorinating agent were rinsed with the water being collected. The containers and caps were not rinsed otherwise. All DBP vials were filled at an angle such that sample water ran down the sides of the vials and were filled headspace-free to prevent volatilization of DBPs. Replicate samples were collected whenever necessary. Samples were stored in coolers containing frozen ice packs. One field blank was also prepared at a randomly selected location for each analyte and at each utility during all samplings, as described in the Control section presented below.

Upon completion of sampling, all samples were shipped on the day of sampling by overnight carrier to the appropriate laboratory. Upon receipt, samples were stored at 4°C (39°F) until analyzed. THM, HAA, UV, and TOC samples were sent to the laboratories of Newport News Waterworks (NNWW), Virginia and the University of North Carolina at Chapel Hill (UNC). NDMA samples were sent to the University of Waterloo, Ont., and AOC and HPC samples were sent to American Water's Quality Control and Research Laboratory, Illinois (American). Analyses performed by each laboratory and utility are presented below.

ANALYTICAL PARAMETERS

The following DBPs were monitored throughout the study:

- All four THMs: chloroform ($CHCl_3$), bromodichloromethane ($CHBrCl_2$), dibromochloromethane ($CHBr_2Cl$), and bromoform ($CHBr_3$)
- All nine HAAs: monochloroacetic acid (ClAA), dichloroacetic acid (Cl_2AA), trichloroacetic acid (Cl_3AA), bromochloroacetic acid (BrClAA), bromodichloroacetic acid ($BrCl_2AA$), dibromochloroacetic acid (Br_2ClAA), monobromoacetic acid (BrAA), dibromoacetic acid (Br_2AA), and tribromoacetic acid (Br_3AA)
- NDMA.

The following general water quality parameters were also monitored:

- Free and total chlorine residuals
- Temperature and pH
- UV absorbance at 254 nm, TOC and AOC
- HPCs
- Ammonia, nitrate and nitrite at chloraminating utilities
- Bromide at the utility where it occurs.

Table 3.17 summarizes the analytical methods used during this study. Immediately after collection, all samples were kept at 4°C or on ice until analysis.

Table 3.17
Monitoring parameters, location of analysis and methods used

	Newport News Waterworks		University of North Carolina		American Water		Utilities	
Parameters	Method	Utilities	Method	Utilities	Method	Utilities	Method	Utilities
THM4	Purge and Trap, and GC; EPA Method 502.2 (USEPA, 1995)	B, C	Liquid-Liquid Extraction and GC/ECD (SM 6232 B)	A, D, E				
HAA9	Liquid-Liquid Extraction and GC; EPA Method 552.2 (USEPA, 1992)	B, C	Liquid-Liquid Extraction and GC/ECD (SM 6251 B with modifications; Brophy et al. 2000; Singer et al. 2001)	A, D, E				
NDMA								
Bromide	EPA 300.1 (USEPA, 1997b)	C			EPA 300.1 (USEPA, 1997b)	A	EPA 300.1 (USEPA, 1997b)	D
TOC	SM 5310 B with modifications	B, C	SM 5310 B	A, D, E				
UV absorbance at 254 nm	SM 5910 B	B, C	SM 5910 B with modifications	A, D, E				
AOC					LeChevallier et al (1993)	All utilities		
Free and total chlorine residual			Onsite (colorimetry, adapted from APHA, 1998)	E			Onsite (colorimetry, adapted from APHA, 1998)	A, B, C, D
Temperature			Onsite	E			Onsite	A, B, C, D

(Continued)

Table 3.17 (Continued)

	Newport News Waterworks		University of North Carolina		American Water		Utilities	
Parameters	Method	Utilities	Method	Utilities	Method	Utilities	Method	Utilities
pH			Onsite	E			Onsite	A, B, C, D
Ammonia	EPA Method 350.1 (USEPA, 1978b)*	C	EPA Method 350.3 (USEPA, 1979b)†	E			HACH Method 8155 for DR-890‡	D
Nitrite	EPA Method 353.2 (USEPA, 1978a)	C	EPA Method 300.0 (USEPA, 1993)	E			HACH Method 8507 for DR-890 (USEPA, 1979)	D
Nitrate	EPA Method 353.2 (USEPA, 1978a)	C	EPA Method 300.0 (USEPA, 1993)	E			EPA Method 300.0 (USEPA, 1993)	D
HPC	SM 9215 C (spread plate on R2A media, 7-day incubation at 25°C)	C			SM 9215 C (spread plate on R2A media, 7-day incubation at 25°C)	A, B, D, E		

*Free and total ammonia
†Free ammonia only
‡Total ammonia only (salicylate method adapted from *Clin. Chim. Acta.*, 14:403, 1966)
SM:Standard Method (APHA 1998)

Disinfection By-Products

Trihalomethanes

Samples were stored at 4°C (39°F) after collection before the extraction was conducted. THM4 were extracted via liquid-liquid solvent extraction based on USEPA Method 551.1 (USEPA 1995) at NNWW, or Standard Method 6232B (APHA 1998) with minor modifications at UNC. The methods used at each laboratory are detailed below.

Newport News Waterworks Laboratory. Pre-cleaned 40-mL amber borosilicate vials with Teflon-lined septa and screw caps were used once only and were similar to those used during the ICR monitoring period. Approximately 3 mg/vial of sodium thiosulfate ($Na_2S_2O_3$) was used as a dechlorinating agent.

NNWW laboratory, THM4 were analyzed using EPA Method 502.2 (USEPA 1995), which involves extracting THM species from the sample matrix by bubbling an inert gas (helium) through a 5-mL aqueous sample. The purged sample components are trapped in a tube containing a suitable sorbent material that is heated and backflushed with helium to desorb trap components onto a capillary gas chromatography (GC) column (Perkin Elmer; Shelton, Conn.). The column is temperature programmed to separate the components, which are detected and measured with a halogen specific detector and photoionization detector (PID) in series (9K PID Detector Autosystem Photoionization and 1000 Hall Detector Autosystem ELCD; Perkin Elmer; Shelton, Conn.).

Prior to the analysis of THM samples at NNWW laboratory, a Laboratory Reagent Blank (LRB) and a Laboratory Fortified Blank (LFB) at a concentration between 0.1 to 5 μg/L were analyzed and the results needed to be within 20 percent of accepted value for each THM species. After every 20th sample and at the end of each analytical batch, a LFB at a mid-calibration concentration was analyzed and the results needed to be within 20 percent of the accepted value. If the percent recovery for the LFB standard was out of the range of 80 to 120 percent, analyses were repeated for any samples analyzed since the last LFB standard that provided acceptable results if sufficient sample volume was available. QCS were analyzed monthly and the results needed to be within 20 percent of the accepted value. Five percent of the analytical batch was analyzed in duplicate, and results needed to be within 20 percent difference. A travel blank consisting of TOC-free distilled water treated through a deionized system with an organic-free cartridge was processed at each participating utility and during each sampling, as described in the Control section below.

University of North Carolina at Chapel Hill Laboratory. Sample vials utilized for DBP analysis were 40-mL clear borosilicate vials with Teflon-lined septa and screw caps. Vials were prepared by soaking them overnight in a bath of tap water and detergent (Alconox, Inc; White Plains, N.Y.), rinsing three times with laboratory grade water (Dracor; Durham, N.C.), immersing the vials overnight in a 10 percent nitric acid bath (Fisher Chemicals; Fair Lawn, N.J.), rinsing three times with laboratory grade water, rinsing three times with methanol (Mallinckrodt Chemicals; Phillipsburg, N.J.,) and oven drying for 24 hours at 180°C. All field blank and batch test bottles were prepared in the same manner. Plastic screw caps and Teflon-lined silicone septa (Laboratory Supply Distributors Corp.; Mt Laurel, N.J.) were soaked overnight in a tap water and detergent batch, rinsed three times with laboratory grade water, rinsed three times with methanol, and left to dry in a hood for several hours. Samples were collected with 20 mg/vial ammonium sulfate and 0.7 g/vial phosphate buffer.

A 25-mL graduated cylinder was rinsed three times with laboratory grade water and once with a small amount of sample. The cylinder was then used to measure a 20-mL aliquot of the 40-mL sample. The rest of the sample was discarded, and the 20-mL aliquot was returned to the 40-mL vial. A 4-mL volume of methyl-*tert*-butyl ether (MtBE, Sigma Aldrich; Milwaukee, Wis.) containing 100 μg/L of 1,2 dibromopropane (Sigma Aldrich, Milwaukee, Wis.) was added to the 20-mL aliquot. The 1,2 dibromopropane served as an internal standard (IS) to account for variations in sample injection volume. Approximately 6 g of sodium sulfate (Mallinckrodt; Paris, Ky.), which had been baked at 400°C for 24 hours to remove organics, was added to the 20-mL aliquot to aid in the extraction of the THMs into the MtBE. The sample was then vortex-mixed (Type 16700 Mixer-MaxiMix I, Thermolyne; Dubuque, Ind.) for 1 minute. The samples were left to sit at least 5 minutes so that the sodium sulfate crystals could settle to the bottom of the vial and solvent layer could separate. One mL or more of MtBE layer was transferred by Pasteur pipette to a 2-mL glass autosampler vial (Laboratory Supply Distributors; Mt. Laurel, N.J.) and capped with an aluminum Teflon-faced seal (Laboratory Supply Distributors; Mt. Laurel, N.J.). During the first few analyses, samples were double-vialed in case re-analysis was required. Both sets of samples were stored in a freezer at -15°C until analysis by GC.

At UNC laboratory, a calibration curve was created for each set of samples processed. THM4 stock standards (5,000 μg/mL of each analyte, Supelco; Bellefonte, Pa.) were diluted several times in methanol and two calibration standards were created. Standard 1 was comprised of 100 μg/mL of $CHCl_3$, 50 μg/mL of $CHBrCl_2$ and $CHBr_2Cl$, each, and 20 μg/mL of $CHBr_3$. Standard 2 was comprised of 4 μg/mL of $CHCl_3$, 2 μg/mL of $CHBrCl_2$ and $CHBr_2Cl$, each, and 0.8 μg/mL of $CHBr_3$. Standards were stored in 5 mL amber glass bottles with open-top screw caps containing PTFE-lined silicone septa. Calibration points were prepared with Standards 1 or 2 and laboratory grade water.

Separation, identification, and quantification of THMs occurred via GC with electron capture detection (ECD) on a Hewlett Packard GC 5890A with an autosampler/autoinject tower and a Hewlett Packard Model ECD (Hewlett-Packard Company; Cary, N.C.), respectively.

At UNC laboratory, reagents and calibration standards were run on the GC prior to their use to verify that no contamination or degradation had occurred. All new batches of MtBE, along with a sample of MtBE and IS, were also run on the GC prior to use. THM species stock standards were discarded and replaced every six months regardless of contamination or degradation. Test mixes of Standards 1 and 2 were generated prior to use and run on the GC to determine if degradation or contamination had occurred. If degradation of more than 20 percent had occurred, the standard was discarded and replaced. Regardless of contamination or degradation, Standards 1 and 2 were discarded every two months and new standards were made.

Check standards and matrix spikes were prepared to evaluate retention time drift on the GC and recovery from the matrix, respectively. Check standards were prepared using the calibration point concentrations and analyzed at a minimum of one standard after every eight samples. A sample of MtBE and IS was injected before or after each check standard to confirm that carry over from sample to sample was not taking place. If the IS area was more than ±10 percent of the average IS area, the sample was removed from consideration. Matrix spike samples were used to check spike recoveries, with no more than ±20 percent variation in the calculated recovery for acceptability of the sample set.

The sample data were analyzed in duplicate to check for reproducibility. If the relative percent difference between duplicate samples was greater than 20 percent, the extraction procedure and analysis were reevaluated to identify possible sources of error. A travel blank consisting of laboratory-grade water sent along with the coolers was processed at each participating utility and during each sampling, as described in the Control section below.

Newport News Waterworks and University of North Carolina. The minimum reporting level (MRL) for all species was 1 μg/L. The results are presented as an average of the duplicate measurements for each sample. When a duplicate sample was not available, the result of the single sample analysis was reported. If the sample and duplicate were both below the MRL, the value was reported as less than the MRL. However, if only one value was below the MRL, one half of the detection limit (0.5 μg/L for THM species) was averaged with the value above the detection limit. If the average was above the detection limit, then the average value was reported. If the average was below the MRL, the value was reported as less than the MRL.

Haloacetic Acids

Samples were stored at 4°C after collection before the extraction was conducted. At UNC laboratory, HAA9 samples were analyzed according to USEPA Method 552 (USEPA 1995), whereas they were extracted using the micro liquid-liquid extraction procedure based on Standard Method 6251B (APHA 1998) with minor variation (Brophy, Weinberg, and Singer 2000) at NNWW laboratory.

Newport News Waterworks Laboratory. Pre-cleaned 60-mL amber glass vials with Teflon-lined screw caps were used once only and were similar to those used during the ICR monitoring period. Approximately 100 mg/vial of ammonium chloride was used as a dechlorinating and preservative agent.

At NNWW laboratory, HAA9 were analyzed using EPA Method 552.2 (USEPA 1992), which involves a liquid-liquid extraction followed by an extraction with MtBE. The HAAs in the MtBE extract are converted to their methyl esters by the addition of acidic methanol followed by a slight heating for 2 hours. The acidic extract is neutralized by a back extraction with a saturated solution of sodium carbonate and the HAA species are identified and measured by capillary column GC/ECD (CP-3800 Gas Chromatograph and CP-8400 Autosampler; Varian; Walnut Creek, Calif.).

Prior to the analysis of HAA samples, a Continuing Calibration Check (CCC) at a low concentration, a LRB and a LFB were analyzed, and results needed to be within 30 percent of accepted value for each HAA species for the results to be considered acceptable. After every 10th sample and at the end of each analytical batch, a CCC at a mid-calibration concentration was analyzed, and the results needed to be within 30 percent of the accepted value. If the percent recovery for the CCC standard was out of the range of 70 to 130 percent, analyses were repeated for any samples analyzed since the last CCC standard that provided acceptable results if sufficient sample volume was available. All standards, samples, and blanks were fortified with a surrogate prior to extraction and an IS prior to injection. Both the IS and surrogate recoveries needed to be within 30 percent of the accepted value. A QCS was analyzed monthly, and the percent recovery for the quality control check standard needed to be within 30 percent of the accepted value. One sample per batch per day was analyzed in duplicate, and results needed to be within 30 percent difference. Data validation included evaluation of a spike sample. All spiked results needed to be within the methods' recommended percent recovery of 30 percent of the expected value, and values outside the acceptance range were reported as suspect of a matrix effect. A travel blank consisting of TOC-free distilled water treated through a deionized system with an organic-free cartridge was processed at each participating utility and during each sampling, as described in the Control section below.

At NNWW laboratory, the MRLs were set according to the ICR guidelines (USEPA 1996). MRL for the HAA species was 1 μg/L, except for the following species: ClAA (2 μg/L), Br_2ClAA (2 μg/L) and Br_3AA (4 μg/L). Anything less than the MRL was reported as 0 μg/L.

University of North Carolina at Chapel Hill Laboratory. Sample vials utilized for DBP analysis were 40-mL clear borosilicate glass with Teflon-lined septa and screw caps. Vials were prepared by soaking them overnight in a bath of tap water and detergent (Alconox, Inc; White Plains, N.Y.), rinsing three times with laboratory grade water (Dracor; Durham, N.C.), immersing the vials overnight in a 10-percent nitric acid bath (Fisher Chemicals, Fair Lawn, N.J.), rinsing three times with laboratory grade water, rinsing three times with methanol (Mallinckrodt Chemicals; Phillipsburg, N.J.), and oven drying for 24 hours at 180°C. All field blank and batch test bottles were prepared in the same manner. Plastic screw caps and Teflon-lined silicone septa (Laboratory Supply Distributors Corp.; Mt Laurel, N.J.) were soaked overnight in a tap water and detergent batch, rinsed three times with laboratory grade water, rinsed three times with methanol, and left to dry in a hood for several hours. Samples were collected with 20 mg/vial ammonium sulfate and 50 μg/vial of an 80-mg/L solution of sodium azide.

Glassware utilized in the generation of the diazomethane was soaked in a 5 N sodium hydroxide bath for a minimum of 24 hours, rinsed with tap water, rinsed three times with laboratory grade water, and rinsed once with methanol. Diazomethane was generated and captured by collecting the diazomethane vapor in MtBE. One 40-mL glass vial, Vial 1, was connected to another 40-mL glass vial, Vial 2, by Teflon tubing, with an inner diameter of 0.8 mm. The tubing was connected to each vial via open-top screw caps which were mounted with Teflon-lined silicone septa. Small holes were drilled in each septum and the tubing was inserted in each hole. A vent was also placed through the septum of Vial 1. A volume of 10 mL of MtBE (Sigma Aldrich; Milwaukee, Wis.) was poured into Vial 1 and the tubing was immersed in the MtBE. Vial 1 was then placed in an ice-water bath.

Reagent 1, comprised of 3.3 g of Diazald (Sigma Aldrich; Milwaukee, Wis.), 5 mL of Carbitol (Sigma Aldrich; Milwaukee, Wis.), and 5 mL of MtBE, was produced in a third 40-mL vial, capped with a screw top and Teflon-lined silicone septum, and gently swirled. Reagent 2, comprised of 6 mL of laboratory grade water, 10 mL of high purity methanol, and 4 mL of potassium hydroxide was produced in a fourth 40-mL vial, capped with a screw top and Teflon-lined silicone septum, and gently swirled. A volume of 6 mL of Reagent 1 and 6 mL of Reagent 2 were combined in Vial 2 and quickly capped to capture the diazomethane gas. The gas traveled through the tubing to Vial 1 and was trapped in the MtBE. Once the MtBE turned a dark yellow color, Vial 1 was quickly capped with a new screw cap and PTFE-lined silicone septum, placed in a double-walled bomb, and stored in an explosion-proof freezer.

A 25-mL graduated cylinder was rinsed three times with laboratory grade water and once with a small amount of sample. The cylinder was then used to measure a 20-mL aliquot of the 40-mL sample. The rest of the sample was discarded and the 20-mL aliquot returned to the

40-mL vial. Each 20-mL sample aliquot received 20 μL of 2,3 dibromopropionic acid (Supelco; Bellefonte, Pa.) followed by 1.5 mL of sulfuric acid. The 2,3 dibromopropionic acid served as an acid surrogate to assess the derivatization efficiency. The addition of the sulfuric acid heated the samples; thus, the samples were placed on ice until they reached room temperature. A volume of 4 mL of MtBE with 100 μg/L of 1,2 dibromopropane (Sigma-Aldrich; Milwaukee, Wis.) were added. The 1,2 dibromopropane was used as an IS, which accounted for any variance in sample volume injected into the GC. Approximately 10 g of sodium sulfate (Mallinckrodt, Ky.), baked at 400 °C for 24 hours, was added to each sample and each sample was vortex-mixed (Type 16700 Mixer-MaxiMix I, Thermolyne; Dubuque, Ind.) for 1 minute. An aliquot of 2 mL of the MtBE layer was transferred to a 2-mL volumetric flask by a Pasteur pipette. Each flask was capped with a screw cap and PTFE-lined silicone septa. Approximately 50 mg of anhydrous, magnesium sulfate (Sigma Aldrich; Milwaukee, Wis.) was added to each volumetric flask to quench any water that might affect the derivatization process. Samples extracted at the beginning of the study were dosed with 225 μL of diazomethane; samples extracted near the end of the study were dosed with 200 μL of diazomethane due to replacement of the micropipetter. Samples were refrigerated for 15 minutes and checked for a yellow color upon removal from the refrigerator. Presence of the yellow color indicated that sufficient diazomethane was present to drive the esterification process. Samples were allowed to warm to room temperature after which approximately 40 mg of silicic acid (J.T. Baker; Phillipsburg, N.J.) was added to the samples to quench any residual diazomethane. Approximately 1 mL of sample was transferred using a Pasteur pipette to a glass GC vial and crimped with an aluminum Teflon cap. Extracted samples were stored at -15°C (5°F) until analysis by the GC, using the same procedure as described for THM.

At UNC laboratory, a calibration curve was created for each set of samples processed. Six calibration points were prepared by using an HAA9 calibration standard and laboratory grade water. The calibration standard was prepared from an HAA6 stock standard (EPA 552 Halogenated Acetic Acids Mix, Supelco; Bellefonte, Pa.) containing 2,000 μg/mL each of ClAA, BrAA, Br_2AA, Cl_2AA, BrClAA, and Cl_3AA and individual stock standards of Br_2ClAA, Cl_2BrAA, and Br_3AA with concentrations of 1,000 μg/mL (Supelco; Bellefonte, Pa.). The HAA9 standard was prepared with MtBE at a concentration of 20 μg/ml of each HAA species and stored in an amber vial with a screw-top and Teflon-lined silicone septum. Three matrix spikes were prepared in duplicate for each set of samples by adding 8 to 10 mL of HAA9 standard into 25 mL of sample.

Quality control procedures (travel blank, check standards, matrix spikes, extraction in duplicate, reporting of average concentrations, etc.) utilized for HAA9 analysis at UNC laboratory were the same as those utilized for THM4 analysis, with the following exceptions. The Br_3AA stock standard was discarded and replaced every 4 months and the Br_2ClAA and Cl_2BrAA stock standards were discarded and replaced every 5 months due to increased rates of degradation of the stock standards. The HAA9 standard was replaced every month regardless of degradation. In addition, if the acid surrogate concentration had a coefficient of variation greater than ±10 percent when compared to the average value for the entire sample set, the extraction procedure and analysis were reevaluated to identify possible sources of error. The MRL was 2 μg/L for all HAA species. Sample data points that were below the MRL were reported as <2 μg/L.

N-Nitrosodimethylamine

NDMA was analyzed at the University of Waterloo. Samples were collected in 1-L amber borosilicate glass bottles with Teflon-lined caps. With the exception of glass syringes used for the preparation of analytical standards and stock solutions, all glassware was cleaned in an automatic dishwasher using potassium hydroxide dishwashing detergent for the cleaning cycle. An acetic acid rinse cycle was also included, followed by a final distilled water rinse. Hamilton glass syringes with Teflon®-tipped plungers (VWR; Mississauga, Ont.), used for transfer of NDMA stock solutions, were triple-rinsed with deionized water, methanol and dichloromethane, and then rinsed five times with the solution to be transferred. Approximately 50 mg ascorbic acid was used per bottle as a dechlorinating and preservative agent.

NDMA was analyzed by a method developed by Taguchi and colleagues (1994) and further modified at the university. It employs Ambersorb® XEN 572 resin adsorption followed by desorption with dichloromethane, and analysis of the isolated NDMA by isotope dilution gas chromatography and mass spectrometry (GC/MS). An initial sample screening was conducted using low-resolution GC/MS to assess the presence of NDMA. If NDMA was present, the samples were analyzed by high resolution GC/MS to confirm the concentration observed.

From the 1-L sample collected, a 500-mL subsample was measured into a separate 1-L amber Boston round bottle with a Teflon liner (VWR; Mississauga, Ont.). The internal standard, the deuterated form of NDMA (d_6-NDMA; CDN Isotopes; Pointe-Claire, Québec), was injected into the dechlorinated water sample at a concentration of 100 ng/L. Approximately 400 μL (200 mg) of the resin beads (Ambersorb® XEN 572; Sigma-Aldrich; Milwaukee, Wis.) were added to the amber bottle, and the bottle was then shaken for approximately 1 hour on a reciprocating shaker table at 180 cycles per minute.

After the bottles were removed from the shaker table, the contents were filtered through Whatman No. 4 filter paper (VWR; Mississauga, Ont.). The carbonaceous beads and the filter paper on which they were retained were placed in an aluminum-weighing dish and allowed to air dry for approximately 30 to 60 minutes in a fume hood. Once the beads were dried enough to be removed from the filter paper and without clumping together, they were placed in a 2-mL autosampler vial with a Teflon liner. NDMA was then desorbed from the beads using 400 μL of dichloromethane. The extracts were either analyzed immediately by GC/MS or refrigerated at 4°C until analysis. Refrigerated samples may be held for up to 50 days.

A Hewlett Packard low resolution 5890 Series II Gas Chromatograph/5971 Mass Selective Detector (GC/MS) (Mississauga, Ont.) was used to analyze the extracts and was operated in selected ion monitoring (SIM) mode. Helium was the carrier gas (1.25 mL/min at 80°C), and the sample components were separated using a J&W Scientific DB-1701 capillary column (30 m x 0.25 mm x 0.25 μm, 14 percent cyanopropylphenyl silicone phase; J&W Scientific; Folsom, Calif.).

Quality control procedures for NDMA analysis at the University of Waterloo included:

- Use of the deuterated form of NDMA (d_6-NDMA) as an internal standard
- Use of solvent blanks, method blanks, check standards, spiked samples (approximately 15 percent of the total sample load)
- Inclusion of field and travel blanks, as well as travel standards

- Duplicate samples (bottles) analyzed (during the December 3, 2002 sampling in Utility A, triplicate samples were anlayzed)
- Regular calibration of GC/MS.

Organic Matter

Organic matter was characterized using TOC, UV absorbance at 254 nm, and AOC analyses.

Total Organic Carbon

TOC samples collected from Utilities B and C were analyzed at NNWW, and TOC samples collected at Utilities A, D, and E were analyzed at UNC. Both laboratories used the same analytical method (Standard Method 5310 B; APHA 1998).

Newport News Waterworks Laboratory. Pre-cleaned 125-mL amber glass bottles were used once only to collect the samples. TOC samples were first measured using a Shimadzu 5000 Total Organic Carbon Analyzer (Shimadzu Corporation; Norcross, Ga.) equipped with an ASI auto-sampler. The 2003 samples were analyzed on a Shimadzu TOC-VCP Total Organic Analyzer and a Shimadzu ASI-V autosampler. The modification involved the use of ACS-certified hydrochloric acid (EMD; Gibbstown, N.J.) for sample acidification instead of phosphoric acid because the Shimadzu 5000 analyzer is sensitive to excess phosphoric acid.

Five-calibration point curves were used to calibrate the TOC analyzer. Calibration points ranged from 0 to 10 mg/L C. Dilutions were made from a prepared 1,000 mg/L C standard (Environmental Resource Associates; Arvada, Colo.). Prior to loading the samples on the analyzer, the calibration was checked by analyzing a low, middle, and high standard (0.7, 4.0, 7.0, 10.0 mg/L) and a QC standard (Environmental Resource Associates; Arvada, Colo.). All results had to be within 10 percent of the expected value; otherwise, the TOC analyzer was recalibrated. Duplicate analyses were run for at least 10 percent of the samples, and 10 percent of the samples were spiked. A calibration check points standard or QC standard was analyzed after every 10th sample and at the end of the analytical batch. A field blank consisting of TOC-free deionized/distilled water was processed at each utility monitored and during each sampling, as described in the Control section below.

University of North Carolina at Chapel Hill Laboratory. Samples were collected in 40-mL borosilicate vials with Teflon-lined septa and screw caps. Glassware used to analyze TOC was rinsed three times with tap water, soaked overnight in a 10-percent nitric acid bath, rinsed three times with tap water, rinsed three times with laboratory grade water, and oven dried overnight at 180°C.

TOC was measured at UNC laboratory following a slightly modified method using a Shimadzu 5000 Total Organic Carbon Analyzer (Shimadzu Corporation; Norcross, Ga.) equipped with an ASI auto-sampler. The modification involved the use of hydrochloric acid (Fisher Scientific; Fairlawn, N.J.) for sample acidification instead of phosphoric acid because the Shimadzu 5000 analyzer is sensitive to excess phosphoric acid.

Four calibration points were developed each time the analyzer was run. Calibration points ranged from 0 to 6 mg/L of TOC. Dilutions were made from a 1,000 mg/L TOC standard. A new standard was made every month by adding 2.128 g of potassium hydrogen phthalate (Fisher Scientific; Fairlawn, N.J.) to 1,000 mL of laboratory grade water.

Prior to loading the samples at UNC laboratory, a mechanical check was performed on the analyzer. Ultra-high purity air (Holox Company; Morrisville, N.C.) was utilized as the carrier gas. TOC vials were rinsed with laboratory grade water, rinsed with a small amount of sample, and then filled with sample to 90 percent capacity. Three drops of hydrochloric acid were added to samples and five drops were added to calibration standards. Duplicate analyses were run for all samples and were checked for more than a 10 percent variation, except in cases when there was no second sample available. A calibration check point was added every 10th sample, and the full calibration curve was run after all the samples had been analyzed to provide quality control. A travel blank consisting of laboratory-grade water sent along with the coolers was processed at each utility monitored and during each sampling, as described in the Control section below.

UV Absorbance at 254 nm

UV absorbance samples collected from Utilities B and C were analyzed at NNWW, and TOC samples collected at Utilities A, D and E were analyzed at UNC. Both laboratories used the same analytical method (Standard Method 5910 B; APHA 1998).

Newport News Waterworks Laboratory. Pre-cleaned 125-mL amber glass bottles were used once only to collect the samples. UV absorbance was measured at 254 nm on a Lambda UV/VIS Spectrophotometer (Perkin Elmer; Shelton, Conn.) using a 5-cm cell. Prior to the analysis of any samples, the spectrophotometer was zeroed with TOC-free deionized/distilled water, and a low, medium, and high UV calibration standards (0.042, 0.102 and 0.304 cm^{-1}) and a QC standard were analyzed. The UV calibration standards were prepared from a 1,000 mg/L Total Carbon standard (Accustandard; New Haven, Conn.) and the QC sample was purchased from Environmental Resource Associates (Arvada, Colo.). Before transferring a sample aliquot to the 5-cm cell, the cell was rinsed with laboratory grade water, rinsed with a small amount of sample, and then filled with sample. Care was taken to ensure that no air bubbles were attached to the walls of the cell.

Duplicate analyses were run for at least 10 percent of the samples. A calibration check or a QC standard was analyzed after every 10th sample and at the end of the analytical batch. A field blank consisting of TOC-free deionized/distilled water was processed at each utility monitored and during each sampling, as described in the Control section below.

University of North Carolina at Chapel Hill Laboratory. Samples were collected in 40-mL borosilicate vials with Teflon-lined septa and screw caps. UV absorbance was measured at 254 nm on a U-2000 UV-Visible spectrophotometer (Hitachi Instruments; Danbury, Conn.) using a 5-cm cell. The spectrophotometer was zeroed with laboratory grade water prior to measuring UV absorbance of the samples. Before transferring a sample aliquot to the 5-cm cell, the cell was rinsed with laboratory grade water, rinsed with a small amount of sample and then filled with sample. Care was taken to ensure that no air bubbles were attached to the walls of the cell.

Duplicate analyses were run for all samples to assure that there was less than 10 percent variation, except in cases when there was no second sample available. A check with laboratory grade water was performed after every 10th sample. A travel blank consisting of laboratory-grade water sent along with the coolers was processed at each utility monitored and during each sampling, as described in the Control section below.

Assimilable Organic Carbon

AOC was measured at American's laboratory using the procedure described by LeChevallier et al (1993). Water samples were collected and analyzed in duplicate in 500-mL Pyrex bottles, which had been muffled at 525°C for 6 hours. Bottle caps were washed in an acid-rinse dishwasher, rinsed with purified low-TOC water, and autoclaved in foil at 121°C for 20 minutes. Bottles were shipped capped.

Samples were neutralized with 700 µL of a sterile solution of 10-percent sodium thiosulfate and pH was adjusted to 6.8 to 7.5. Samples were pasteurized at 70°C for 30 minutes to kill endogenous bacteria. Approximately 20 mL of each sample was poured into five pre-cleaned 20-mL sample vials (Scientific Specialties, Inc.; Lodi, Calif.; Kaplan and LeChevallier 1993), identified as follows:

- Method Blank consisted of negative controls and received no additional treatment beyond the pasteurization step.
- *Pseudomonas fluorescens* P-17 Matrix Spike vials received 50 µL of a 40-mg/L solution of sodium acetate, for a final concentration of approximately 100 µg/L acetate AOC. This spike also received 50 µL refrigerated P-17 culture, or approximately $2x10^4$ CFU.
- *Spirillum* NOX Matrix Spike vials received the same sodium acetate spike and 50 µL refrigerated NOX culture or approximately $6x10^4$ CFU.
- P-17 and NOX Experimental Vials received separate inoculations of approximately 10^4 CFU.

Spiking and bacterial inoculations were performed through the vial septa with sterile 3-mL syringes bearing 21G 1.5" needles. During the incubation period (at room temperature), the bacteria were allowed to multiply to full cell yield at stationary phase using nutrients available in the water sample. From each of the five vials, 1-mL aliquots were tested each day over a three-day period. On Days 3, 4, and 5, post-inoculation P-17 and NOX Experimental Vials were agitated and subjected to serial tenfold dilution in phosphate buffered water (Standard Method 9050 C; APHA 1998). Aliquots of 0.1 mL of the appropriate dilution was spread onto R2A agar plates for colony enumeration from approximately 30 to 300 CFU per plate. The conversion factors from cell yield into AOC concentration were those determined by van der Kooij (1990 and 1992). AOC concentrations were reported as three-day means of plate counts. Total AOC values for sampled water were expressed as the sum of P-17 and NOX mean AOC values.

Outlying AOC measurements were defined as outside the range +/-0.5 times the mean and were removed from further calculations. Outliers were also identified on the basis of high Method Blank values or other procedural anomalies. On rare occasions, two or more data points were considered outliers. In these instances, the sample result was considered invalid.

General Water Quality Parameters

Chlorine residual, temperature, and pH were measured onsite at the time of sample collection using the methods presented in Table 3.17. Free and total chlorine were measured with a Hach® Pocket Colorimeter™ Analysis System (Hach; Loveland, Colo.).

Nitrogen Species

Nitrogen species (ammonia, nitrite, and nitrate) were only measured in the systems using chloramines (Utilities C, D and E). Methods used and responsible laboratories are presented in Table 3.17.

For the sampling conducted on February 28, 2003, at Utility E, sample analysis could not be completed within 48 hours of sample collection; therefore, sulfuric acid was added to the samples as a preservative. The sulfuric acid addition allowed for the analysis of ammonia, but not nitrite and nitrate.

Bromide

Samples were collected in 125-mL polyethylene bottles. At NNWW laboratory, bromide was analyzed using USEPA Method 300.1 (USEPA 1997b), which involves introducing a sample into an ion chromatograph. The anions of interest are separated and measured using a system consisting of a guard column, analytical column, suppressor device, and conductivity detector (DX-500 Dionex; Sunnyvale, Calif.).

Prior to the analysis of ions by this method, a CCC at a low concentration, an LRB, and an LFB were analyzed, and the results needed to be within 10 percent for each ion species. After every 10th sample and at the end of each analytical batch, a CCC at a mid-calibration concentration was analyzed, and the results needed to be within 10 percent of the accepted value. If the percent recovery for the CCC standard was out of the range of 90 to 110 percent, analyses were repeated for any samples analyzed since the last CCC standard provided acceptable results if sufficient sample volume was available. A QCS was analyzed monthly, and the percent recovery for the quality control check standard needed to be within 25 percent of the accepted value. Five percent of the analytical batch was analyzed in duplicate, and results needed to be within 10 percent difference. Data validation included evaluation of a spike sample. All spiked results needed to be within the methods' recommended percent recovery of 25 percent of the expected value, and values outside the acceptance range were reported as suspect of a matrix effect.

Heterotrophic Bacteria

At both laboratories where heterotrophic bacteria were enumerated (NNWW and American), HPCs were analyzed by the spread plate method in accordance with Standard Method 9215 C (APHA 1998). Samples were diluted and spread onto R2A medium within 48 hours of sample collection and incubated for 7 days at 25°C. A field blank consisting of sterile buffered water (Standard Methods 9050 C.1a; APHA 1998) was processed at each participating utility and during each sampling.

CONTROLS

Depending on the type of parameter analyzed, a variety of controls and standards were processed during each sampling. Field control blanks were included with each shipment to ensure that the sample quality was maintained during sample collection and transportation. These blanks were filled onsite at the same time than the samples were collected. Travel control blanks were also processed to ensure that the sample quality was maintained during sample transportation. A method control blank was run with each set of samples to ensure that contamination from glassware, materials, or reagents did not interfere with accurate sample analysis. For this method blank, all sample handling and analytical operations using the specified materials and reagents were performed in the absence of sample substrate. Some parameters required an analysis control blank without sample handling and processing. This blank was analyzed during instrument calibration, as well as during sample analysis. All control blanks were processed and analyzed like all other samples. They had to provide results below the method detection limit (MDL) to validate the samples collected during the same experiment.

Matrix spikes and standards were also analyzed along with each batch of samples. The methodology used for these particular samples was detailed in the above sections.

Inter-Laboratory Comparison Experiments

Inter-laboratory comparison experiments for THM4 and HAA9 were conducted at the beginning of the project and then quarterly (approximately every three months) to ensure comparability of results. Further information and results are presented in Appendix E.

CHAPTER 4
CHANGES IN DISINFECTION BY-PRODUCTS AND WATER QUALITY IN FULL-SCALE DISTRIBUTION SYSTEMS

THM4, HAA9, NDMA, and a variety of water quality parameters were monitored in five full-scale distribution systems. The water quality parameters analyzed included chlorine residual, HPC, TOC concentration, UV absorbance at 254 nm, AOC concentration, and nitrification indicators. This chapter examines the changes in DBP concentration, DBP speciation and water quality with increasing water age, including effects of seasons, in two systems receiving free chlorine as the final disinfectant, and two systems receiving chloramines as the final disinfectant. The same effects were examined in a system using chloramines as the final disinfectant, but switching to free chlorine every year to limit nitrification occurrence. In this system, results obtained during the chloramination periods were also compared with those obtained during the free chlorination period.

Batch incubations in bottles were conducted during several samplings to distinguish the effect of pipes from the effect of water age on disinfectant decay and DBP formation. These results are presented in Appendix B. Linear correlations were calculated between DBP concentrations and water age or selected water quality parameters, in an attempt to identify one or more parameters that could be used as surrogate(s) of the DBPs examined. These results are presented in Appendix C. When selecting the sampling locations within each system, consideration was given, as much as possible, to select locations with similar water age and pipe diameter but different pipe material, as well as locations with similar water age and pipe material but different pipe diameter. This approach allowed the examination of the effect of pipe material and diameter on DBP concentration; these results are presented in Appendix D. In Utility A, samples were collected throughout the distribution system, as well as before and after a storage reservoir and a booster chlorination station to allow examination of the effect of these facilities on DBP concentrations. These results are also presented in Appendix D.

In this chapter, illustrative graphs and tables of the findings are presented. A complete set of graphs and tables for all data collected is presented on the enclosed CD-Rom. THM4, HAA9, TOC and UV absorbance were analyzed at two different laboratories: the University of North Carolina at Chapel Hill analyzed samples collected from Utilities A, D and E, and Newport News Waterworks analyzed samples collected from Utilities B and C. Because of this, only the trends observed at each utility are compared. Quality control results observed for inter-laboratory comparison samples analyzed by both laboratories are presented in Appendix E.

TRIHALOMETHANES AND HALOACETIC ACIDS

The discussion that follows is organized by type of disinfectant residual used at the utilities studied. The results discussed below have been interpreted in consideration of the limitations of the analytical methods in terms of MDL of each DBP species, relative percent difference (RPD) between replicate samples, matrix spike and standard recovery, results of method and analysis control blanks, as well as travel and field blanks.

Free Chlorinated Systems

Two of the distribution systems studied used free chlorine as the final disinfectant, and are designated as Utilities A and B.

Trends in THM Concentrations with Increasing Water Age in the Free Chlorinated Systems

Figures 4.1 and 4.2 present representative results of THM4 concentration, HAA9 concentration and chlorine residual as a function of water age in Utility A and Utility B, respectively. Tables 4.1 and 4.2 present all THM4 data obtained from Utility A and Utility B, respectively. THM4 concentrations increased as water age increased and chlorine residual decreased, a trend observed for all samplings at both free chlorinated utilities. This pattern is consistent with theoretical expectations and the literature (LeBel et al. 1996, Harms and Loogenga 1977, Brett and Calverly 1979, Levine et al. 1987, Koch et al. 1991, Meyer et al. 1993, Williams et al. 1994, Vasconcelos et al. 1996, Chen and Weisel 1998, Obolensky and Frey 2002).

In Utility A, THM4 concentrations at the POE were generally the lowest, while THM4 concentrations at the MRTL (water age of 323 hours) were the highest (maximum concentrations reached of 97 to 269 μg/L). The increases in THM4 concentration from the POE to the MRTL were important, averaging 452 percent and ranging from 153 to 1,118 percent in Utility A. In Utility B, THM4 concentrations also increased from the POE to the MRTLs (maximum concentrations reached of 60 to 161 μg/L), although variations were observed in the increases in THM4 concentration with water age. Increases in THM4 concentration from the POE to the MRTL of Utility B (water age of 47 hours) were much lower than in Utility A and averaged 58 percent, ranging from 23 to 80 percent. Most likely the higher chlorine residuals and lower water ages measured in Utility B compared to Utility A were responsible for the differences observed.

Results indicate that THM4 concentrations were very high in both distribution systems, particularly at the MRTLs. Using the current MCL of 80 μg/L for THM4 as a marker, individual THM4 concentrations measured at the MRTL of Utility A (water age of 323 hours) exceeded this concentration during all samplings, and THM4 concentrations above 160 μg/L (two times the MCL marker) were measured during five out of the seven samplings. During three out of the seven samplings conducted in Utility A, only the POE THM4 concentration was below 80 μg/L. Discussions with this utility's personnel indicated that DBP concentrations measured in 2003 were much higher than normal due to higher precipitation and runoff, as was also shown by DBP regulatory compliance data. In Utility B, individual THM4 concentrations exceeded 80 μg/L at various sampling locations during five out of the eight samplings conducted.

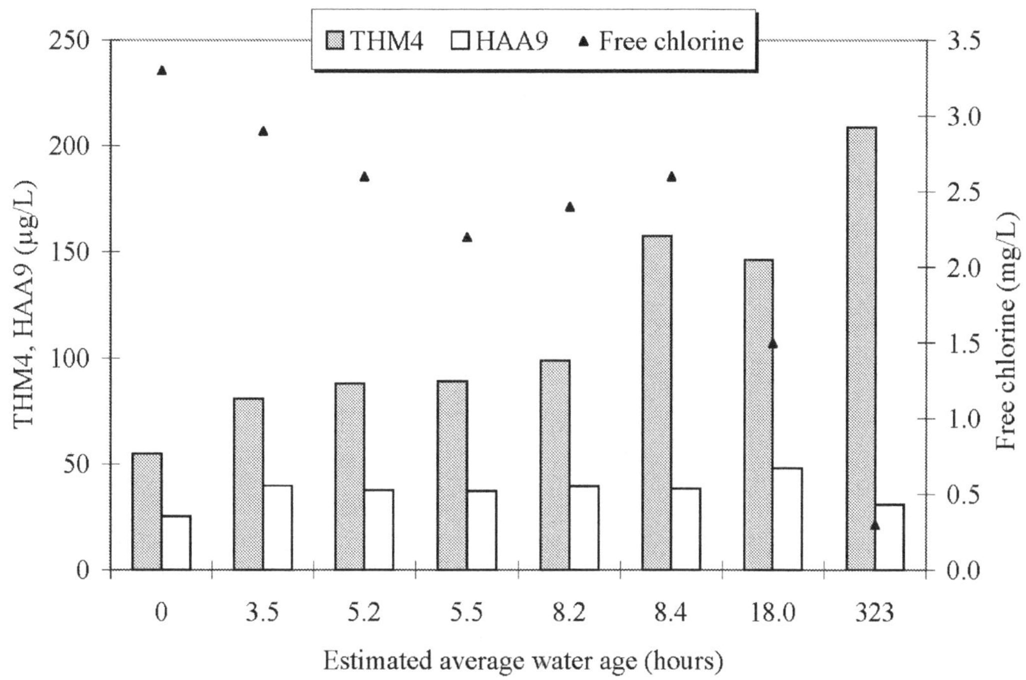

Figure 4.1 THM4 concentration, HAA9 concentration and free chlorine residual as a function of water age measured during the September 23, 2003 sampling in Utility A (free chlorinated system) (water temperature at the POE was 16.6°C, 62°F)

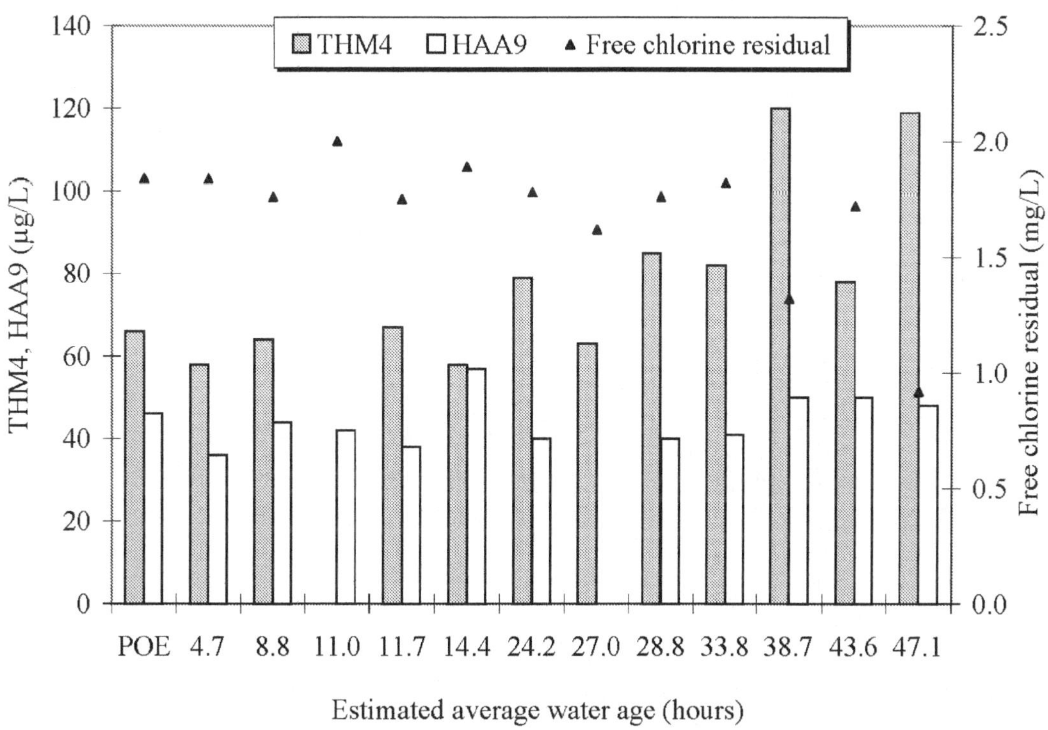

Figure 4.2 THM4 concentration, HAA9 concentration and free chlorine residual as a function of water age measured during the August 12, 2003 sampling in Utility B (free chlorinated system) (water temperature at the POE was 27°C, 81°F)

Table 4.1
THM4 concentrations (µg/L) measured during each sampling in Utility A (free chlorinated system), with water temperature at the POE (°C, °F)

Water age (hours)	October 16, 2002 (13.3°C, 56°F)	December 3, 2002 (5.6°C, 42°F)	January 7, 2003 (6.1°C, 43°F)	February 25, 2003 (4.0°C, 39°F)	June 10, 2003 (15.0°C, 59°F)	August 12, 2003 (20.1°C, 68°F)	September 23, 2003 (16.6°C, 62°F)
0 (POE)	51	13	13	38	50	54	55
3.5	88	23	23	35	73	163	81
5.2	81	25	27	64	81	97	88
5.5	85	21	26	63	68	95	89
8.2	88	47	32	66	80	137	99
8.4	107	22	31	67	80	106	158
18.0	120	44	66	62	97	157	146
323	194	163	108	97	161	269	209

Table 4.2
THM4 concentrations (µg/L) measured during each sampling in Utility B (free chlorinated system), with water temperature at the POE (°C, °F)

Water age (hours)	October 15, 2002 (26°C, 79°F)	November 18, 2002 (17°C, 63°F)	January 7, 2003 (11°C, 52°F)	February 25, 2003 (12°C, 54°F)	April 29, 2003 (22°C, 72°F)	June 10, 2003 (26°C, 79°F)	August 12, 2003 (27°C, 81°F)	October 21, 2003 (22°C, 72°F)
0 (POE)	51	49	39	48	41	76	66	47
4.7	50	59	41	46	51	76	58	48
8.8	57	52	36	51	43	88	64	47
11.0	49	52	44	50	55	83	NA	59
11.7	53	63	67	51	42	81	67	49
14.4	58	66	44	52	51	84	58	61
19.3	64	56	NA	NA	NA	NA	NA	NA
24.2	89	59	50	52	47	92	79	60
27.0	57	62	51	50	41	78	63	49
28.8	65	61	48	53	60	98	85	64
33.8	58	57	NA	NA	NA	NA	NA	NA
33.8	71	65	50	55	57	99	82	59
38.7	59	61	62	60	62	126	120	NA
43.6	68	56	49	53	51	99	78	63
47.1	87	86	69	59	60	119	119	66

NA: Not analyzed

Table 4.3
THM4 LRAA values calculated in Utilities A and B (free chlorinated systems)

Utility A		Utility B	
Water age of sampling location (hours)	THM4 LRAAs (μg/L)	Water age of sampling location (hours)	THM4 LRAAs (μg/L)
0 (POE)	44.3	0 (POE)	50.6
3.5	77.6	4.7	51.0
5.2	73.3	8.8	51.9
5.5	70.6	11.0	56.8
8.2	**83.6**	11.7	63.8
8.4	**89.8**	14.4	56.2
18.0	**106.5**	24.2	64.0
323	**179.2**	27.0	57.0
		28.8	63.3
		33.8	64.5
		38.7	76.5
		43.6	62.7
		47.1	**82.9**

Concentrations in bold characters are greater than the 80 μg/L MCL

An exercise was conducted during which the data obtained at each of the sampling locations in this study were used to calculate RAA and LRAA values. RAAs were calculated by first averaging all THM4 concentrations measured at all locations during each sampling, leading to one average value per sampling. These sampling averages were then combined by season, and averages were calculated (seasonal averages). The resulting four seasonal averages were then averaged together into a single RAA value for each system studied. THM4 RAA values were 90.6 and 61.5 μg/L in Utility A and Utility B, respectively. Similarly, the LRAA values for each sampling location were calculated by first determining an average THM4 concentration for each season for that sampling location (locational seasonal averages). The resulting four locational seasonal averages were then averaged into LRAA values for each sampling location (Table 4.3). The results suggest that Utility A would be out of compliance with the proposed Stage 2 DBPR at all locations with a water age greater than about 8 hours, whereas Utility B would be in compliance for all locations with a water age less than about 45 hours. Although this exercise is not entirely consistent with the Stage 1 and Stage 2 DBPR requirements because many more sampling locations were used in this study, results lead to very high values, questioning the ability of these utilities to meet the proposed regulatory requirements.

THM Speciation in the Free Chlorinated Systems

Figures 4.3 and 4.4 present representative THM speciation results obtained from Utility A and Utility B, respectively. Table 4.4 summarizes the weight fractions of $CHCl_3$ and chlorinated THM species (i.e., $CHCl_3$ and $CHBrCl_2$) for both free chlorinated systems studied. Results show that the chlorinated species comprised most of the THM concentrations because of the lower bromide concentrations in the source water of these systems. In Utility A, where plant influent bromide concentrations were 0.03 mg/L or lower, the percentage of chlorinated THM species were similar throughout the year. In Utility B, however, these percentages seemed higher in warm water than in colder water. Bromide concentration was not measured at the influent of Utility B's treatment plant, although historical bromide levels ranged from <0.1-0.5 mg/L.

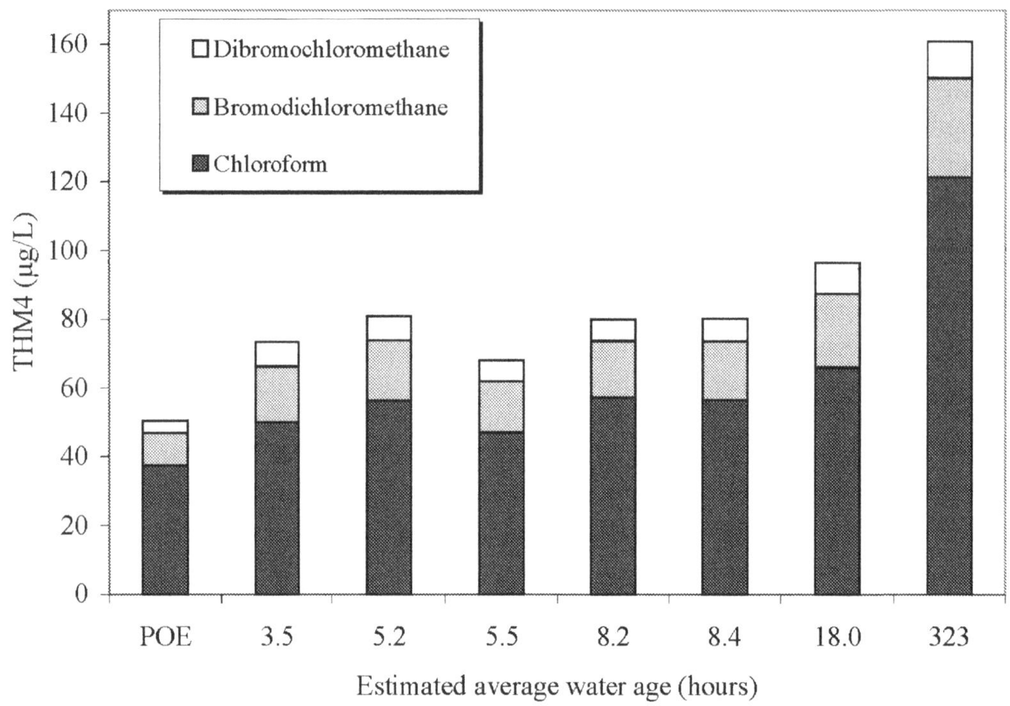

Figure 4.3 THM4 speciation as a function of water age measured during the June 10, 2003 sampling in Utility A (free chlorinated system) (water temperature at the POE was 15.0°C, 59°F; bromoform was not detected)

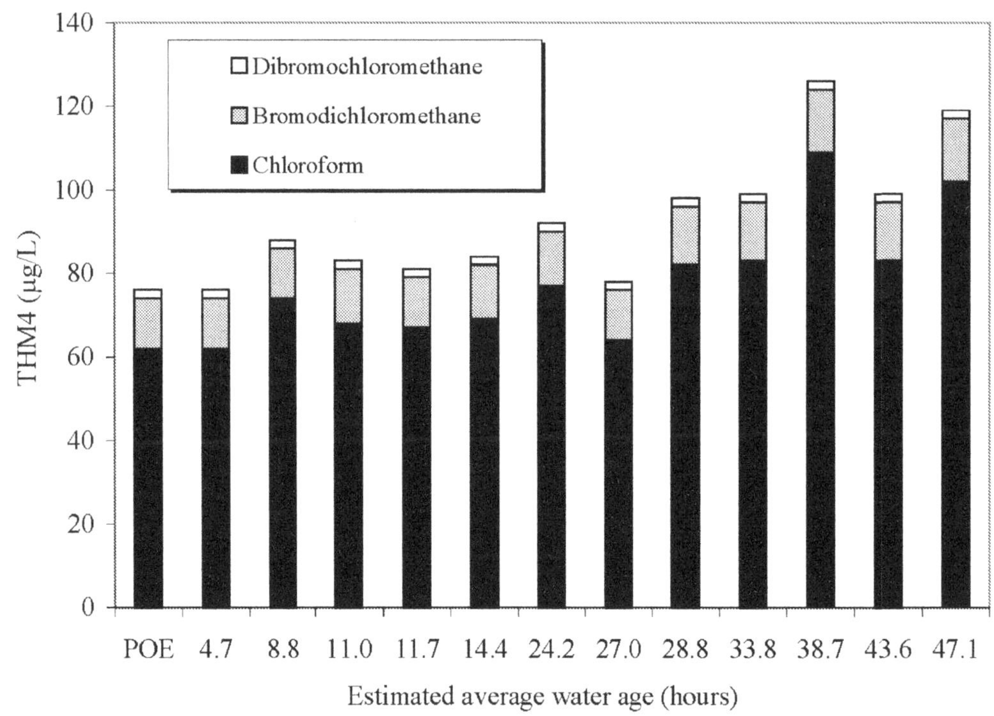

Figure 4.4 THM4 speciation as a function of water age measured during the June 10, 2003 sampling in Utility B (free chlorinated system) (water temperature at the POE was 26°C, 79°F; bromoform was not detected)

The weight fraction of chlorine-containing THMs (i.e., $CHCl_3$ and $CHCl_2Br$) decreased slightly (<12 percent decrease from the POE to the MRTL) with increasing water age during some of the samplings conducted in Utility A, or remained stable. As expected, the decreases in chlorine-containing THMs were associated with increases in bromine-containing THMs. The opposite was observed in Utility B; the weight fraction of chlorine-containing THMs increased slightly (up to 7 percent increase from the POE to the MRTL) with increasing water age, consistent with a slight decrease in bromine-containing THMs. The later trend is consistent with expectations as the bromine-containing DBP species are known to form faster than the chlorinated species. As such, the fraction of chlorine-containing THMs should increase with increasing water age in the distribution system, as observed in Utility B but not in Utility A.

Table 4.4
Mass ratios of $CHCl_3$ to THM4 and chlorine-containing THMs to THM4 calculated in the free chlorinated systems (Utilities A and B)

	Utility A			Utility B		
	Minimum	Average	Maximum	Minimum	Average	Maximum
October 2002:	Water temperature at POE: 13.3°C, 56°F			Water temperature at POE: 26°C, 79°F		
$CHCl_3$/THM4	60%	66%	74%	51%	56%	65%
Cl-THM/THM4	82%	88%	94%	80%	83%	88%
November/December 2002;	Water temperature at POE: 5.6°C, 42°F			Water temperature at POE: 17°C, 63°F		
$CHCl_3$/THM4	65%	69%	75%	61%	62%	67%
Cl-THM/THM4	89%	92%	94%	89%	90%	91%
January 2003:	Water temperature at POE: 6.1°C, 43°F			Water temperature at POE: 11°C, 52°F		
$CHCl_3$/THM4	48%	53%	68%	64%	68%	74%
Cl-THM/THM4	80%	84%	89%	92%	93%	95%
February 2003:	Water temperature at POE: 4.0°C, 39°F			Water temperature at POE: 12°C, 54°		
$CHCl_3$/THM4	64%	73%	82%	71%	73%	77%
Cl-THM/THM4	88%	94%	97%	94%	94%	95%
April 2003:				Water temperature at POE: 22°C, 72°F		
$CHCl_3$/THM4				75%	78%	82%
Cl-THM/THM4				95%	96%	97%
June 2003:	Water temperature at POE: 15.0°C, 59°F			Water temperature at POE: 26°C, 79°F		
$CHCl_3$/THM4	68%	71%	75%	82%	83%	87%
Cl-THM/THM4	90%	92%	93%	97%	98%	98%
August 2003:	Water temperature at POE: 20.1°C, 68°F			Water temperature at POE: 27°C, 81°F		
$CHCl_3$/THM4	83%	84%	87%	71%	76%	83%
Cl-THM/THM4	95%	96%	97%	93%	95%	98%
September/October 2003:	Water temperature at POE: 16.6°C, 62°F			Water temperature at POE: 22°C, 72°F		
$CHCl_3$/THM4	66%	70%	79%	65%	67%	76%
Cl-THM/THM4	86%	90%	95%	89%	91%	94%

Cl-THM: Total amount of chlorine incorporated into THM4 including $CHCl_3$ and $CHBrCl_2$

Trends in HAA Concentrations with Increasing Water Age in the Free Chlorinated Systems

Figures 4.1, 4.2, 4.5 and 4.6 show representative HAA9 concentrations as water age increased in Utilities A and B's distribution systems. Tables 4.5 and 4.6 present all HAA5 and HAA9 data obtained from Utility A and Utility B, respectively. HAA9 concentrations generally increased or remained constant as water age increased and chlorine residual decreased, except at the MRTLs, where a decrease in HAA9 concentrations was measured (mainly in Utility A).

During six of the seven samplings conducted at Utility A, the HAA concentrations decreased at the MRTL (water age of 323 hours) most likely due to HAA biodegradation at this location. This biodegradation was more pronounced when water temperature was high and the free chlorine residual was low (<0.5 mg/L Cl_2). The difference between the maximum HAA9 concentration and the concentration measured at the MRTL averaged 41 percent, and ranged from 0 percent (when the free chlorine residual was 1.3 mg/L) to 92 percent in Utility A. HAA biodegradation at MRTLs has been observed by other authors in a number of distribution systems, typically at locations with low disinfectant residuals and high HPCs (Koch et al. 1991; Jammes et al. 1994; Williams et al. 1994, 1995; LeBel et al. 1995; Chen and Weisel 1998; Obolensky and Frey 2002; Singer et al. 2002). In Utility A, losses of the XAAs and X_2AAs were more pronounced than losses of the X_3AAs, which is consistent with the literature (van der Ploeg et al. 1991; Williams et al. 1996, 1997, 1998; Zhou and Xie 2002; Xie and Zhou 2002).

HAA biodegradation in Utility B was not as pronounced as in Utility A, although lower HAA9 concentrations were measured at the MRTL (water age of 47 hours) during two of the eight samplings conducted (mainly in warmer water). When lower HAA concentrations were measured at the MRTL, the decreases in HAA concentration from POE to MRTL were 11 and 26 percent. The higher chlorine residuals and lower water ages measured in Utility B were most likely responsible for the lower HAA biodegradation observed.

Using the current MCL of 60 μg/L for HAA5 as a marker, results suggest that individual HAA5 concentrations never exceeded this MCL marker in Utility A. In Utility B, however, HAA5 concentrations measured at some of the sampling locations exceeded this concentration during two of the eight samplings conducted. As explained above for THM4, if all sampling locations are used, HAA5 calculations using all data obtained would lead to RAA values of 22.6 and 36.0 μg/L in Utility A and Utility B, respectively. LRAA calculations would lead to the values shown in Table 4.7. Thus both Utilities A and B would be in compliance with both the Stage 1 and Stage 2 DBPR.

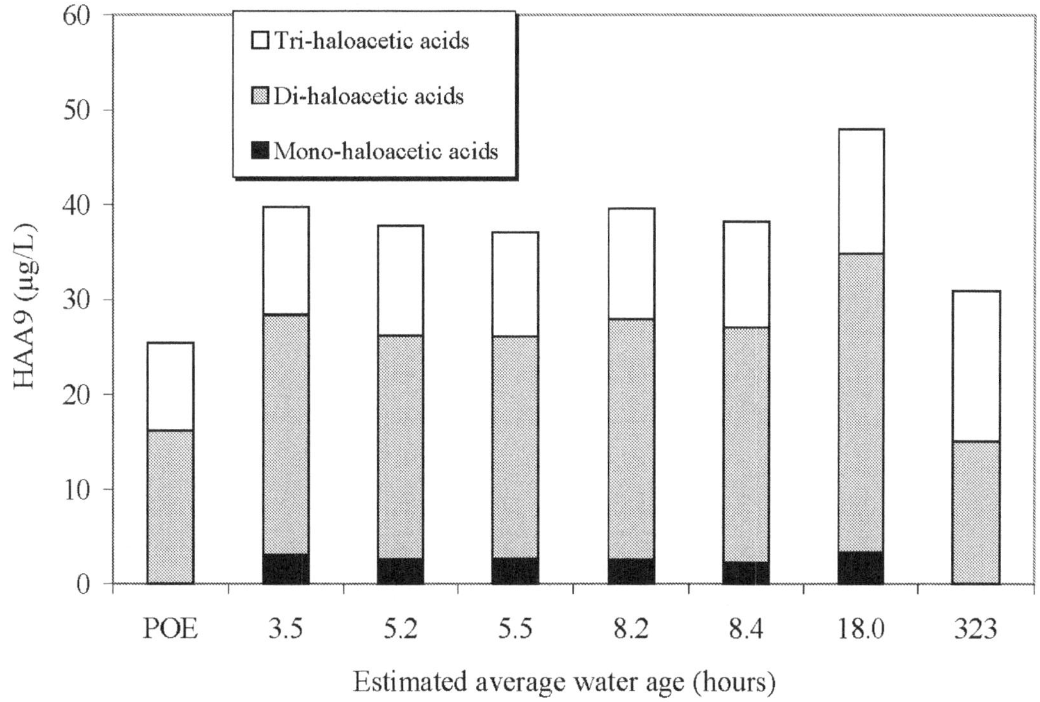

Figure 4.5 HAA9 speciation as a function of water age measured during the September 23, 2003 sampling in Utility A (free chlorinated system) (water temperature at the POE was 16.6°C, 62°F)

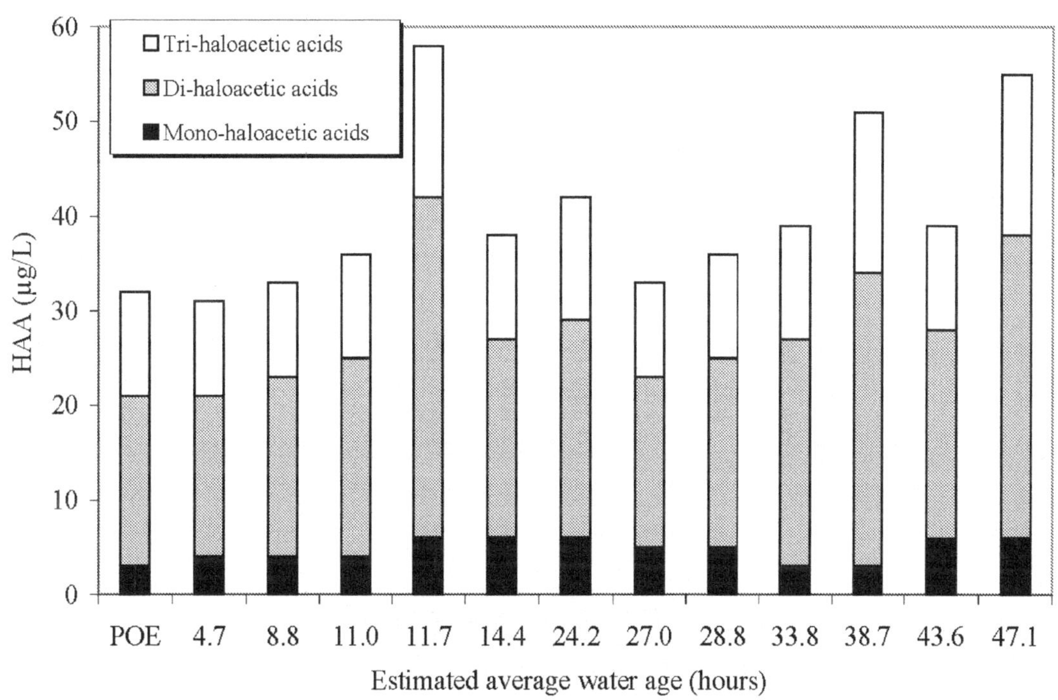

Figure 4.6 HAA9 speciation as a function of water age measured during the January 7, 2003 sampling in Utility B (free chlorinated system) (water temperature at the POE was 11°C, 52°F)

Table 4.5
HAA5 and HAA9 concentrations (µg/L) measured during each sampling in Utility A (free chlorinated system), with water temperature at the POE (°C, °F)

Water age (hours)	October 16, 2002 (13.3°C, 56°F)	December 3, 2002 (5.6°C, 42°F)	January 7, 2003 (6.1°C, 43°F)	February 25, 2003 (4.0°C, 39°F)	June 10, 2003 (15.0°C, 59°F)	August 12, 2003 (20.1°C, 68°F)	September 23, 2003 (16.6°C, 62°F)
0 (POE)	33 and 39	5 and 5	8 and 13	31 and 33	23 and 32	17 and 28	16 and 25
3.5	31 and 40	9 and 9	7 and 13	39 and 46	26 and 36	0 and 0	27 and 40
5.2	28 and 36	10 and 15	8 and 13	37 and 44	23 and 30	25 and 40	26 and 38
5.5	26 and 34	10 and 12	8 and 13	35 and 43	29 and 39	24 and 34	25 and 37
8.2	30 and 38	12 and 16	10 and 17	34 and 41	24 and 31	32 and58	27 and 40
8.4	27 and 37	9 and 9	11 and 18	34 and 41	27 and 36	25 and 36	26 and 38
18.0	25 and 34	10 and 14	13 and 22	23 and 32	26 and 36	30 and 42	34 and 48
323	0 and 3	3 and 6	21 and 32	23 and 33	22 and 29	23 and 33	21 and 31

Table 4.6
HAA5 and HAA9 concentrations (µg/L) measured during each sampling in Utility B (free chlorinated system), with water temperature at the POE (°C, °F)

Water age (hours)	October 15, 2002 (26°C, 79°F)	November 18, 2002 (17°C, 63°F)	January 7, 2003 (11°C, 52°F)	February 25, 2003 (12°C, 54°F)	April 29, 2003 (22°C, 72°F)	June 10, 2003 (26°C, 79°F)	August 12, 2003 (27°C, 81°F)	October 21, 2003 (22°C, 72°F)
0 (POE)	28 and 38	47 and 61	25 and 32	39 and 50	25 and 31	60 and 69	37 and 46	17 and 23
4.7	36 and 49	43 and 55	24 and 31	37 and 45	29 and 36	44 and 51	29 and 36	18 and 24
8.8	39 and 55	38 and 49	27 and 33	37 and 46	30 and 37	68 and 76	36 and 44	24 and 32
11.0	31 and 41	41 and 52	29 and 36	35 and 42	26 and 32	46 and 53	34 and 42	20 and 28
11.7	36 and 50	60 and 78	48 and 58	33 and 43	30 and 37	45 and 51	30 and 38	19 and 26
14.4	38 and 53	60 and 80	31 and 38	44 and 54	24 and 30	41 and 48	47 and 57	17 and 24
19.3	29 and 41	60 and 78	NA	NA	NA	NA	NA	NA
24.2	45 and 62	35 and 45	33 and 42	32 and 40	25 and 31	57 and 64	32 and 40	22 and 30
27.0	29 and 39	55 and 69	26 and 33	37 and 45	26 and 32	70 and 79	NA	17 and 23
28.8	34 and 46	51 and 65	29 and 36	32 and 40	31 and 38	56 and 63	32 and 40	19 and 27
33.8	25 and 35	44 and 56	NA	NA	NA	NA	NA	NA
33.8	29 and 39	53 and 69	31 and 39	35 and 43	36 and 43	45 and 52	33 and 41	21 and 29
38.7	34 and 48	80 and 110	42 and 51	46 and 58	37 and 43	59 and 66	42 and 50	26 and 34
43.6	30 and 43	55 and 70	32 and 39	36 and 44	33 and 40	45 and 52	41 and 50	23 and 31
47.1	43 and 59	41 and 54	46 and 55	30 and 37	36 and 43	64 and 72	39 and 48	18 and 24

NA: Not analyzed

Table 4.7
HAA5 LRAA values calculated in Utilities A and B (free chlorinated system)

Utility A		Utility B	
Water age of sampling location (hours)	HAA5 LRAAs (μg/L)	Water age of sampling location (hours)	HAA5 LRAAs (μg/L)
0 (POE)	21.7	0 (POE)	34.0
3.5	22.4	4.7	32.1
5.2	23.7	8.8	36.9
5.5	24.2	11.0	32.2
8.2	25.6	11.7	36.6
8.4	24.2	14.4	36.0
18.0	24.6	24.2	34.0
323	14.8	27.0	40.3
		28.8	35.0
		33.8	35.6
		38.7	44.5
		43.6	36.5
		47.1	39.9

HAA Speciation in the Free Chlorinated Systems

The proportion of HAA9 represented by HAA5 is summarized in Table 4.8 and illustrated on Figures 4.7 and 4.8 for Utility A and Utility B, respectively. Results indicate that the weight ratios of HAA5 to HAA9 were similar in both free chlorinated systems studied, ranging from 52 to 100 percent (averaging 72 percent) in Utility A, and from 65 to 89 percent (averaging 79 percent) in Utility B. These results suggest that the non-regulated HAAs comprise a significant proportion of the HAAs in the systems studied. In general, the HAA5-to-HAA9 weight ratios decreased slightly or remained constant with increasing water age in Utility A, whereas they remained stable with increasing water age in Utility B. Decreases in the HAA5-to-HAA9 ratio with increasing water age in a distribution system where HAA biodegradation occurs was expected as the XAAs and X_2AAs are the HAA species that are generally biodegraded, as mentioned above. The XAA and X_2AAs comprise four of the five HAAs in HAA5, whereas the less biodegradable X_3AAs mainly constitute the majority of the remaining non-regulated HAAs.

Figures 4.7 and 4.8 show representative HAA results obtained from Utility A and Utility B, respectively. For most of the samplings, the dominant HAA species in both systems was X_2AA, followed by X_3AA, consistent with observations made by other authors.

The weight ratio of chlorine-containing HAA species (i.e., ClAA, Cl_2AA, Cl_3AA and $BrCl_2AA$) measured in the free chlorinated systems is summarized in Table 4.8. In general, the ratio of chlorinated HAAs decreased slightly or remained constant with increasing water age in Utility A. In Utility B, the weight ratio of chlorinated HAAs increased slightly with increasing water age during most samplings, particularly in warm water. These trends are consistent with those observed with the THM4. As mentioned above, increases in the ratio of chlorine-containing HAAs with increasing water age would be expected as the brominated DBP species are known to form faster than the chlorinated species.

Table 4.8
Mass ratios of chlorine-containing HAAs to HAA9 and HAA5 to HAA9 calculated in the free chlorinated systems (Utilities A and B)

	Utility A			Utility B		
	Minimum	Average	Maximum	Minimum	Average	Maximum
October 2002	Water temperature at POE: 13.3°C			Water temperature at POE: 26°C		
Cl-HAA/HAA9	81%	88%	100%	57%	64%	71%
HAA5/HAA9	72%	77%	85%	70%	73%	76%
November/ December 2002	Water temperature at POE: 5.6°C			Water temperature at POE: 17°C		
Cl-HAA/HAA9	82%	94%	100%	69%	72%	74%
HAA5/HAA9	52%	81%	100%	72%	77%	79%
January 2003	Water temperature at POE: 6.1°C			Water temperature at POE: 11°C		
Cl-HAA/HAA9	73%	79%	83%	69%	75%	82%
HAA5/HAA9	55%	59%	66%	75%	79%	82%
February 2003	Water temperature at POE: 4.0°C			Water temperature at POE: 12°C		
Cl-HAA/HAA9	81%	91%	100%	75%	77%	79%
HAA5/HAA9	68%	81%	93%	77%	81%	83%
April 2003				Water temperature at POE: 22°C		
Cl-HAA/HAA9				77%	82%	91%
HAA5/HAA9				80%	82%	86%
June 2003	Water temperature at POE: 15.0°C			Water temperature at POE: 26°C		
Cl-HAA/HAA9	77%	85%	89%	84%	88%	93%
HAA5/HAA9	72%	75%	78%	85%	88%	90%
August 2003	Water temperature at POE: 20.1°C			Water temperature at POE: 27°C		
Cl-HAA/HAA9	63%	76%	84%	63%	80%	90%
HAA5/HAA9	55%	66%	72%	65%	80%	84%
September/ October 2003	Water temperature at POE: 16.6°C			Water temperature at POE: 22°C		
Cl-HAA/HAA9	73%	76%	81%	57%	63%	77%
HAA5/HAA9	64%	68%	71%	70%	74%	77%

Cl-HAA: Sum of ClAA, Cl_2AA, Cl_3AA and $BrCl_2AA$

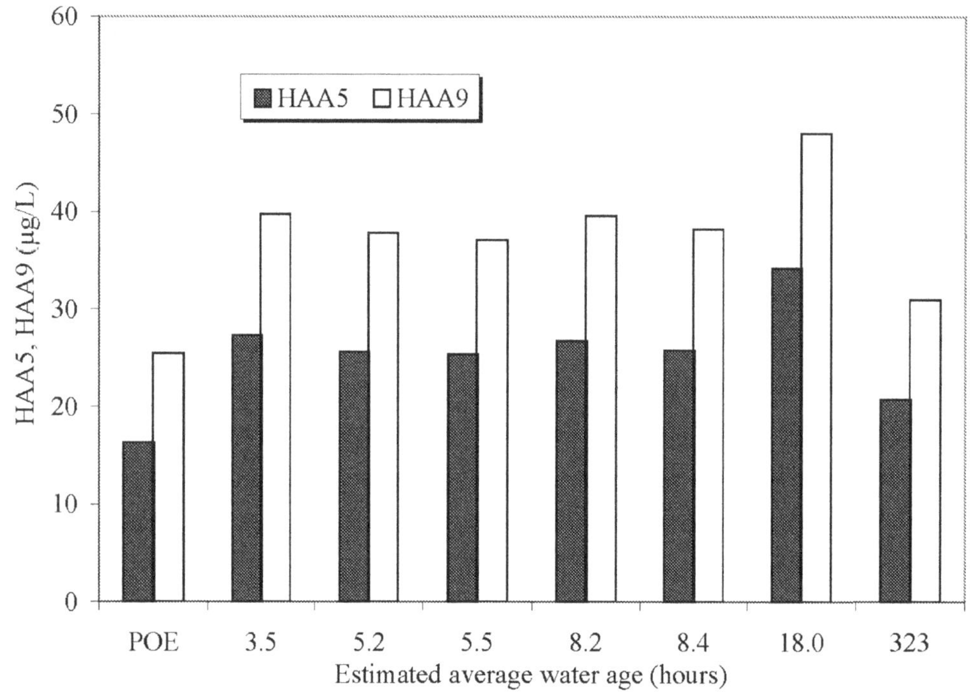

Figure 4.7 HAA5 and HAA9 concentrations as a function of water age measured during the September 23, 2003 sampling at Utility A (free chlorinated system) (water temperature at the POE was 16.6°C, 62°F)

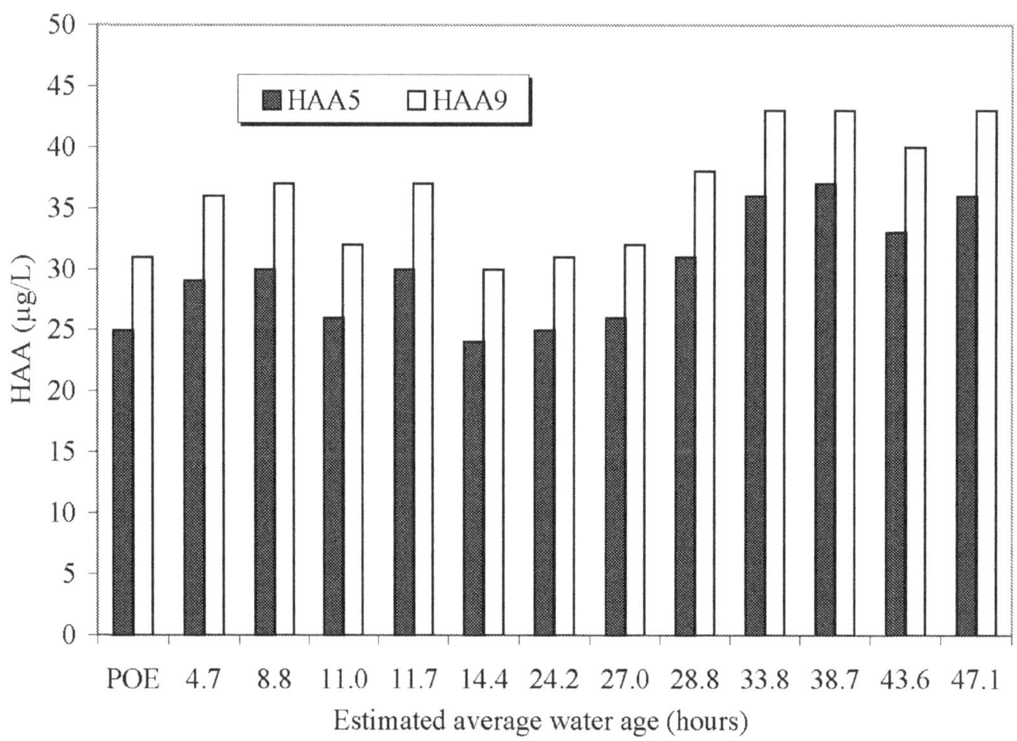

Figure 4.8 HAA5 and HAA9 concentrations as a function of water age measured during the April 29, 2003 sampling at Utility B (free chlorinated system) (water temperature at the POE was 22°C, 72°F)

Chloraminated Systems

Two of the distribution systems studied used chloramines as the final disinfectant, and are designated here in as Utilities C and D.

Trends in THM Concentrations with Increasing Water Age in the Chloraminated Systems

Figures 4.9 and 4.10 present representative results of THM4 concentrations, HAA9 concentrations and chlorine residuals as a function of water age in Utility C and Utility D, respectively. Tables 4.9 and 4.10 present all THM4 data obtained from Utility C and Utility D, respectively. In Utility C, THM4 concentrations were relatively stable as water age increased, a trend that was expected. Surprisingly occasional decreases in concentration were observed at the MRTL (water age of >120 hours) in warmer water and when the total chlorine residual was lower (<2.0 mg/L). Decreases in THM4 concentration of up to 23 percent were measured between the sampling location with the maximum THM4 concentration and the MRTL.

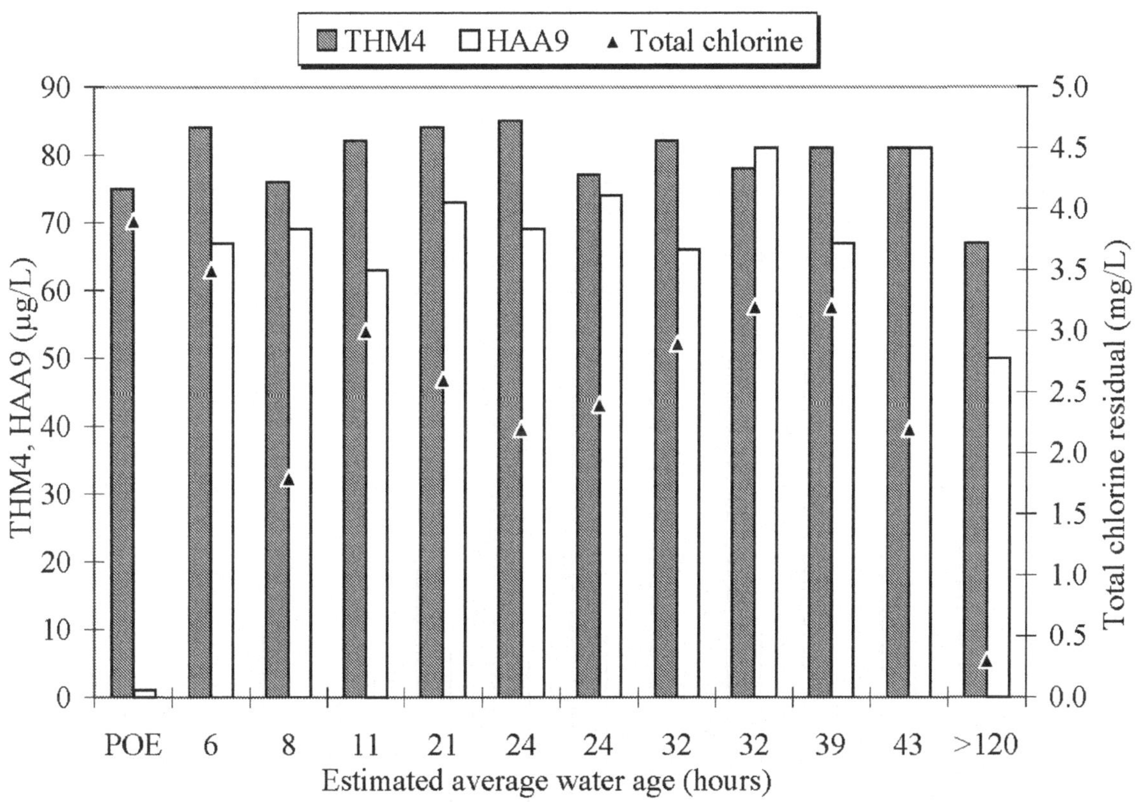

Figure 4.9 THM4, HAA9 and total chlorine residual as a function of water age measured during the June 11, 2003 sampling in Utility C (chloraminated system) (water temperature at the POE was 25°C, 77°F)

a) Both POEs were used (April 8, 2003) (water temperature at the POEs was 18°C, 64°F)

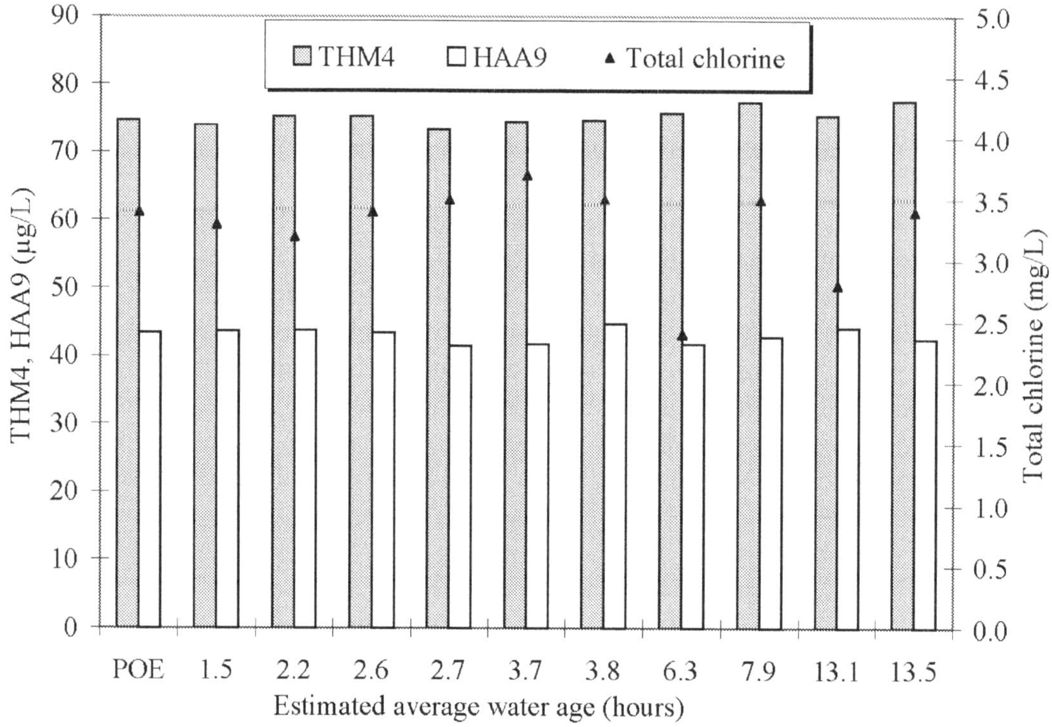

b) One POE was used (June 10, 2003) (water temperature at the POE was 17.8°C, 64°F)

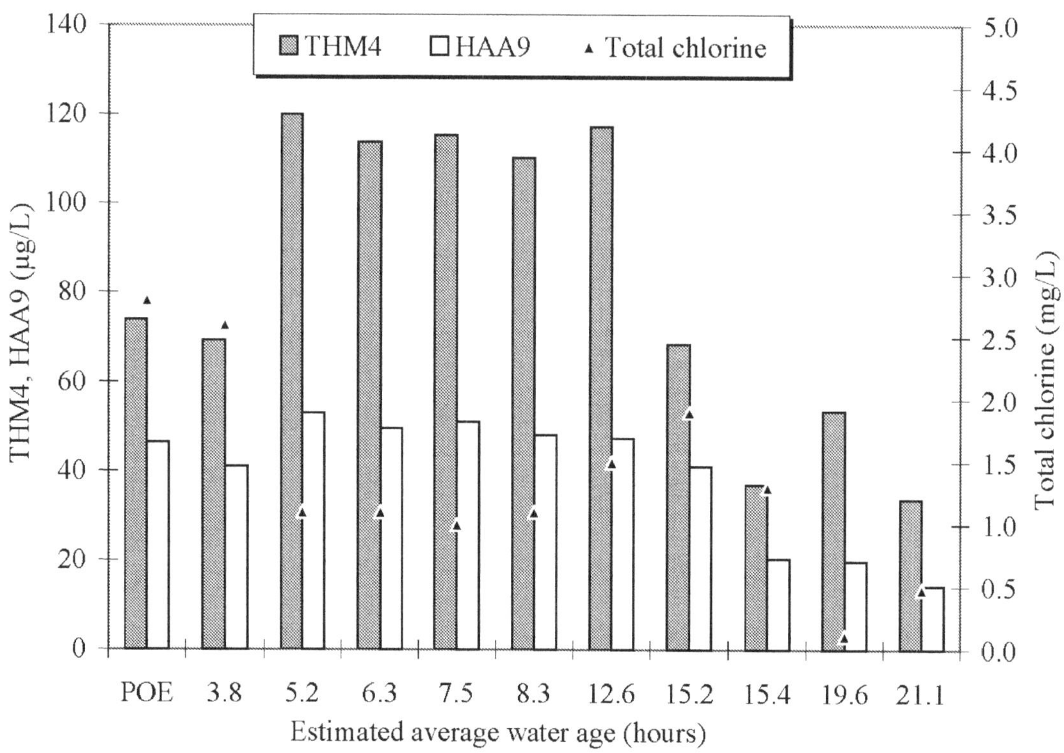

Figure 4.10 THM4, HAA9 and total chlorine residual as a function of water age measured in Utility D (chloraminated system)

Table 4.9
THM4 concentrations (µg/L) measured during each sampling in Utility C (chloraminated system), with water temperature at the POE (°C, °F)

Water age (hours)	October 15, 2002 (22.0°C, 72°F)	November 18, 2002 (16.0°C, 61°F)	January 7, 2003 (10.5°C, 51°F)	February 25, 2003 (8.0°C, 46°F)	April 29, 2003 (20.0°C, 68°F)	June 11, 2003 (25.0°C, 77°F)	August 12, 2003 (27.0°C, 81°F)	October 21, 2003 (19.0°C, 66F)
0 (POE)	56	43	NA	49	72	75	88	72
6	70	46	61	52	72	84	73	77
8	65	51	61	57	82	76	86	76
11	68	43	64	57	79	82	80	77
21	69	39	65	55	80	84	85	75
24	72	52	61	55	79	85	88	73
24	60	46	54	65	81	77	82	76
32	76	52	65	56	93	82	91	82
32	70	43	62	57	76	78	88	71
39	72	44	59	55	75	81	82	78
43	73	51	51	57	78	81	79	80
>120	73	52	61	50	78	67	72	77

NA: Not analyzed

Table 4.10
THM4 concentrations (μg/L) measured during each sampling in Utility D (chloraminated system), with water temperature at the POE (°C, °F)

Site identification	Water age (hours)	November 18, 2002 (21.1°C, 70°F)	January 7, 2003 (15.6°C, 60°F)	April 8, 2003 (18.3°C, 64°F)	June 10, 2003 (17.8°C, 64°F)	July 22, 2003 (21.1°C, 70°F)	September 23, 2003 (21.1°C, 70°F)
No.1 - POE	0	94	79	75	74	74	80
No.2	1.6 – 5.2	46	77	75	120	72	82
No.3	2.4 – 6.3	69	76	73	114	72	87
No.4	3.3 – 7.5	62	75	75	115	73	79
No.5	3.7 – 7.3	67	76	74	110	74	84
No.6	6.3 – 12.6	58	74	76	117	74	80
No.7	10.9 – 18.4	71	75	75	68	75	79
No.8-A	50.0	69	53				
No.8-B	7.9 – 15.4			77	37	73	80
No.9-A	15.3	52	51				
No.9-B	12.7 – 21.1			78	34	75	80
No.10-A	8.3	26	44				
No.10-B	2.2 – 3.8			75	69	74	83
No.11-A	37.2	36	43				
No.11-B	1.5			74			
No.11-C	10.0 – 19.6				53	74	90
No.12-A	11.1	39	44				
No.12-B - POE	0					73	82

The decreases in THM4 concentrations at the MRTLs were surprising. There is no evidence in the literature suggesting biodegradation of THMs as there is for HAAs, except under anoxic conditions (Singer et al. 1993), which is not likely to occur in full-scale distribution systems. However recent studies by Wahman et al. (2003) have reported THM biodegradation via co-metabolism by species of nitrifying bacteria that have been isolated from drinking water distribution systems, such as *Nitrosomonas europaea* and *oligotropha*. In Utility C, nitrification was observed at the MRTL as shown by increases in nitrite, nitrate and HPC concentrations, and decreases in total chlorine residual and total ammonia concentration. Further information on the occurrence of nitrification in the distribution systems studied is presented at the end of this chapter. Although microorganism identification was not conducted during this study, it is possible that nitrifying bacteria may have degraded THMs at MRTLs.

In Utility D, the trends in THM4 concentrations as a function of water age varied from one sampling event to another. It appears that these trends might have been related to the different hydraulic scenarios that occurred in this distribution system over the course of the study. When both POEs were used (during the samplings conducted in April, July, and September 2003), water ages were lower and THM4 concentrations remained constant with increasing water age (Figure 4.10a). The lower water ages most likely helped limit THM variations throughout this distribution system. However when only one POE was used (samplings conducted in November 2002, January 2003, and June 2003), irregardless of the POE (whether it was the main or secondary POE), water ages were higher throughout the distribution system, and THM concentrations decreased at the sampling locations with higher water ages and low chlorine residuals (Figure 4.10b). Decreases in THM4 concentration of 46 to 62 percent from the POE to some of the MRTLs were observed during these samplings. Signs of nitrification were obvious at the sampling locations that showed lower THM4 concentrations during these sampling events, as described at the end of this chapter. As explained above, it is possible that THMs were biodegraded via co-metabolism by nitrifying bacteria (Wahman et al. 2003).

It appears that THM4 concentrations measured in Utility D's distribution system were much more affected by relatively small increases in water age (>13 hours). This is actually not the case as Utility D is a consecutive system, and the water ages presented here are only for the section of the distribution system that was studied in this project, without consideration for the travel time in the wholesaler's distribution system prior to reaching Utility D's POE.

Although Utility C uses chloramines as the final disinfectant, individual high THM4 concentrations were measured and exceeded the MCL marker of 80 μg/L during the four samplings conducted in warmer water. Similarly, individual THM4 concentrations measured in Utility D exceeded 80 μg/L during three of the six samplings conducted. The concentrations measured in these systems during this study were consistent with historical data. If all sampling locations were used to calculate RAA and LRAA values, as explained above, RAA values for THM4 would be 71.1 and 68.7 μg/L in Utility C and Utility D, respectively. LRAA values are summarized in Table 4.11. These results suggest that although these utilities would be in compliance with the Stage 1 DBPR, Utility D would be out of compliance with the Stage 2 DBPR if the sampling locations selected for this study were used for compliance purposes.

Table 4.11
THM4 LRAA values calculated in Utilities C and D (chloraminated systems)

Utility C		Utility D	
Water age of sampling locations (hours)	THM4 LRAAs (μg/L)	Water age of sampling locations (hours)	THM4 LRAAs (μg/L)
0 (POE)	64.5	0 (POE 1)	**80.8**
6	66.2	1.6 to 5.2	69.2
8	70.8	2.4 to 6.3	74.6
11	69.2	3.3 to 7.5	72.4
21	70.0	3.7 to 8.3	74.2
24	71.1	6.3 to 12.6	71.5
24	70.5	10.9 to 18.4	76.1
32	78.6		
32	68.9		
39	69.0		
43	68.8		
>120	67.4		

Concentrations in bold characters are greater than the 80 μg/L MCL marker

THM Speciation in the Chloraminated Systems

Utility C was experiencing drought conditions when sampling began in October 2002. As a result, raw water bromide concentrations were higher during the first two samplings of October and November 2002 (370 to 400 μg/L) than normal (32 to 66 μg/L during the subsequent samplings). Results from the first two samplings indicate that the treated water contained mainly the brominated THM species (i.e., $CHClBr_2$ and $CHBr_3$), whereas subsequent samplings were dominated by $CHCl_3$ and $CHCl_2Br$ (Figure 4.11 and Table 4.12). Raw water bromide concentrations were relatively high in Utility D (up to 0.36 mg/L), resulting in a high proportion of brominated THMs (Figure 4.12 and Table 4.12).

In general, the weight fraction of chlorine-containing THMs (i.e., $CHCl_3$ and $CHCl_2Br$) decreased slightly with increasing water age or remained stable in Utility C. In a chloraminated system, changes in THM speciation with water age are not expected as chloramines form only limited amounts of these DBPs. The bromine-containing THMs were conserved throughout the distribution system. Also, the distribution of the THM species remained essentially the same at the locations with high and low THM4 concentrations; i.e., the loss in THMs was not attributed to any particular species. These results are surprising as the bromine-containing THM species would be expected to degrade faster than the chlorine-containing species (Wahman et al. 2003).

In Utility D, the weight fraction of chlorine-containing THMs (i.e., $CHCl_3$ and $CHCl_2Br$) did not follow any trends with increasing water age, as expected. During most of the samplings, the variations in these fractions were too small to show a clear trend. Only during the January 2003 sampling did the weight fraction of chlorine-containing THMs significantly increased with increasing water age. As observed in Utility C, the losses of THMs could not be attributed to any specific species in Utility D: the chlorine-containing THMs were the species that were lost at times (June 2003 sampling), while at other times (November 2002 and January 2003 samplings), it was the bromine-containing species that were lost.

a) High raw water bromide concentration (October 15, 2002) (water temperature at POE of 22°C, 72°F)

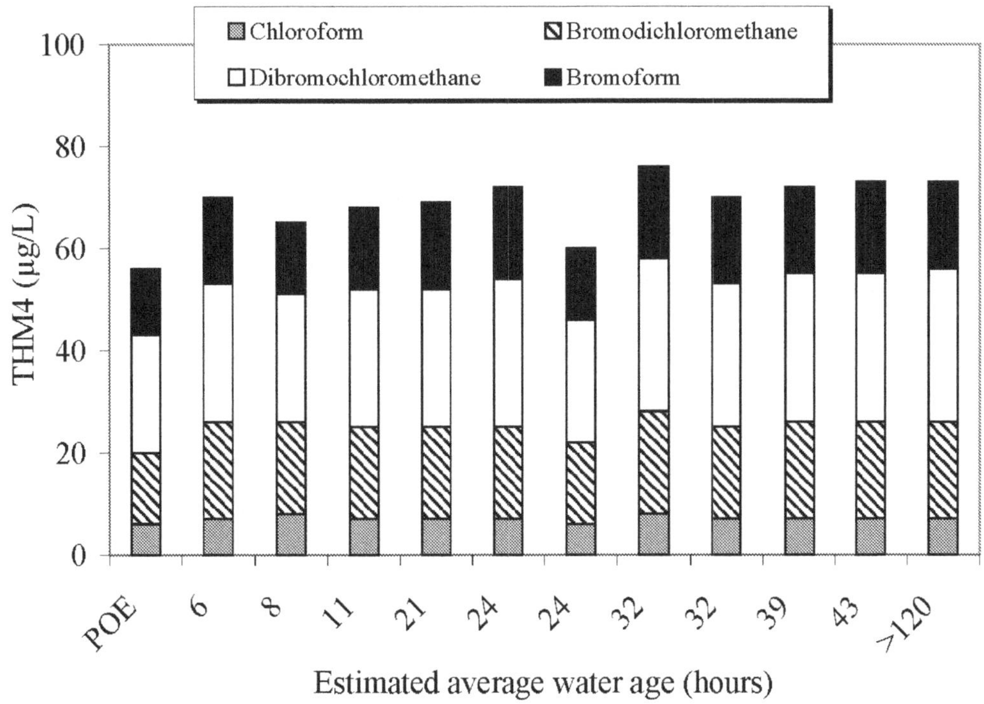

b) Low raw water bromide concentration (October 21, 2003) (water temperature at POE of 19°C, 66°F)

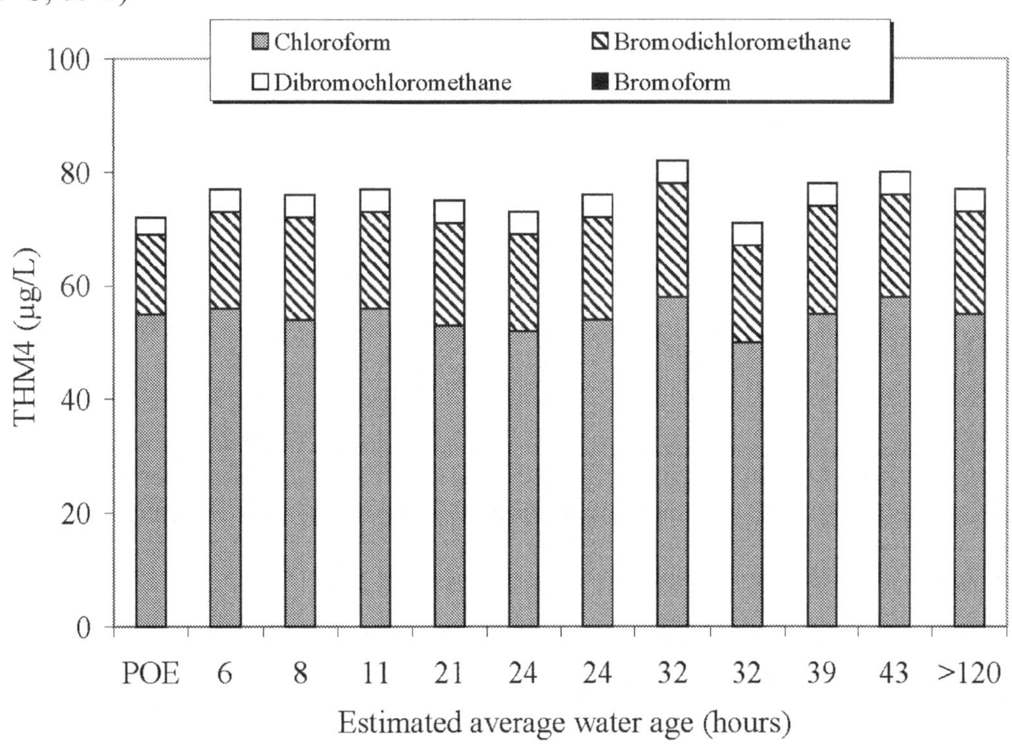

Figure 4.11 THM speciation as a function of water age measured in Utility C (chloraminated system)

Table 4.12
Mass ratios of $CHCl_3$ to THM4, chlorine-containing THMs to THM4, and bromine-containing THMs to THM4 calculated in the chloraminated systems (Utilities C and D)

	Utility C			Utility D		
	Minimum	Average	Maximum	Minimum	Average	Maximum
October 2002 :	Water temperature: 22°C, 72°F					
$CHCl_3$/THM4	9.6%	10%	12%			
Cl-THM/THM4	35%	36%	40%			
Br-THM/THM4	60%	64%	65%			
November/December 2002:	Water temperature: 16°C, 61°F			Water temperature: 21.1°C, 70°F		
$CHCl_3$/THM4	14%	20%	23%	13%	19%	29%
Cl-THM/THM4	42%	54%	58%	36%	41%	51%
Br-THM/THM4	42%	46%	58%	49%	59%	64%
January 2003:	Water temperature: 10.5°C, 51°F			Water temperature: 15.6°C, 60°F		
$CHCl_3$/THM4	63%	77%	79%	10%	14%	19%
Cl-THM/THM4	93%	96%	97%	39%	43%	50%
Br-THM/THM4				50%	57%	61%
February 2003:	Water temperature: 8°C, 46°F					
$CHCl_3$/THM4	63%	66%	79%			
Cl-THM/THM4	92%	93%	97%			
April 2003:	Water temperature: 20°C, 68°F			Water temperature: 18.3°C, 64°F		
$CHCl_3$/THM4	79%	81%	82%	6.7%	6.9%	7.4%
Cl-THM/THM4	97%	98%	99%	35%	36%	36%
Br-THM/THM4				64%	64%	65%
June 2003:	Water temperature: 25°C, 77°F			Water temperature: 17.8°C, 64°F		
$CHCl_3$/THM4	80%	81%	83%	6.3%	10%	15%
Cl-THM/THM4	97%	98%	99%	37%	39%	42%
Br-THM/THM4				58%	61%	63%
July/August 2003:	Water temperature: 27°C, 81°F			Water temperature: 21.1°C (70°F) at POE 1, and 23.3°C at POE 2 (74°F)		
$CHCl_3$/THM4	86%	89%	91%	14%	14%	15%
Cl-THM/THM4	99%	99%	100%	42%	43%	44%
Br-THM/THM4				56%	57%	58%
September/October 2003:	Water temperature: 19°C, 66°F			Water temperature: 21.1°C (70°F) at POE 1, and 22.2°C at POE 2 (72°F)		
$CHCl_3$/THM4	70%	72%	76%	15%	16%	18%
Cl-THM/THM4	94%	95%	96%	44%	45%	46%
Br-THM/THM4				54%	55%	56%

Cl-THM: Chlorinated THM4 species including $CHCl_3$ and $CHBrCl_2$
Br-THM: Brominated THM4 species including $CHBr_3$ and $CHBr_2Cl$

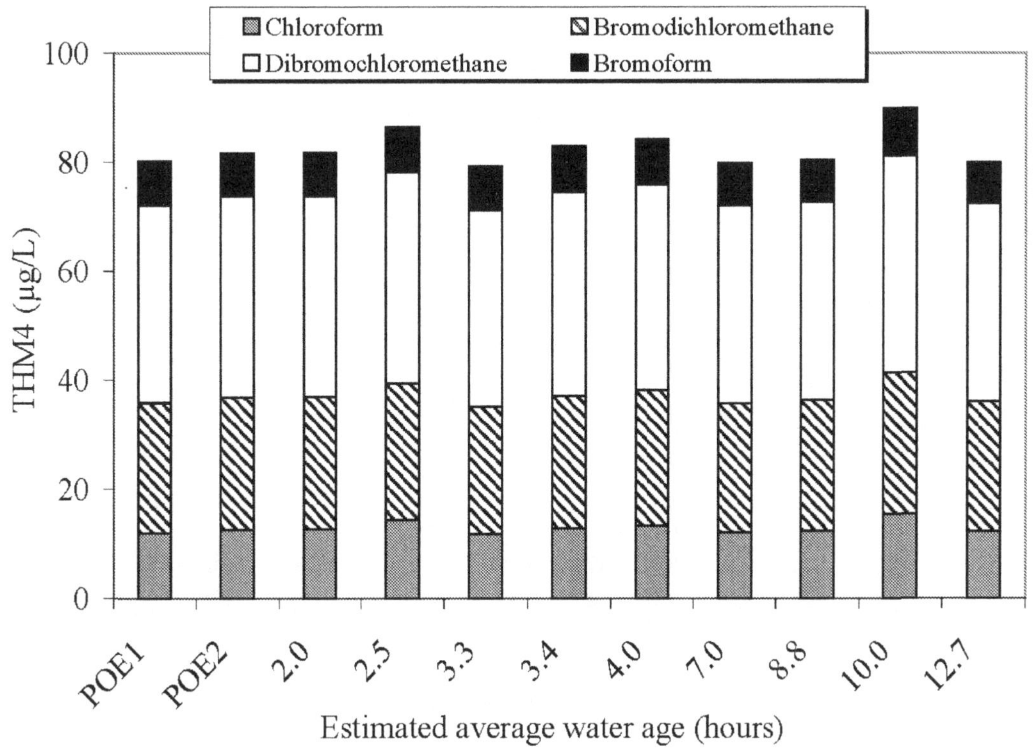

Figure 4.12 THM4 speciation as a function of water age measured during the September 23, 2003 sampling in Utility D (chloraminated system) (water temperature at the POEs was 21-22°C, 70-72°F)

Trends in HAA Concentrations with Increasing Water Age in the Chloraminated Systems

Figures 4.9, 4.10, 4.13 and 4.14 present representative HAA9 concentrations measured in Utility C's distribution system. Tables 4.13 and 4.14 present all THM4 data obtained from Utility C and Utility D, respectively. Results indicate that HAA9 concentrations were relatively constant throughout the distribution system, with the exception of the MRTL (water age of >120 hours), where a decrease in HAA concentrations occurred during six out of the eight samplings conducted. When a decrease in HAA concentration was noticed, the differences between the highest HAA9 concentration and the concentration measured at the MRTL ranged from 23 to 85 percent. The MRTL also showed lower total chlorine residuals (<0.5 mg/L Cl_2) and higher HPCs (1 to 2.7 log higher than at other locations) during most samplings, which suggests that HAA biodegradation was occurring. These conditions are consistent with observations made during previous studies (Koch et al. 1991; Jammes et al. 1994; Williams et al. 1994, 1995; LeBel et al. 1995; Chen and Weisel 1998; Obolensky and Frey 2002; Singer et al. 2002). The decreases in HAA concentrations at MRTLs were generally associated with losses in XAAs and X_2AAs; X_3AAs remained relatively conserved. This is consistent with the literature which reports that microorganisms tend to degrade the XAA and X_2AA species but not the X_3AA species (van der Ploeg et al. 1991; Williams et al. 1996, 1997, 1998; Zhou and Xie 2002; Xie and Zhou 2002).

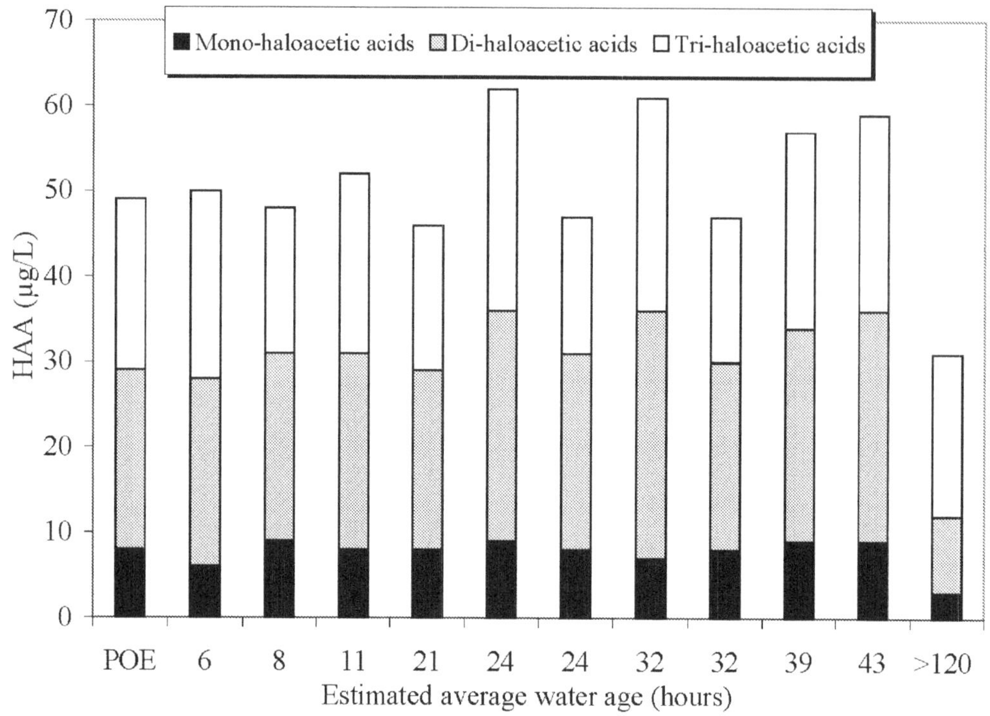

Figure 4.13 HAA9 speciation as a function of water age measured during the January 7, 2003 sampling in Utility C (chloraminated system) (water temperature at POE was 10.5°C, 51°F)

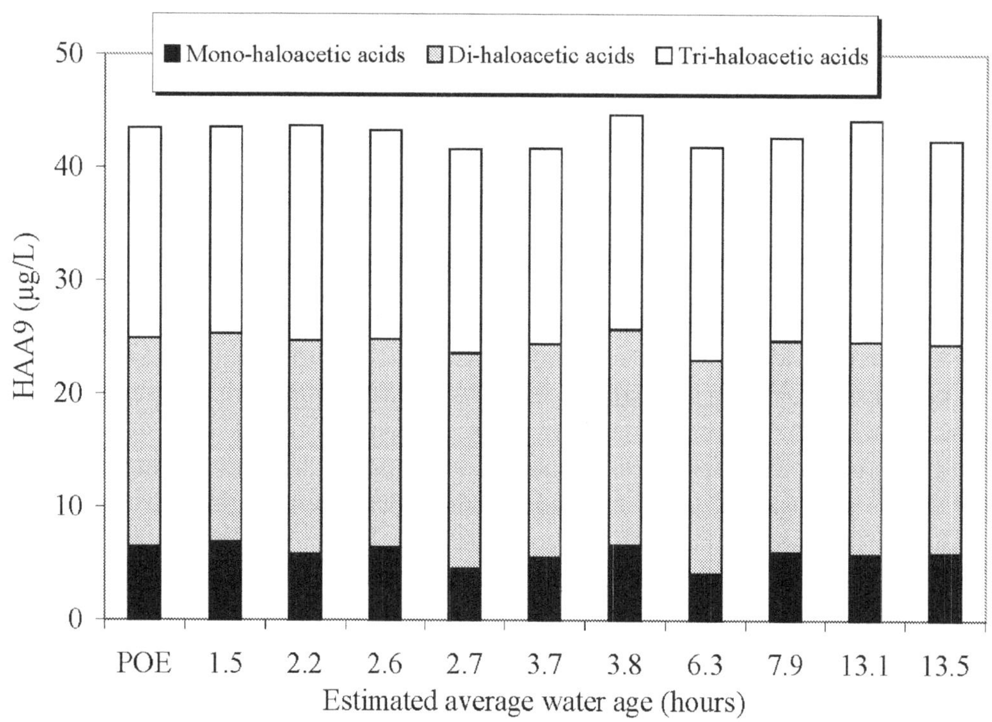

Figure 4.14 HAA9 speciation as a function of water age measured during the April 8, 2003 sampling in Utility D (chloraminated system) (water temperature at POE was 18.3°C, 64°F)

Table 4.13
HAA5 and HAA9 concentrations (µg/L) measured during each sampling in Utility C (chloraminated system), with water temperature at the POE (°C, °F)

Water age (hours)	October 15, 2002 (22.0°C, 72°F)	November 18, 2002 (16.0°C, 61°F)	January 7, 2003 (10.5°C, 51°F)	February 25, 2003 (8.0°C, 46°F)	April 29, 2003 (20.0°C, 68°F)	June 11, 2003 (25.0°C, 77°F)	August 12, 2003 (27.0°C, 81°F)	October 21, 2003 (19.0°C, 66°F)
0 (POE)	26 and 47	47 and 80	39 and 50	48 and 56	74 and 84	NA	80 and 88	50 and 61
6	29 and 63	39 and 65	38 and 50	46 and 54	69 and 77	59 and 67	74 and 82	47 and 58
8	27 and 63	43 and 70	38 and 49	58 and 68	65 and 73	65 and 69	78 and 85	49 and 61
11	22 and 53	50 and 91	40 and 52	47 and 55	66 and 74	56 and 63	83 and 93	66 and 80
21	23 and 57	44 and 75	36 and 47	55 and 66	68 and 76	64 and 73	85 and 93	41 and 51
24	34 and 70	36 and 60	48 and 62	53 and 64	70 and 78	61 and 69	72 and 80	56 and 69
24	29 and 64	44 and 73	38 and 48	49 and 57	NA	65 and 74	81 and 89	64 and 79
32	28 and 60	37 and 61	47 and 61	51 and 61	72 and 80	59 and 66	72 and 80	65 and 82
32	25 and 56	39 and 66	37 and 48	62 and 74	59 and 66	71 and 81	72 and 80	47 and 58
39	32 and 80	38 and 63	44 and 57	58 and 69	71 and 79	59 and 67	74 and 83	52 and 65
43	33 and 63	51 and 89	46 and 59	55 and 65	59 and 66	71 and 81	74 and 81	49 and 61
>120	23 and 48	40 and 73	22 and 31	55 and 64	49 and 57	43 and 50	64 and 72	10 and 12

NA: Not analyzed

Table 4.14
HAA5 and HAA9 concentrations (µg/L) measured during each sampling in Utility D (chloraminated system), with water temperature at the POE (°C, °F)

Site identification	Water age (hours)	November 18, 2002 (21.1°C, 70°F)	January 7, 2003 (15.6°C, 60°F)	April 8, 2003 (18.3°C, 64°F)	June 10, 2003 (17.8°C, 64°F)	July 22, 2003 (21.1°C, 70°F)	September 23, 2003 (21.1°C, 70°F)
No.1 - POE	0	19 and 43	11 and 36	21 and 43	22 and 47	17 and 41	19 and 46
No.2	1.6 – 5.2	11 and 23	14 and 40	21 and 43	22 and 53	16 and 40	18 and 43
No.3	2.4 – 6.3	15 and 32	16 and 42	20 and 42	21 and 50	17 and 41	19 and 46
No.4	3.3 – 7.5	14 and 29	17 and 45	22 and 45	21 and 51	17 and 40	20 and 47
No.5	3.7 – 7.3	17 and 33	16 and 43	21 and 42	20 and 48	17 and 41	19 and 47
No.6	6.3 – 12.6	12 and 27	18 and 45	19 and 42	20 and 47	17 and 41	20 and 48
No.7	10.9 – 18.4	16 and 35	14 and 42	21 and 44	19 and 41	16 and 40	19 and 48
No.8-A	50.0	17 and 41	2 and 15				
No.8-B	7.9 – 15.4			21 and 43	6 and 20	16 and 42	21 and 47
No.9-A	15.3	6 and 20	2 and 17				
No.9-B	12.7 – 21.1			21 and 43	3 and 14	16 and 39	20 and 46
No.10-A	8.3	2 and 11	2 and 15				
No.10-B	2.2 – 3.8			21 and 44	16 and 41	17 and 42	19 and 47
No.11-A	37.2	3 and 9	2 and 15				
No.11-B	1.5			22 and 44			
No.11-C	10.0 – 19.6				6 and 20	17 and 41	20 and 50
No.12-A	11.1	3 and 9	2 and 16				
No.12-B - POE	0					9 and 35	19 and 49

As observed with the THM data measured in Utility D, HAA9 concentrations followed either one of two trends with increasing water age, depending on the hydraulic scenario: concentrations remained constant when both POEs were used (Figure 4.10a), or remained constant and then decreased at MRTLs (where chlorine residual was lower, which suggests that biodegradation was occurring) when only one POE was used (Figure 4.10b). Consistently, the lower HAA concentrations were observed at the locations where lower THM4 concentrations were measured. The decreases in HAA concentrations were essentially due to the loss in XAAs and X_2AAs, the most biodegradable species. The X_3AAs were the only remaining HAA species at the MRTLs.

During five out of the eight samplings conducted in Utility C, some locations exceeded the HAA5 MCL marker of 60 μg/L. In Utility D, individual HAA5 concentrations never exceeded this concentration. In fact, HAA5 concentrations remained below 23 μg/L during all of the samplings and at all locations. However HAA5 concentrations represented less than 50 percent of HAA9 concentrations measured in Utility D, as mentioned above. If all sampling locations used were considered in the calculations, HAA5 RAA values would be 55.1 and 14.8 μg/L in Utility C and Utility D, respectively, and LRAA values at both utilities could be summarized as per Table 4.15. Thus both Utilities C and D would be in compliance with both the Stage 1 and Stage 2 DBPR.

Table 4.15
HAA5 LRAA values calculated in Utilities C and D (chloraminated systems)

Utility C		Utility D	
Water age of sampling locations (hours)	HAA5 LRAAs (μg/L)	Water age of sampling locations (hours)	HAA5 LRAAs (μg/L)
0 (POE)	59.6	0 (POE 1)	17.7
6	54.0	1.6 to 5.2	16.3
8	56.0	2.4 to 6.3	17.4
11	56.3	3.3 to 7.5	17.7
21	56.0	3.7 to 8.3	18.4
24	57.3	6.3 to 12.6	17.1
32	57.5	10.9 to 18.4	17.4
32	54.3		
39	57.3		
43	56.6		
>120	41.3		

HAA Speciation in the Chloraminated Systems

The proportion of HAA9 represented by HAA5 is summarized in Table 4.16 and illustrated on Figures 4.15 and 4.16 for Utility C and Utility D, respectively. Results indicate that the HAA5-to-HAA9 weight ratios were much higher in Utility C, ranging from 40 to 94 percent (averaging 77 percent), than in Utility D where they ranged from 12 to 50 percent (averaging 39 percent). These results can be explained by the concentration of bromide present in the source water of these utilities. In the source water of Utility C, bromide concentrations were 370 to 401 μg/L during the first two samplings, and <10 to 66 μg/L for the remaining samplings, whereas bromide concentrations ranged from 0.12 to 0.18 mg/L in the source water of Utility D. Greater bromide concentrations lead to lower HAA5-to-HAA9 ratios as the bromine-containing HAA species are not well represented by the HAA5. Consistent with the high raw water bromide concentrations measured in Utility C at the beginning of the study, the HAA5-to-HAA9 weight ratios calculated during the first two samplings were much lower than the ratios calculated during the subsequent samplings (Table 4.16). In both Utilities C and D, the weight ratios of HAA5-to-HAA9 remained unchanged or decreased slightly with increasing water age. As explained above, this trend was expected as the biodegradable HAA species (XAAs and X_2AAs) mainly compose the HAA5, whereas the less biodegradable X_3AAs are included in the HAA9 calculations.

In general, X_2AAs and X_3AAs accounted for most HAA concentrations measured in Utilities C and D (Figures 4.13 and 4.14), consistent with observations made by others.

Because of the high raw water bromide concentration occurring in Utility C at the beginning of the study, samples collected during the first two samplings, October and November 2002, contained significant concentrations of brominated HAAs (i.e., BrAA, Br_2AA, Br_3AA and Br_2ClAA; Table 4.16). Subsequent samplings were dominated by the chlorinated HAAs (i.e., ClAA, Cl_2AA, Cl_3AA and $BrCl_2AA$). In Utility D, the relatively high bromide concentration normally present lead to similar fractions of brominated and chlorinated HAAs throughout the study. The weight fraction of chlorine-containing HAAs followed various trends with increasing water age in the distribution systems of both chloraminated utilities: they sometimes increased, sometimes decreased, and remained stable at other times. Such trend was expected as chloramines form only limited amounts of HAAs.

Chloraminated System Periodically Switching to Free Chlorine

Utility E uses chloramines as the final disinfectant, but switches to free chlorine every year for a four-week period starting on March 1 to limit nitrification. Two samplings were conducted during chloramination prior to the switch to free chlorine, two samplings during the free chlorination period (on Day 11 and 31 of free chlorination), and two sampling after the return to chloramination (on Days 29 and 50 following the return to chloramines).

Utility E's distribution system is fed by two treatment plants, receiving a constant blend of the two water sources over the course of a year. Both POEs to the distribution system were sampled during this study. As explained in the Materials and Methods section, it appeared that Site 2 was the dominant source water for sampling Sites 3, 4, and 7, while all of the remaining locations receive predominantly Site 1 source water. Figures 4.17 and 4.18 show THM4 and HAA9 concentrations for all water ages categorized by POE water. The POE water did not have any discernable effect on THM4 or HAA9 concentrations observed in the distribution system.

Table 4.16
Mass ratios of chlorine-containing HAAs to HAA9, bromine-containing HAAs to HAA9 and HAA5 to HAA9 calculated in the chloraminated systems (Utilities C and D)

	Utility C			Utility D		
	Minimum	Average	Maximum	Minimum	Average	Maximum
October 2002	Water temperature at POE: 22°C, 72°F					
Cl-HAA/HAA9	30%	34%	38%			
Br-HAA/HAA9	40%	44%	48%			
HAA5/HAA9	40%	46%	55%			
November/December 2002	Water temperature at POE: 16°C, 61°F			Water temperature at POE: 21.1°C, 70°F		
Cl-HAA/HAA9	33%	37%	40%	33%	43%	68%
Br-HAA/HAA9	42%	44%	51%	32%	44%	54%
HAA5/HAA9	54%	58%	61%	22%	40%	50%
January 2003	Water temperature at POE: 10.5°C, 51°F			Water temperature at POE: 15.6°C, 60°F		
Cl-HAA/HAA9	70%	73%	77%	35%	43%	49%
Br-HAA/HAA9	14%	17%	20%	44%	49%	53%
HAA5/HAA9	71%	77%	79%	12%	27%	40%
February 2003	Water temperature at POE: 8°C, 46°F					
Cl-HAA/HAA9	80%	82%	84%			
Br-HAA/HAA9	8.8%	10%	12%			
HAA5/HAA9	83%	85%	86%			
April 2003	Water temperature at POE: 20°C, 68°F			Water temperature at POE: 18.3°C, 64°F		
Cl-HAA/HAA9	87%	89%	90%	41%	44%	45%
Br-HAA/HAA9	4.6%	5.8%	8.3%	41%	42%	44%
HAA5/HAA9	86%	89%	90%	46%	48%	49%
June 2003	Water temperature at POE: 25°C, 77°F			Water temperature at POE: 17.8°C, 64°F		
Cl-HAA/HAA9	85%	87%	91%	25%	38%	44%
Br-HAA/HAA9	6.0%	8.0%	9.9%	42%	46%	59%
HAA5/HAA9	86%	89%	94%	21%	39%	47%
July/August 2003	Water temperature at POE: 27°C, 81°F			Water temperature: 21.1°C (70°F) at POE 1 and 23.3°C at POE 2 (74°F)		
Cl-HAA/HAA9	87%	90%	91%	19%	35%	38%
Br-HAA/HAA9	3.2%	5.3%	7.1%	44%	47%	59%
HAA5/HAA9	89%	90%	92%	26%	39%	42%
Cl-HAA/HAA9 (without POE2)				35%	36%	38%
Br-HAA/HAA9 (without POE2)				44%	46%	48%
HAA5/HAA9 (without POE2)				38%	41%	42%
September/October 2003	Water temperature at POE: 19°C, 66°F			Water temperature: 21.1°C (70°F) at POE 1 and 22.2°C at POE 2 (72°F)		
Cl-HAA/HAA9	71%	76%	83%	42%	42%	43%
Br-HAA/HAA9	9.8%	17%	22%	40%	41%	41%
HAA5/HAA9	79%	81%	83%	39%	42%	45%

Cl-HAA: Sum of ClAA, Cl_2AA, Cl_3AA and $BrCl_2AA$; Br-HAA: Sum of BrAA, Br_2AA, Br_3AA and Br_2ClAA

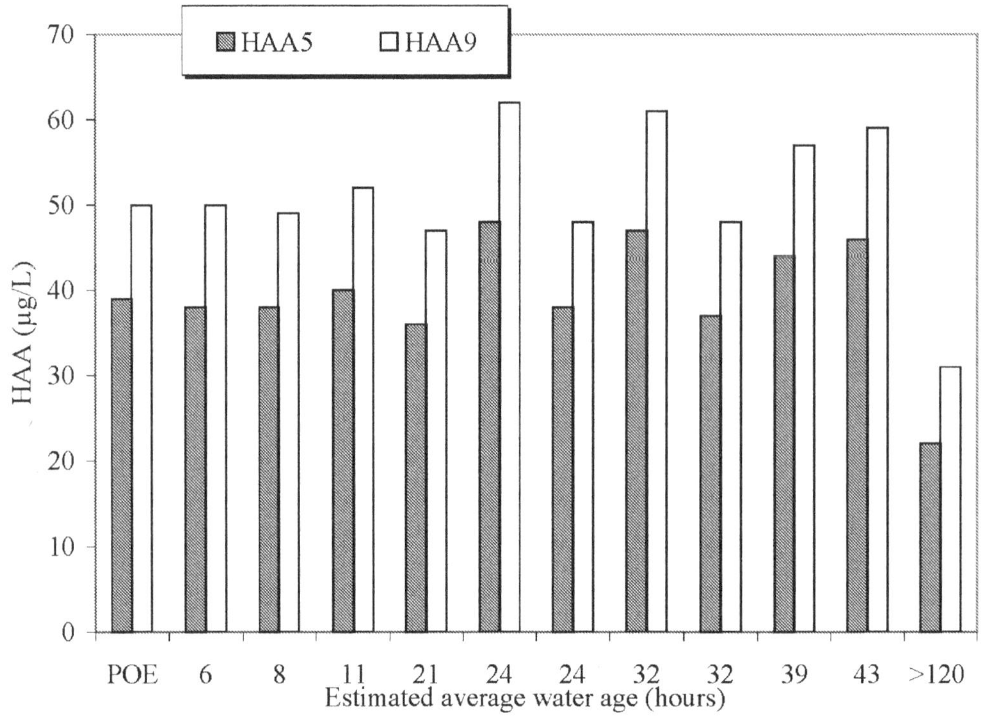

Figure 4.15 HAA5 and HAA9 concentrations as a function of water age measured during the January 7, 2003 sampling at Utility C (chloraminated system) (water temperature at POE was 10.5°C, 51°F)

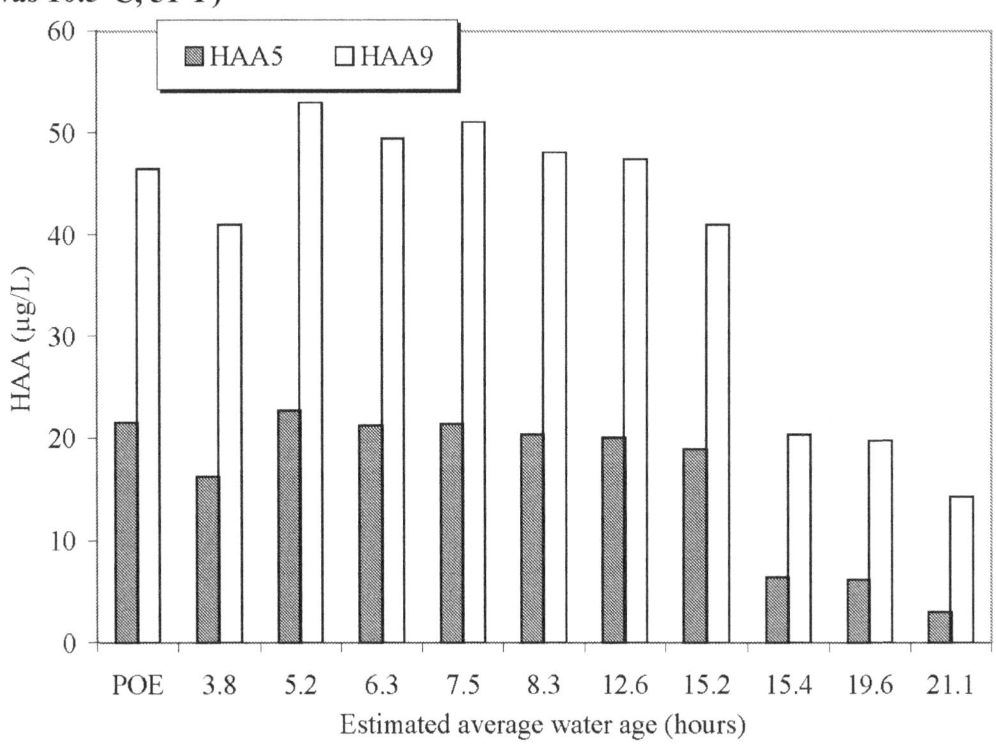

Figure 4.16 HAA5 and HAA9 concentrations as a function of water age measured during the June 10, 2003 sampling at Utility D (chloraminated system) (water temperature at POE was 17.8°C, 64°F)

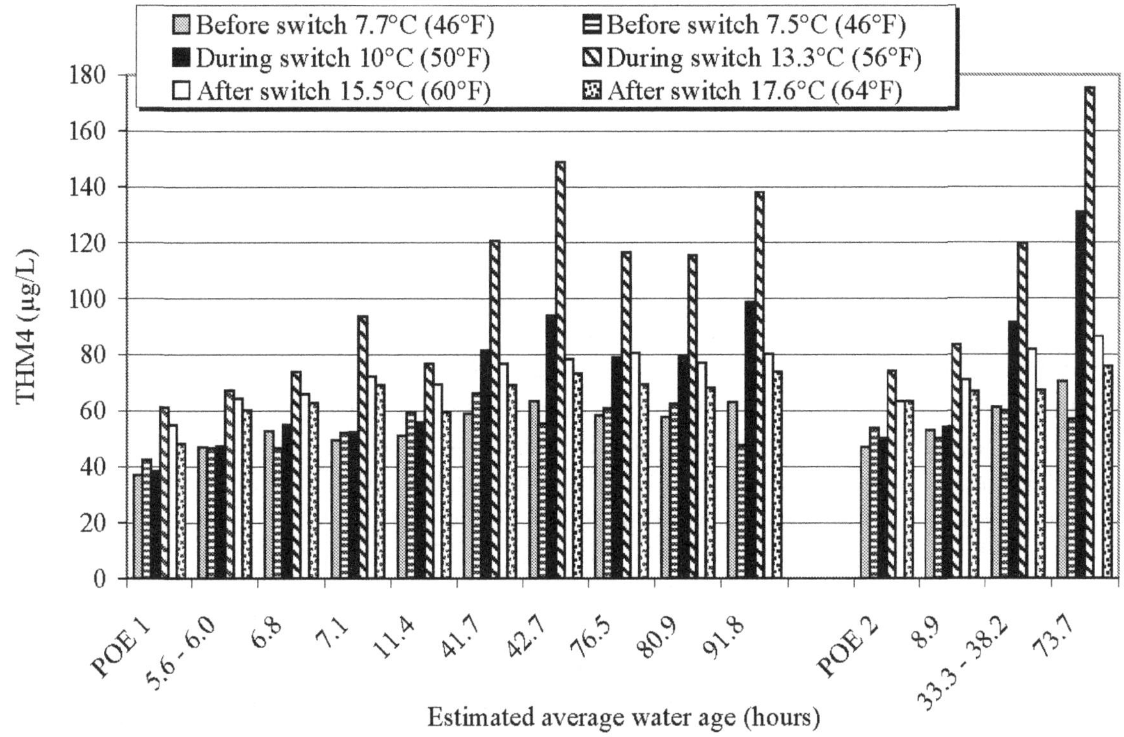

Figure 4.17 THM4 concentration as a function of water age for all samplings sorted by POE water in Utility E

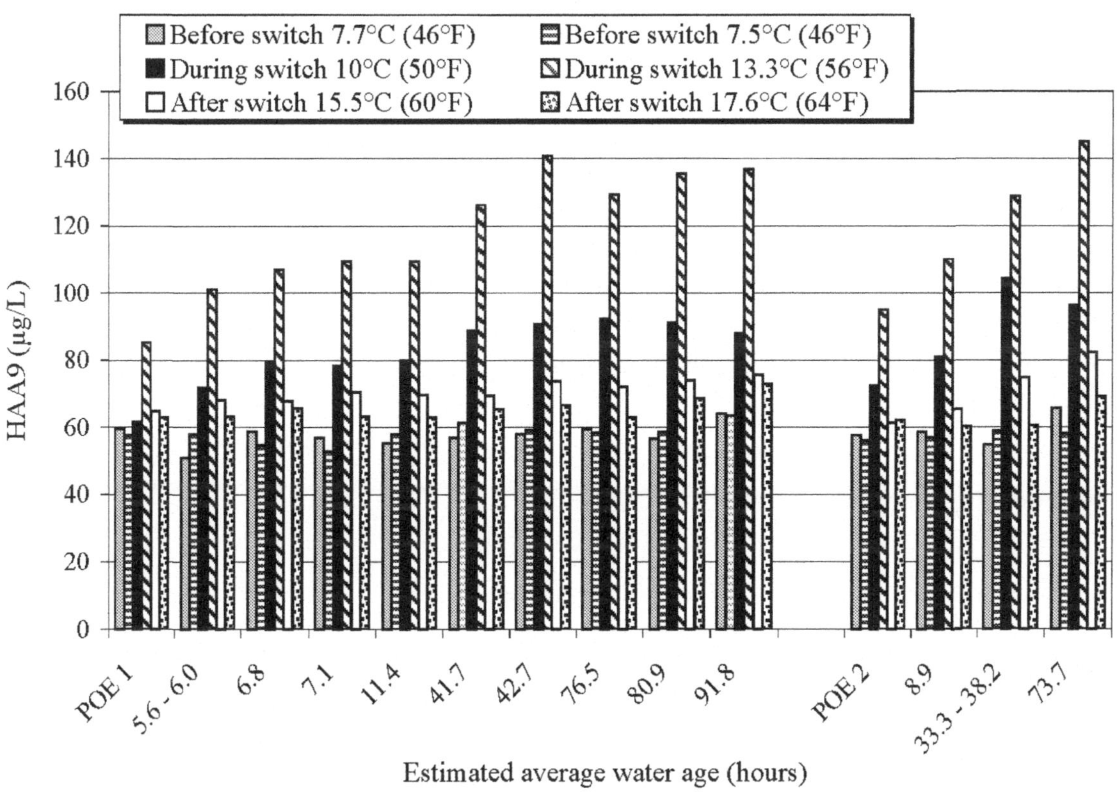

Figure 4.18 HAA9 concentrations as a function of water ages for all samplings sorted by POE water at Utility E

Trends in THM Concentrations with Increasing Water Age in the Chloraminated System Periodically Switching to Free Chlorine

Figure 4.19 presents THM4 concentrations and chlorine residuals as water age increased for samplings conducted before, during and after the switch to free chlorine at Utility E. During both chloramination periods, THM4 concentrations increased slightly with increasing water age. Surprisingly, the peak THM4 concentration was not observed at the MRTL (water age of 92 hours), but rather at an intermediate location (water age of 74 hours). Increases in THM4 concentration from the POE to the location with the maximum THM4 concentration varied from 20 to 50 percent (average increase of 32 percent) during the chloramination periods, with maximum THM4 concentrations measured prior to the switch to free chlorine of 66 and 70 µg/L. During the free chlorination period, however increases in THM4 concentrations were much greater: 135 and 161 percent, with peak THM4 concentrations of 131 and 175 µg/L. When this system returned to chloramines, maximum THM4 concentrations were 87 µg/L. These results were expected as chloramination is known to limit THM formation when compared with free chlorination.

THM4 concentrations were higher for the last two samplings conducted during chloramination (April and May 2003) than for the first two samplings conducted during the original chloramination period (February 2003). This difference most likely results from the increased water temperature during the last two samplings (13 to 24°C, 55 to 75°F) compared to the first two samplings (6 to 12°C, 43 to 54°F). Increased THM formation is expected at higher temperature due to the increased rate of reaction between free chlorine and organic material (Stevens et al. 1976, Kavanaugh et al. 1980, Oliver 1980, Grulois et al. 1989, Krasner et al. 1989, Hooper et al. 1994, Summers et al. 1996, Obolensky and Frey 2002).

During the free chlorination period, THM4 concentrations were lower for the first set of samples collected on March 11, 2003 than for the second set of samples collected on March 31, 2003 (Figure 4.19). This difference can be due to the fact that on March 11, Utility E was still adjusting chlorine doses to maintain a free chlorine residual throughout the entire distribution system (more information is presented below). Therefore, the samples collected at the MRTLs on March 11 had been treated with a lower chlorine dose than the samples that were collected at lower water age locations. The increased chlorine dose over time resulted in increased THM4 concentrations throughout the distribution system.

a) During chloramination, before the switch to free chlorine (water temperature at the POEs was 6.4 to 9.0°C [44 to 48°F] on February 11, and 7.0 to 8.0°C [45 to 46°F] on February 28, 2003)

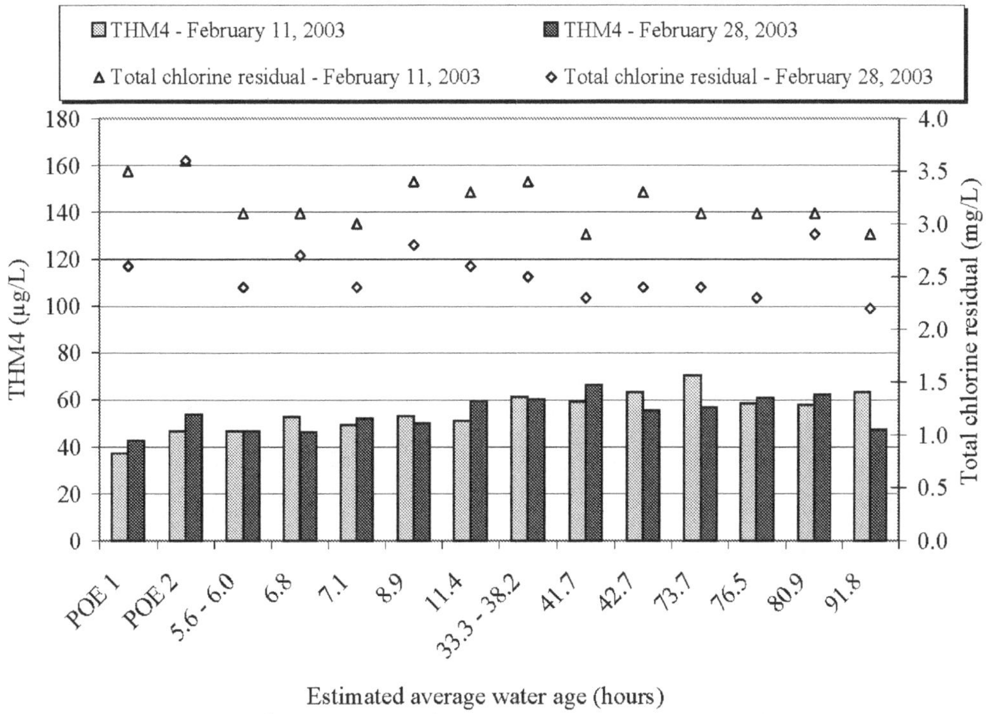

b) During the free chlorination period (water temperature at the POEs was 9.9 to 10.2°C [50°F] on March 11, and 13.3°C [56°F] on March 31, 2003)

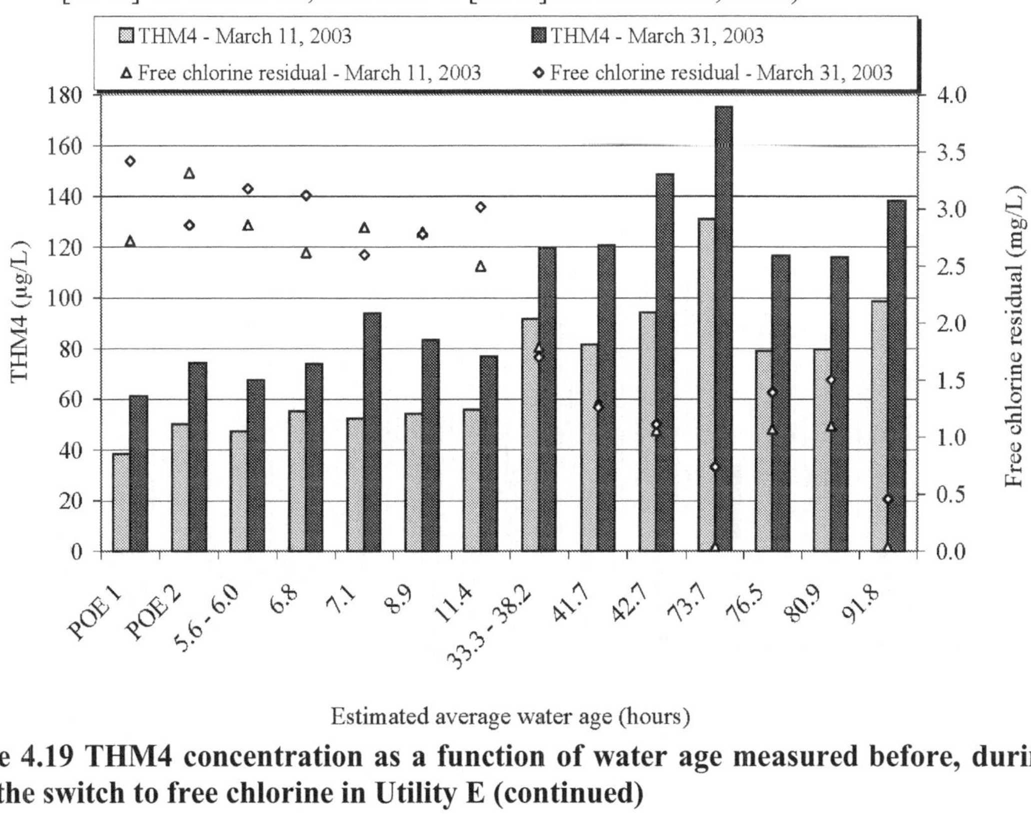

Figure 4.19 THM4 concentration as a function of water age measured before, during and after the switch to free chlorine in Utility E (continued)

Figure 4.19 (continued)

c) Following the return to chloramines (water temperature at the POEs was 15.5 to 15.6°C [60°F] on April 29, and 17.5 to 17.7°C [64°F] on May 20, 2003)

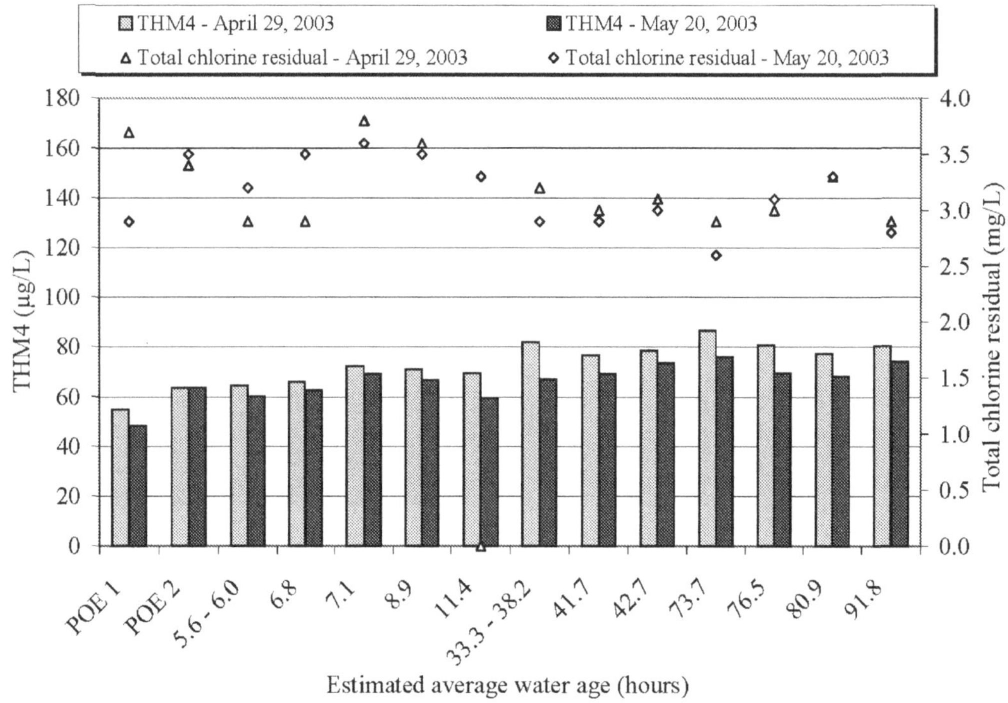

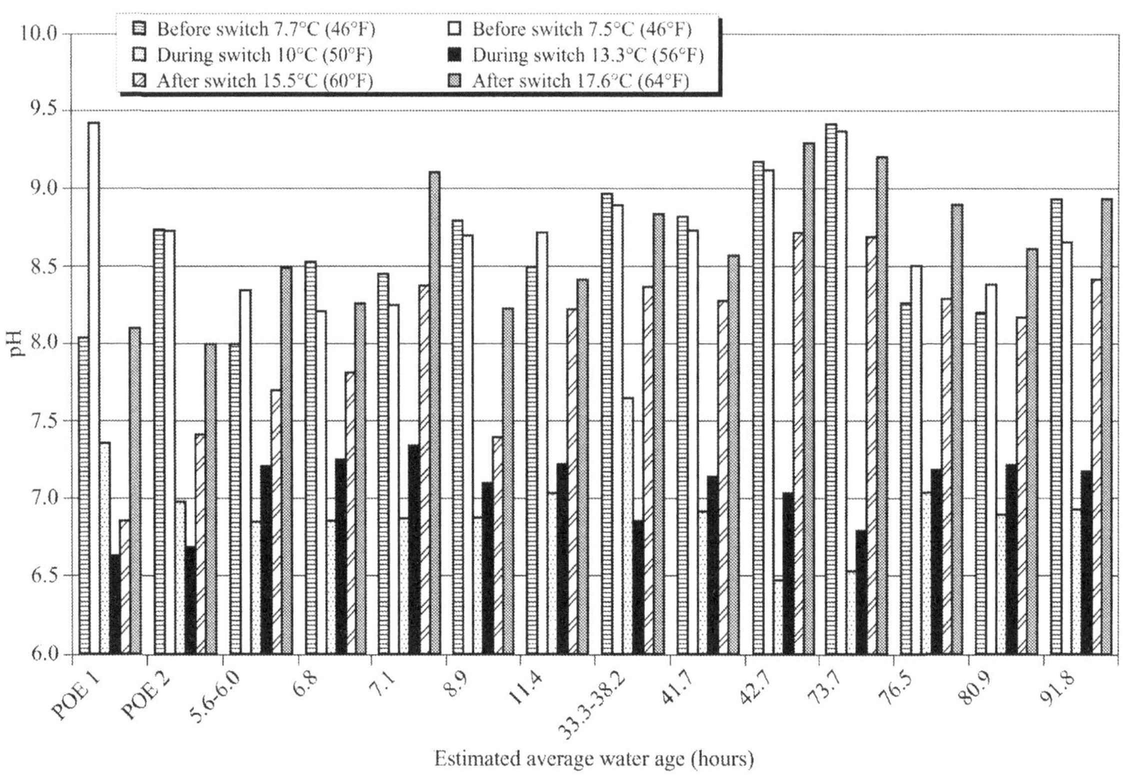

Figure 4.20 pH as a function of water age measured in Utility E

Using a concentration of 80 μg/L as a marker to characterize the THM4 concentrations measured during this study, results show that individual THM4 concentrations did not exceed this concentration during the chloramination period prior to the switch to free chlorine. Maximum THM4 concentrations measured prior to the switch to free chlorine were 66 and 70 μg/L. THM4 concentrations however slightly exceeded the marker at two locations after the free chlorination period, following the return to chloramination, with maximum THM4 concentrations up to 87 μg/L. During the free chlorination period, most sampling locations had THM4 concentrations above 80 μg/L, with peak THM4 concentrations of 131 and 175 μg/L (more than twice the regulatory limit) during both samplings conducted during the free chlorination period. These results confirmed that the final disinfection with chloramines is an effective technology to limit THM4 formation in the distribution system and help meet regulatory requirements. It worth noting that pH was significantly lower during the free chlorination period than during the chloramination periods, as shown on Figure 4.20. As a result, the lower pH may have help limit THM formation during the free chlorination period. Otherwise, even higher concentrations may have been measured.

THM Speciation in the Chloraminated System Periodically Switching to Free Chlorine

The weight fraction of $CHCl_3$ comprised most of the THM4 measured in Utility E (Table 4.17 and Figure 4.21). These results were expected considering the low bromide concentration in Utility E's raw water. The remaining THMs consisted of $CHBrCl_2$ only. The speciation of THM4 was not affected by disinfectant type.

The $CHCl_3$ weight fractions increased with increasing water age in the distribution system during half of the samplings (a trends that was expected, as explained above), and remained stable with water age during the remaining samplings, which included both samplings conducted during the free chlorination period.

Table 4.17
Mass ratios of $CHCl_3$ to THM4 calculated in Utility E. Water temperatures were measured at Site 1 POE and Site 2 POE, respectively

	Water temperature (°C, °F)	Minimum	Average	Maximum
Chloramines (prior to switch):				
February 11, 2003	6.4°C (44°F) at POE 1 9.0°C (48°F) at POE 2	86%	88%	89%
February 28, 2003	7.0°C (45°F) at POE 1 8.0°C (46°F) at POE 2	85%	87%	88%
Free chlorine:				
March 11, 2003	9.9°C (50°F) at POE 1 10.2°C (50°F) at POE 2	83%	86%	90%
March 31, 2003	13.3°C (56°F) at POEs 1 and 2	89%	91%	93%
Chloramines (after switch):				
April 29, 2003	15.6°C (60°F) at POE 1 15.5°C (60°F) at POE 2	89%	91%	93%
May 20, 2003	17.7°C (64°F) at POE 1 17.5°C (64°F) at POE 2	89%	91%	93%

a) During the original chloramination period, prior to the switch to free chlorine (February 11, 2003) (water temperature at the POEs was 6.4 and 9.0°C, 44 and 48°F)

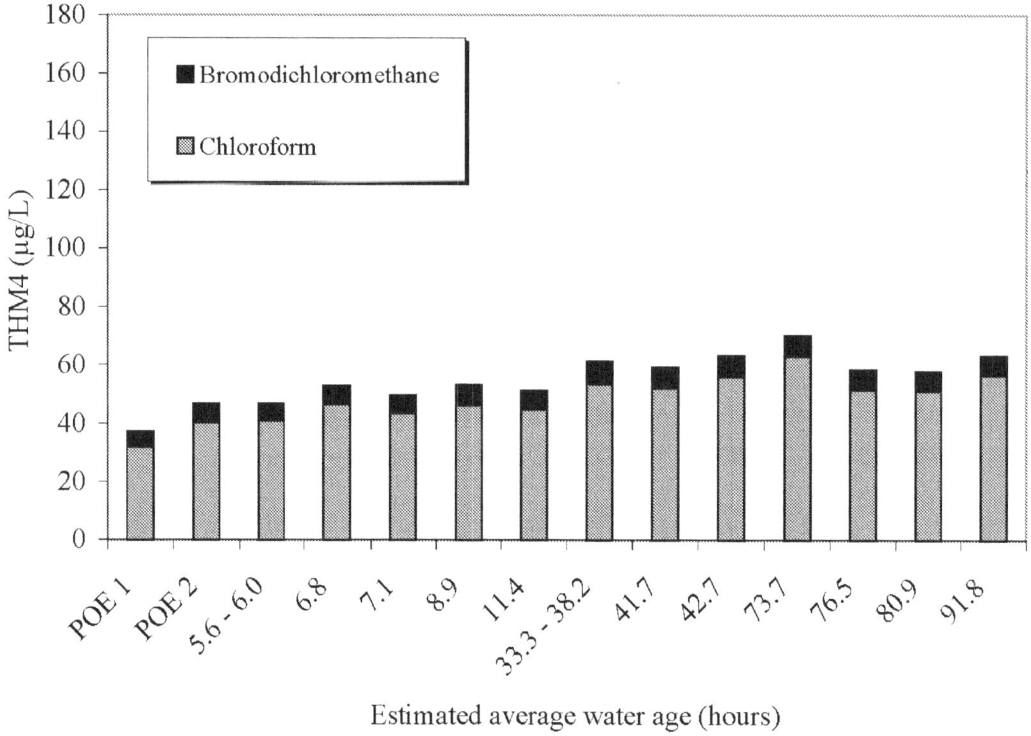

b) During the free chlorination period (March 31, 2003) (water temperature at the POEs was 13.3°C, 56°F)

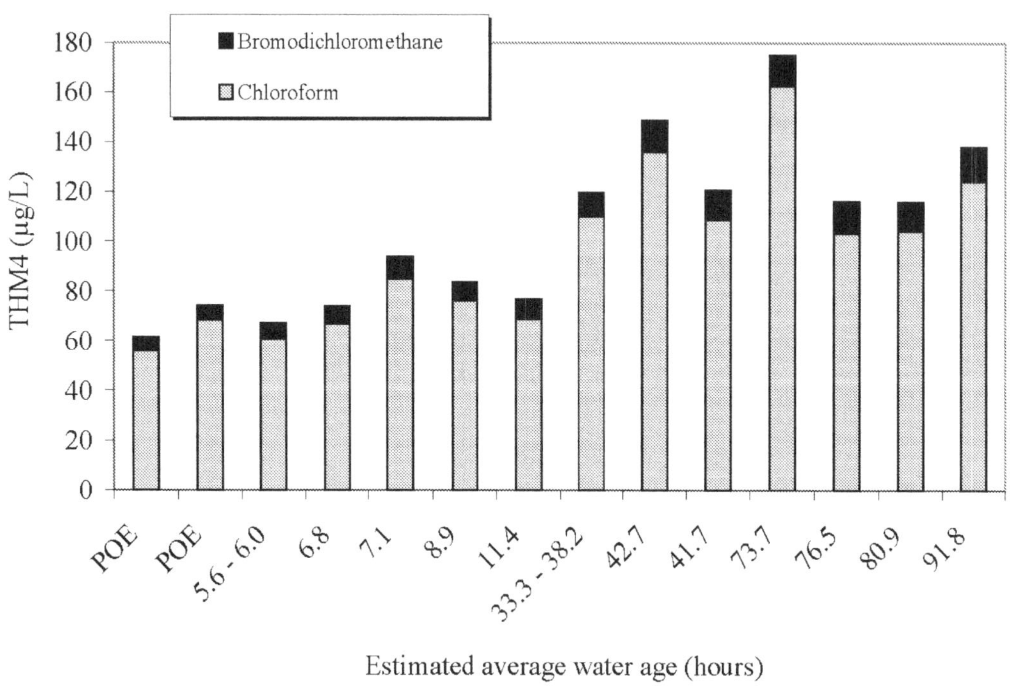

Figure 4.21 THM4 speciation as a function of water age measured in Utility E. $CHBr_2Cl$ and $CHBr_3$ were not detected during any of these samplings (continued)

Figure 4.21 (continued)

c) Following the return to chloramination (April 29, 2003) (water temperature at the POEs was 15.5 and 15.6°C, 60°F)

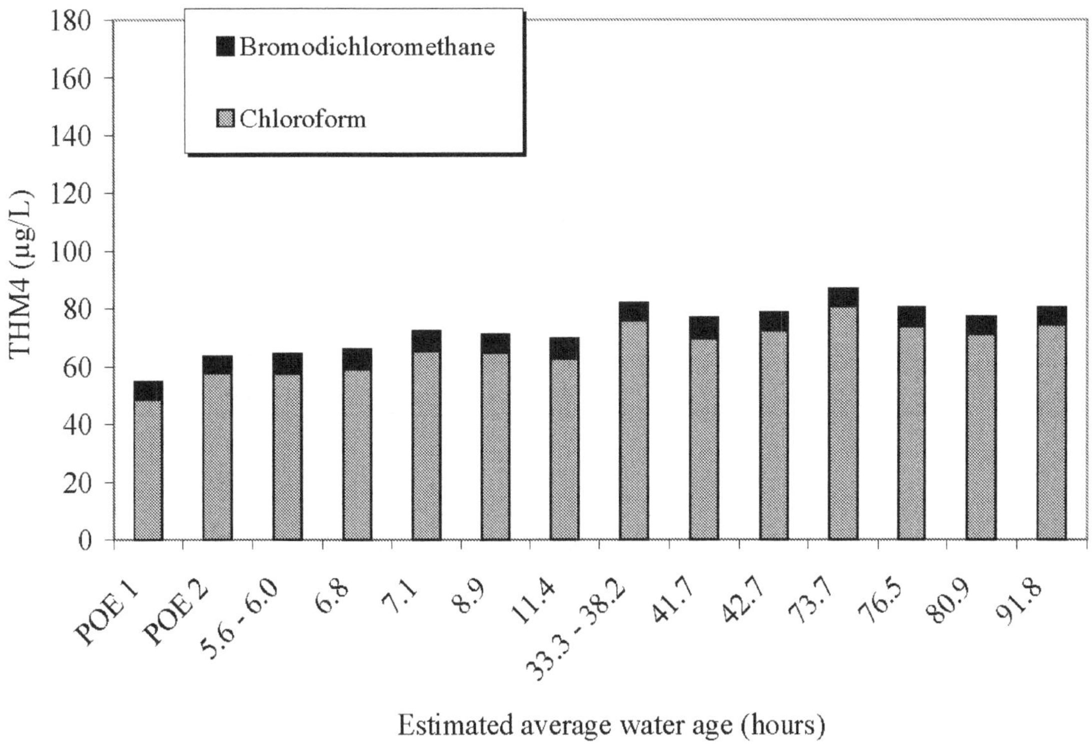

Trends in HAA Concentrations with Increasing Water Age in the Chloraminated System Periodically Switching to Free Chlorine

Figure 4.22 presents HAA9 concentrations and chlorine residuals as water age increases for samplings conducted before, during and after the switch to free chlorine at Utility E. As expected, HAA9 concentrations remained relatively stable with increasing water age during both chloramination periods; only minor increases in HAA9 concentrations were noticed (maximum concentrations of 64 to 66 µg/L). During the free chlorination period however, HAA9 concentrations increased significantly with water age, with peak concentrations of 104 to 145 µg/L. The magnitude of the increase in HAA9 concentration during the free chlorination period (44 and 53 percent from the POE to the location with the maximum HAA9 concentration) was less than the increase observed for THM4 (135 and 161 percent from the POE to the location with the maximum THM4 concentration). This trend is common when examining the fate of THMs and HAAs in distribution systems. Following the return to chloramination, HAA9 concentrations peaked at 73 to 82 µg/L. No clear decrease in HAA concentration was observed at the MRTLs, even though the maximum water age was relatively high (92 hours), suggesting that HAA biodegradation did not occur in Utility E's distribution system over the course of this study. This was most likely due to the lower water temperature when these samples were collected (further information is presented below).

As explained for the THMs, the higher HAA9 concentrations observed during the last two samplings (April and May 2003) most likely resulted from the higher water temperature compared to the first two samplings conducted during the original chloramination period before the switch to free chlorine (February 2003). As explained above for the THM4, the increase in chlorine dose during the free chlorination period resulted in higher HAA9 concentrations during the second free chlorine sampling (March 31, 2003) compared to the first free chlorine sampling (March 11, 2003).

a) During chloramination, prior to the switch to free chlorine (water temperature at the POEs was 6.4 to 9.0°C [44 to 48°F] on February 11, and 7.0 to 8.0°C [45 to 46°F] on February 28, 2003)

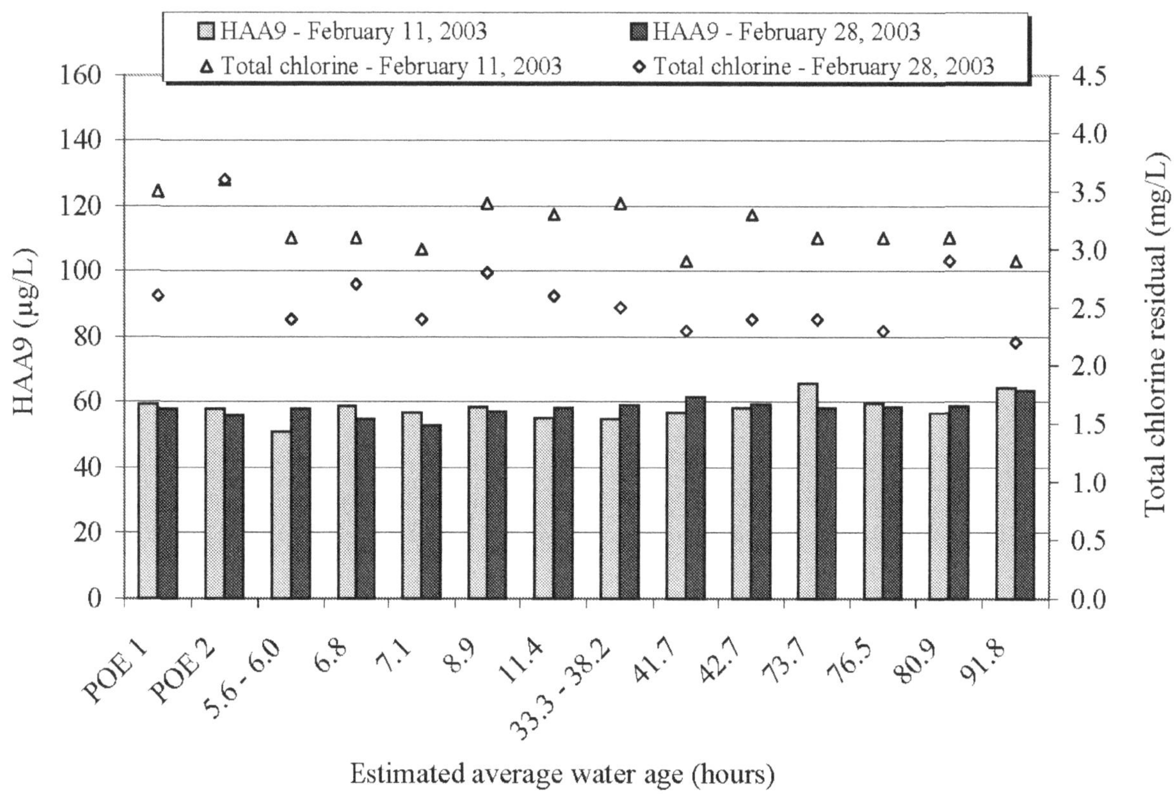

Figure 4.22 HAA9 concentration as a function of water age measured before, during and after the switch to free chlorine in Utility E (continued)

Figure 4.22 (continued)

b) During the free chlorination period (water temperature at the POEs was 9.9 to 10.2°C [50°F] on March 11, and 13.3°C [56°F] on March 31, 2003)

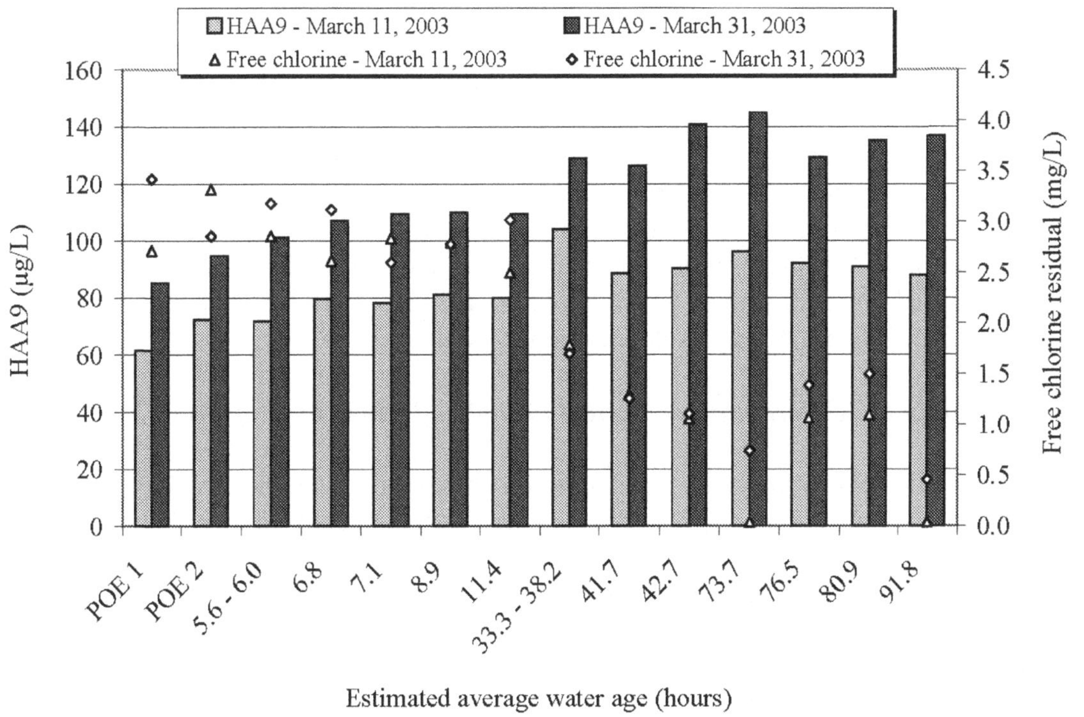

c) Following the return to chloramines (water temperature at the POEs was 15.5 to 15.6°C [60°F] on April 29, and 17.5 to 17.7°C [64°F] on May 20, 2003)

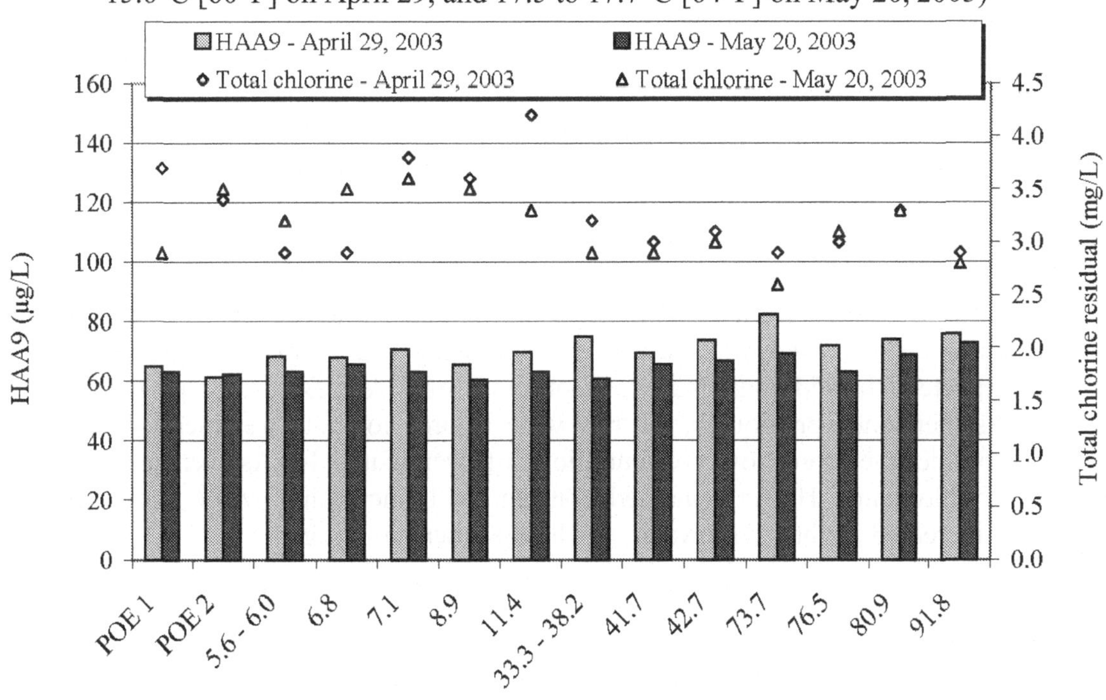

Individual HAA5 concentrations were close to the MCL marker of 60 μg/L during chloramination prior to the switch to free chlorine, and exceeded this concentration at one of the sampling locations. During the free chlorination period, HAA5 concentrations were greater than 60 μg/L at all distribution system locations, including the POEs. HAA5 concentrations as high as 138 μg/L were reached and concentrations measured during the free chlorination period were approximately double the concentrations observed during chloramination. The last two samplings conducted in warmer water (~15.5°C, 60°F, on April 29, 2003, and ~17.6°C, 64°F, on May 20, 2003) following the return to chloramination, also had HAA5 concentrations that exceeded 60 μg/L at several distribution system locations. However, the maximum HAA5 concentration reached (76 μg/L) was much lower than those measured during the free chlorination period. Although chloramines limited HAA5 formation when compared to free chlorine, additional measures may be required to achieve HAA5 concentrations below the regulatory requirements in Utility E's distribution system.

HAA Speciation in the Chloraminated System Periodically Switching to Free Chlorine

Because of the low bromide concentration measured in the source water of Utility E, HAA5 represented almost all of the HAA9 calculated (>84 percent, Table 4.18 and Figure 4.23). The mass ratios of HAA5-to-HAA9 were similar with both disinfectant types. This ratio did not show a specific trend with increasing water age in Utility E: the ratios remained constant during some of the samplings, or increased slightly with water age at other times. These results are consistent with the fact that biodegradation was not observed in Utility E's distribution system over the course of the study.

Figure 4.24 presents each class of HAA9 as a function of water age before, during and after the switch to free chlorine. Results suggest that X_3AAs were slightly more dominant than X_2AAs for both disinfectant residuals used; X_3AAs represented 41 to 57 percent of HAA9 concentrations (average of 49 percent), whereas X_2AAs represented 36 to 52 percent of HAA9 concentrations (average of 42 percent). Other researchers have observed higher concentrations of X_3AA than X_2AA during chlorination (Cowman and Singer 1996; Symons et al. 1998; Zhang et al. 2000), although X_3AAs should not be produced from combined chlorine. The high percentage of X_3AA observed during chloramination may be due in part to the use of free chlorine as primary disinfectant throughout the study.

When using chloramine as the final disinfectant, Cl_2AA and Cl_3AA were the predominant HAA species formed, and were approximately equal in concentration (Table 4.18). A slightly higher concentration of Cl_3AA was measured while using free chlorine compared to Cl_2AA, which was expected. In addition, small concentrations of ClAA, BrClAA, and $BrCl_2AA$ were measured. As mentioned previously, the raw water source for both treatment plants contains very little bromide; therefore, low concentrations of brominated HAAs were expected. The proportion of chlorinated HAAs remained constant with increasing water age during some samplings, or increased slightly with water age during other samplings.

Table 4.18
Mass ratios of chlorine-containing HAAs to HAA9 and HAA5 to HAA9 measured in Utility E

	Minimum	Average	Maximum
Chloramines (prior to switch):			
February 11, 2003:	Water temperature at POEs: 6.4 and 9.0°C (44 and 48°F)		
Cl-HAA/HAA9	97%	100%	100%
HAA5/HAA9	89%	92%	94%
February 28, 2003:	Water temperature at POEs: 7.0 and 8.0°C (45 and 46°F)		
Cl-HAA/HAA9	92%	95%	96%
HAA5/HAA9	84%	88%	90%
Free chlorine:			
March 11, 2003:	Water temperature at POEs: 9.9 and 10.2°C (50°F)		
Cl-HAA/HAA9	96%	96%	97%
HAA5/HAA9	88%	89%	91%
March 31, 2003:	Water temperature at POEs: 13.3°C (56°F)		
Cl-HAA/HAA9	97%	98%	100%
HAA5/HAA9	91%	93%	95%
Chloramines (after switch):			
April 29, 2003:	Water temperature at POEs: 15.6 and 15.5°C (60°F)		
Cl-HAA/HAA9	97%	98%	100%
HAA5/HAA9	90%	93%	95%
May 20, 2003:	Water temperature at POEs: 17.7 and 17.5°C (64°F)		
Cl-HAA/HAA9	100%	100%	100%
HAA5/HAA9	92%	93%	95%

Cl-HAA: Sum of ClAA, Cl_2AA, Cl_3AA and $BrCl_2AA$

a) During the original chloramination period, prior to the switch to free chlorine (February 28, 2003) (water temperature at the POEs was 7.0 and 8.0°C, 45 and 46°F)

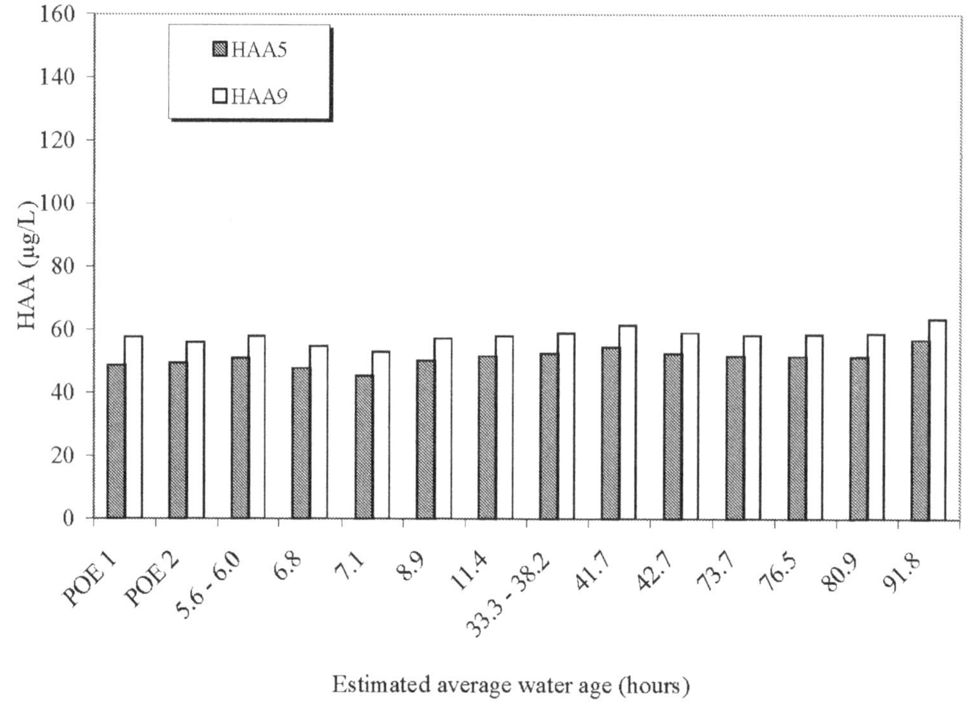

b) During the free chlorination period (March 31, 2003) (water temperature at the POEs was 13.3°C, 56°F)

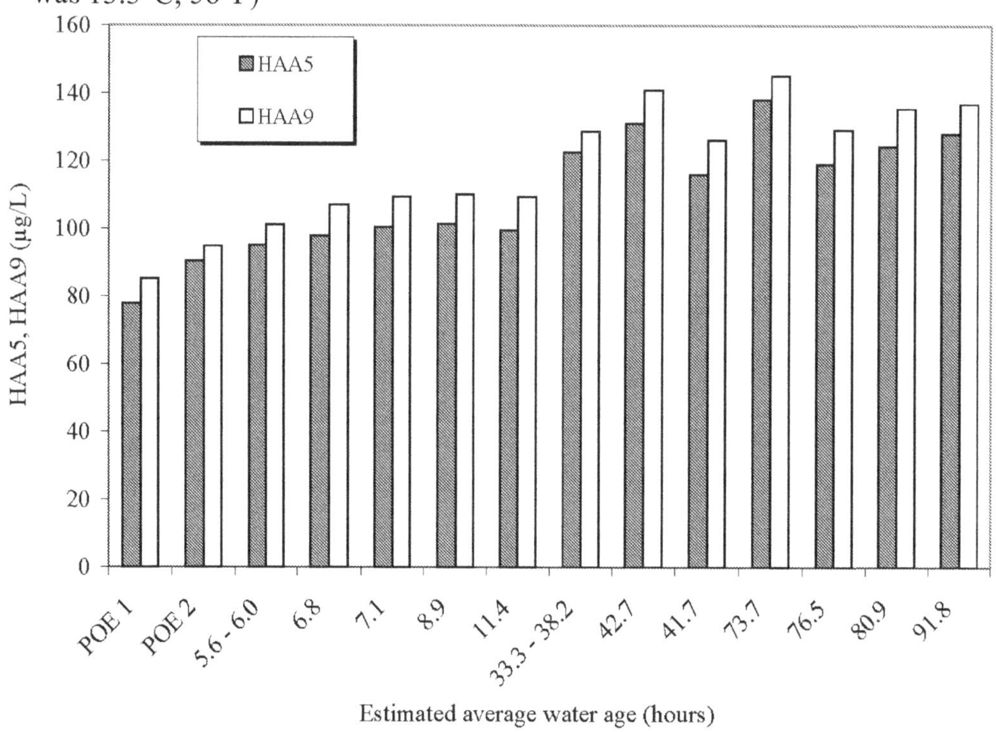

Figure 4.23 HAA5 and HAA9 concentrations as a function of water age in Utility E (continued)

Figure 4.23 (continued)

c) Following to return to chloramination (May 20, 2003) (water temperature at the POEs was 17.7 and 17.5°C, 64°F)

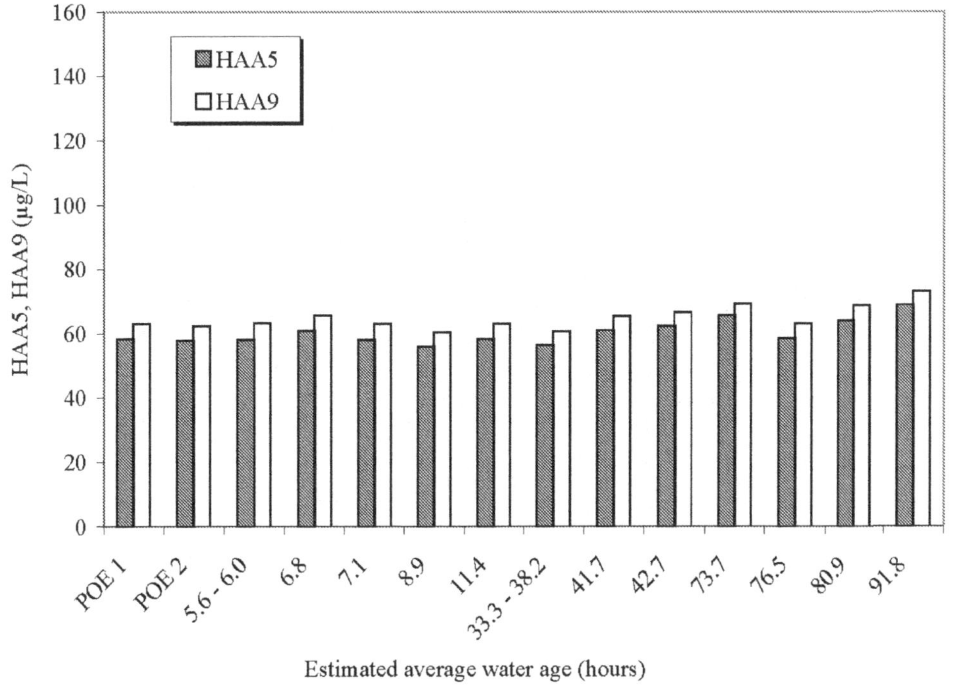

a) During the original chloramination period, prior to the switch to free chlorine (February 28, 2003) (water temperature at the POEs was 7.0 and 8.0°C, 45 and 46°F)

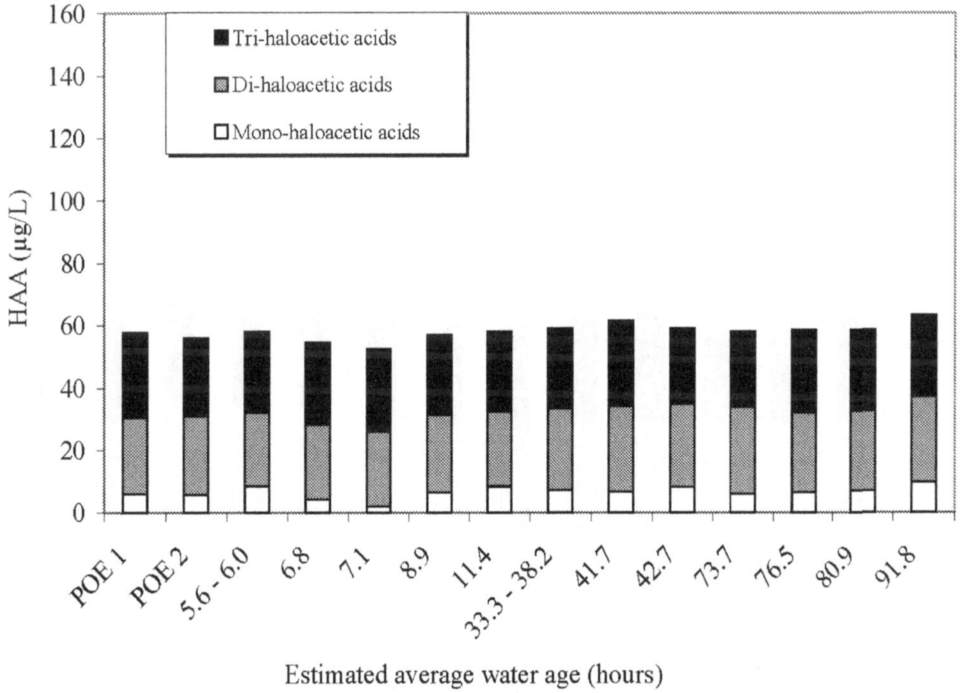

Figure 4.24 HAA9 speciation as a function of water age measured in Utility E (continued)

Figure 4.24 (continued)

b) During the free chlorination period (March 31, 2003) (water temperature at the POEs was 13.3°C, 56°F)

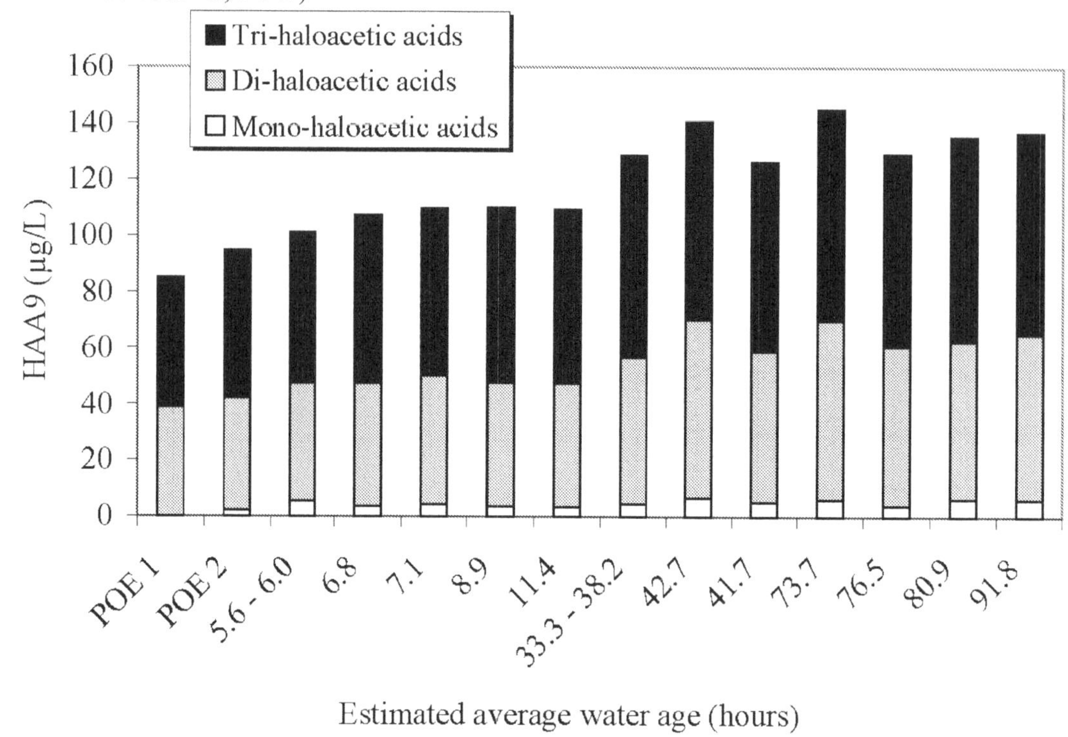

c) Following the return to chloramination (May 20, 2003) (water temperature at the POEs of 17.7 and 17.5°C, 64°F)

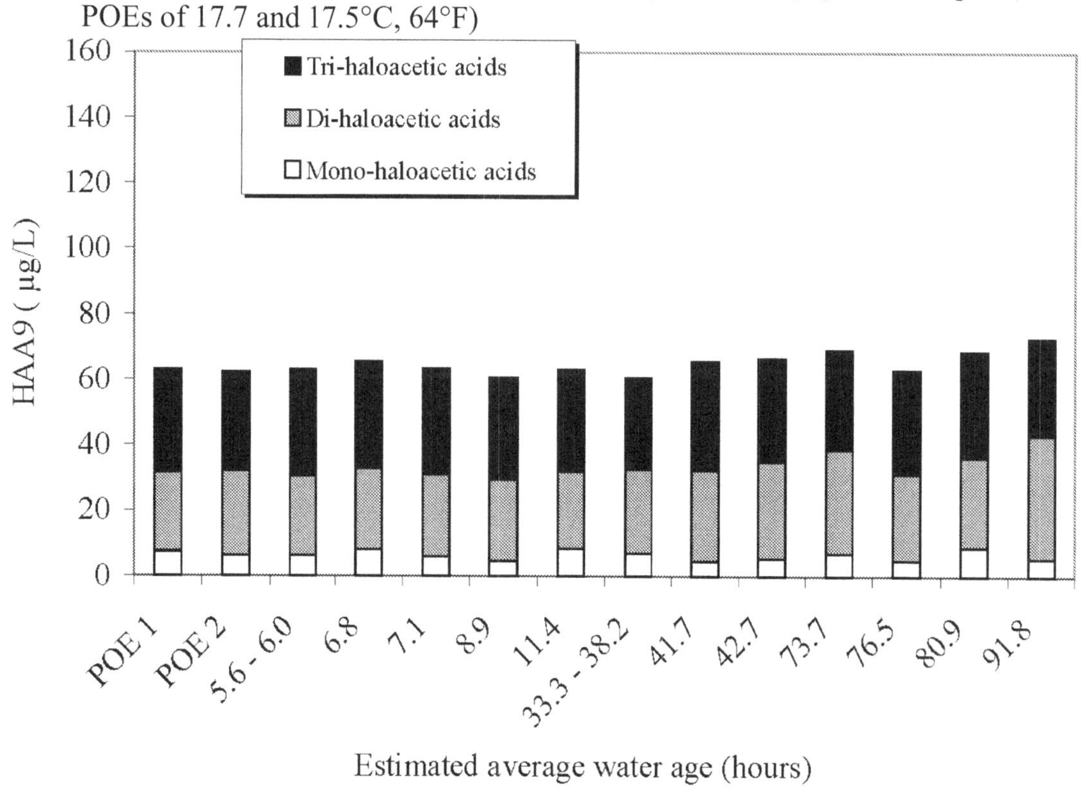

Effect of Seasons on THMs and HAAs

As expected, THM4 and HAA9 concentrations were lowest in the colder months and highest in the warmer months in Utility A (Figure 4.25, where clear bars represent summer samplings, and darker bars represent winter samplings). Similarly, peak concentrations for both classes of DBPs occurred in the summer in Utility B (Figure 4.26). These trends are consistent with the literature (Stevens et al. 1976, Kavanaugh et al. 1980, Oliver 1980, Grulois, Matejka and Parinet 1989, Krasner et al. 1989, Hooper et al. 1994, Summers et al. 1996, Obolensky and Frey 2002). In one of the chloraminated systems (Utility C), THM4 and HAA9 concentrations were generally lower in colder water and higher in warmer water, although the trends with HAA9 were not as clear as with THM4 due to the higher variability of HAA concentrations between the various sampling locations (Figure 4.27). In Utility D, it is likely that the change in water temperature was not important enough to see a difference in DBP concentration (15.6 to 21.1°C, 60 to 70°F).

N-NITROSODIMETHYLAMINE

NDMA concentrations were measured during three samplings conducted in cold water and three samplings conducted in warm water in a chloraminated system (Utility C). It was monitored in Utility E before, during and after the switch to free chlorine to address the effect of a change in disinfectant on this DBP. NDMA was also measured in a free chlorinated system (Utility A) as a control. The results discussed below have been interpreted in consideration of the limitations of the analytical method for NDMA. The average measurement of 10-ng/L matrix spikes was 10.4 ± 2.6 ng/L (n=24), while the average measurement of 50-ng/L standards was 47.4 ± 3.2 ng/L (n=30).

Chloraminated System

Six NDMA samplings were conducted in Utility C. Figure 4.28 presents representative results of NDMA concentration and chlorine residual as a function of water age. No particular trend could be identified between NDMA concentrations and water age, which suggests that NDMA was not forming in the distribution system. High concentrations were measured on two occasions (40 and 46 ng/L). All other sampling locations had concentrations below 20 ng/L, with the majority of samples having concentrations below 10 ng/L. Note that the California DHS has an action level of 10 ng/L, and the Ontario MOE has set a maximum acceptable concentration of 9 ng/L.

Figure 4.29 summarizes data obtained from the samplings conducted under the coldest and warmest conditions, one of the samplings was omitted from this figure as the water temperature was lukewarm at that time (clear bars represent summer samplings, and darker bars represent winter samplings). Results show slightly greater NDMA concentrations in warm water than in colder water, although this trend was not evident at each sampling location. Also, very high concentrations were measured during one of the cold water samplings.

a) THM4 concentrations

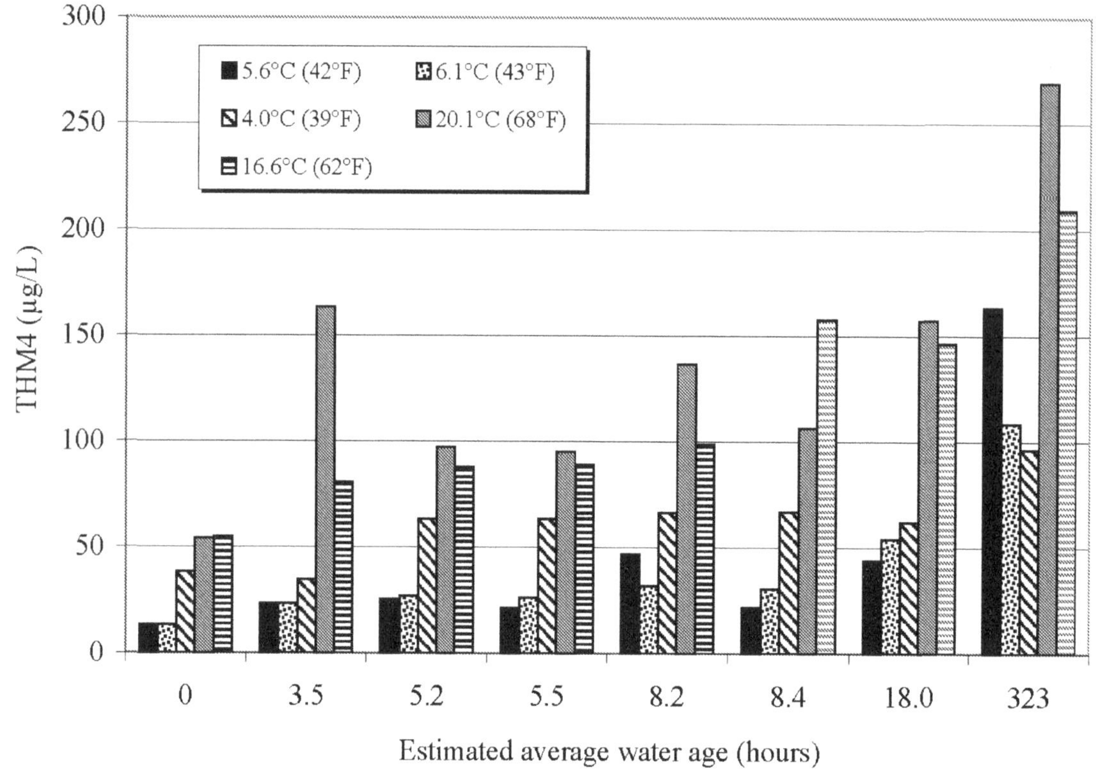

b) HAA9 concentrations

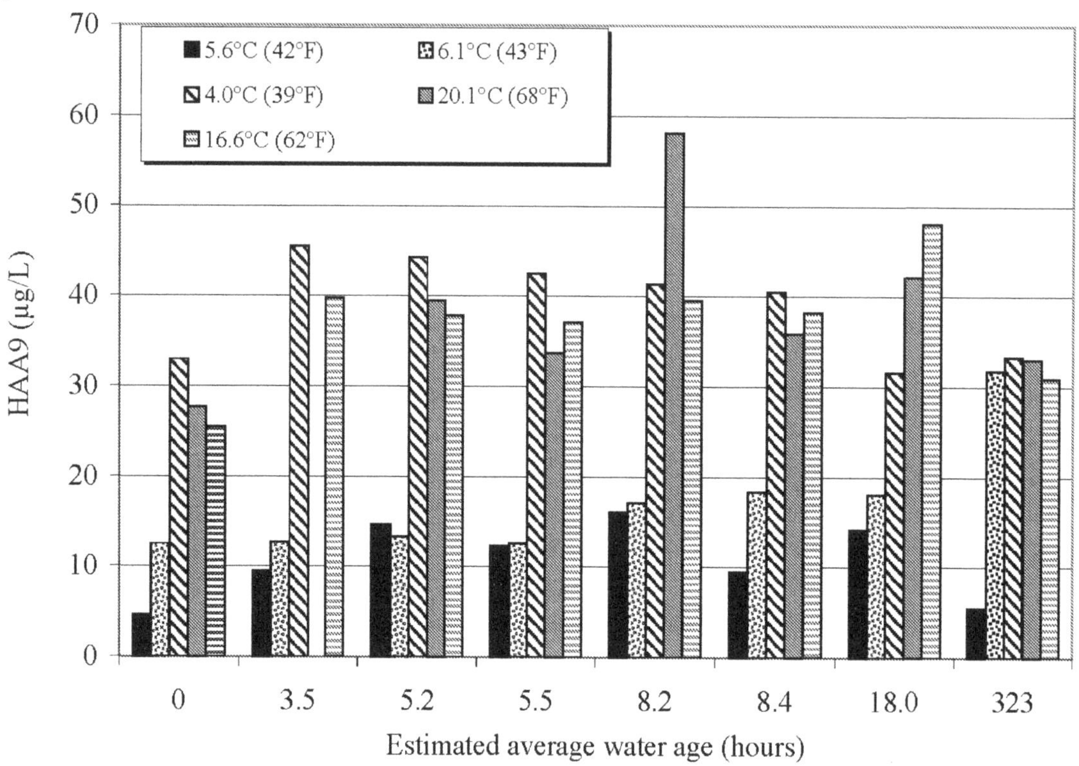

Figure 4.25 DBP concentrations as a function of water age measured in Utility A (free chlorinated system)

a) THM4 concentrations

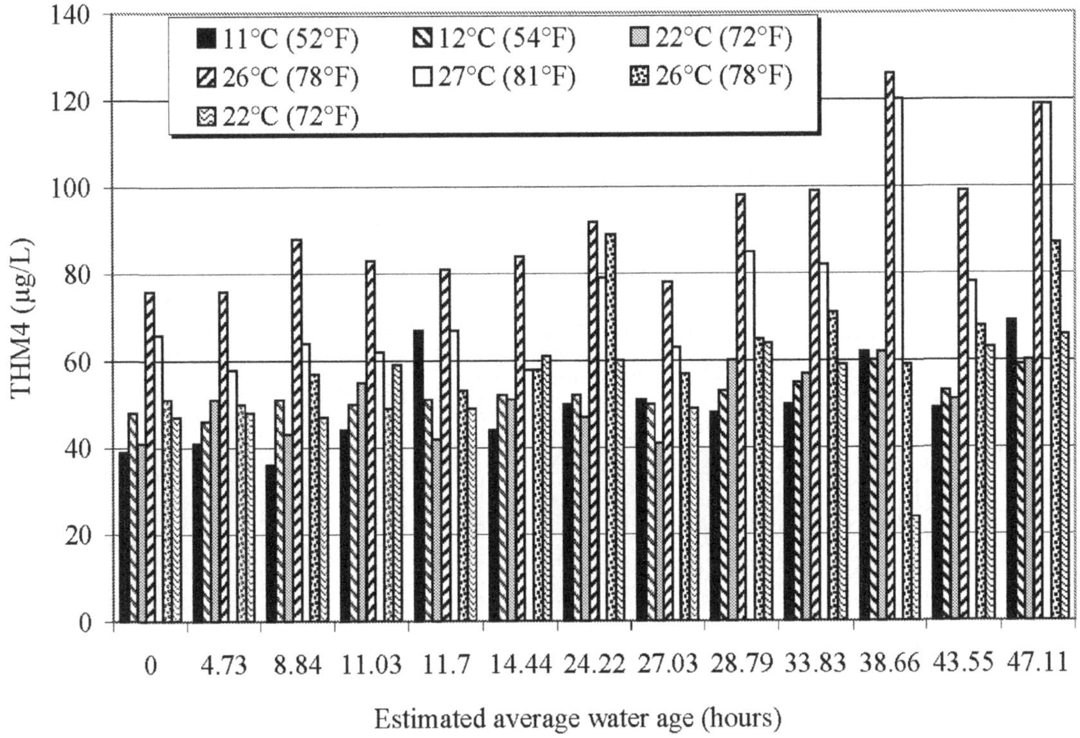

b) HAA9 concentrations

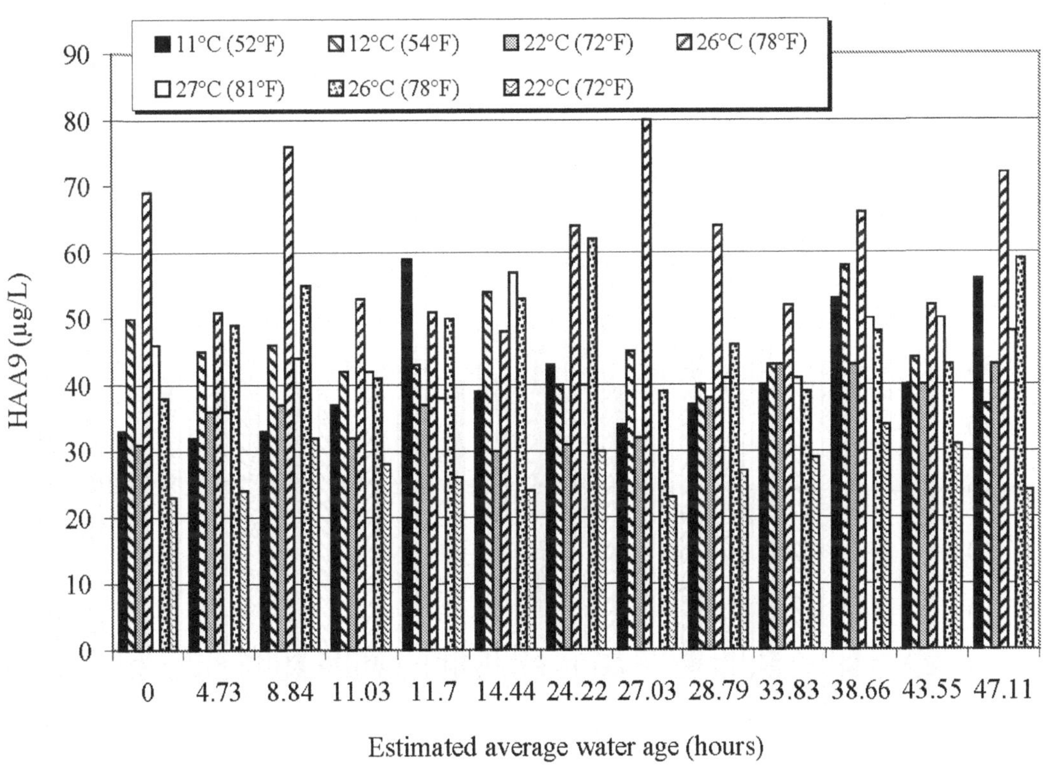

Figure 4.26 DBP concentrations as a function of water age measured in Utility B (free chlorinated system)

a) THM4 concentrations

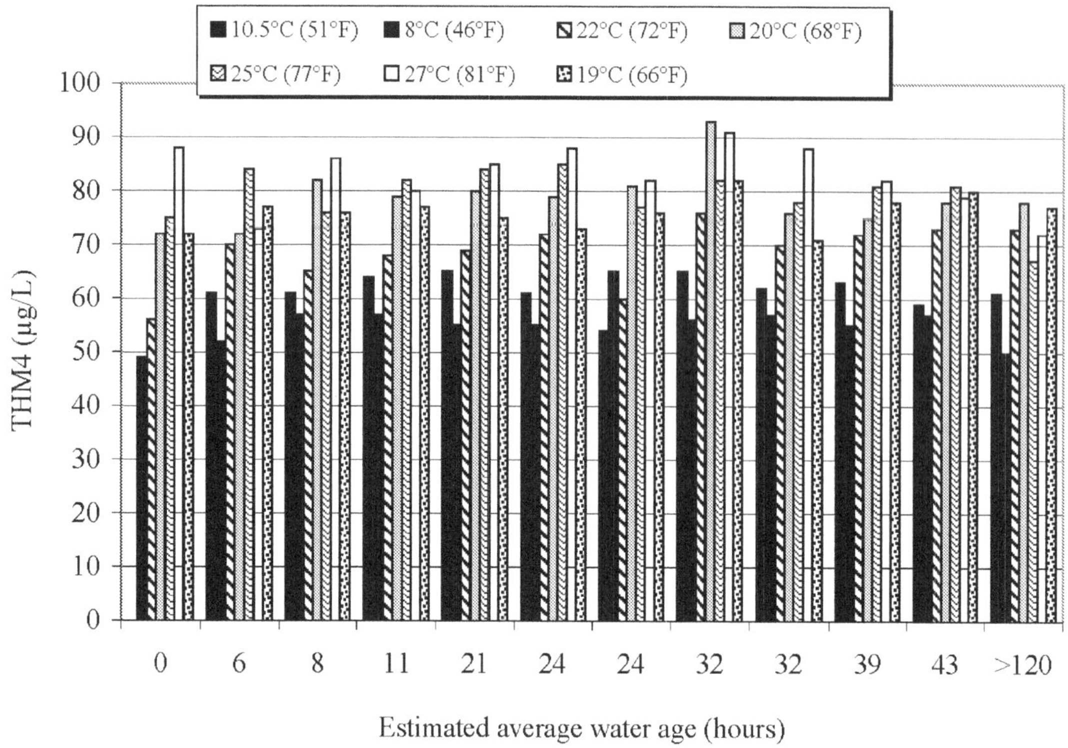

b) HAA9 concentrations

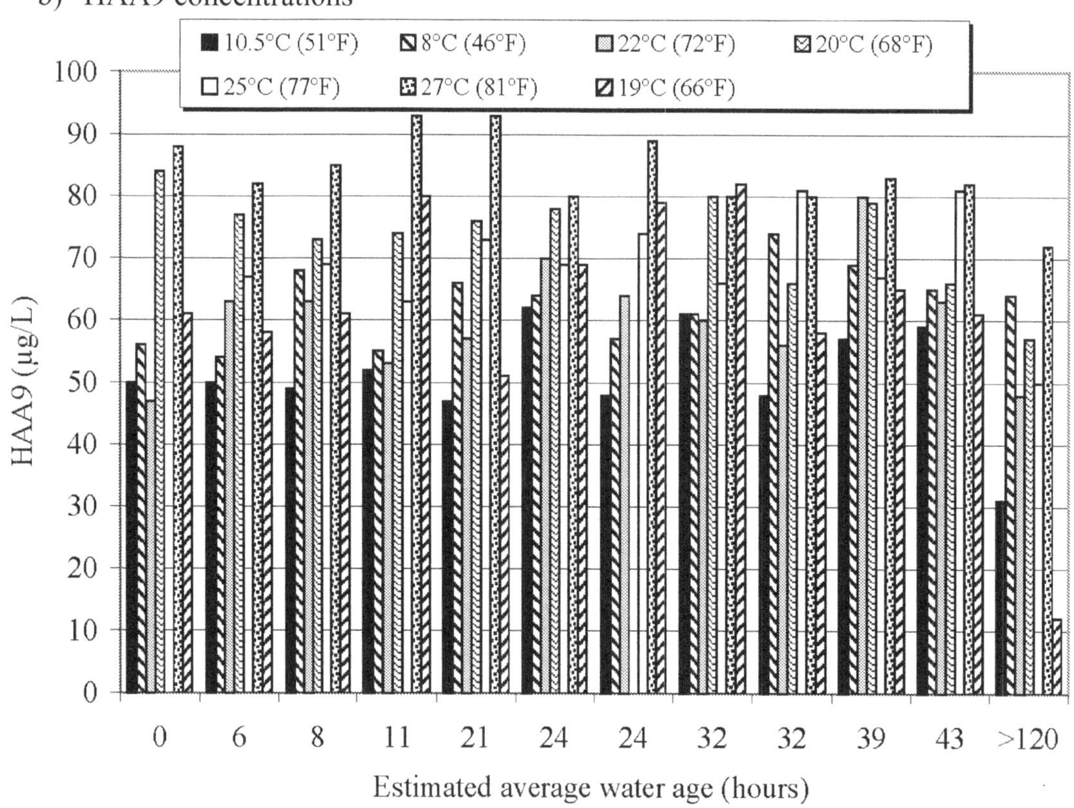

Figure 4.27 DBP concentrations as a function of water age measured in Utility C (chloraminated system)

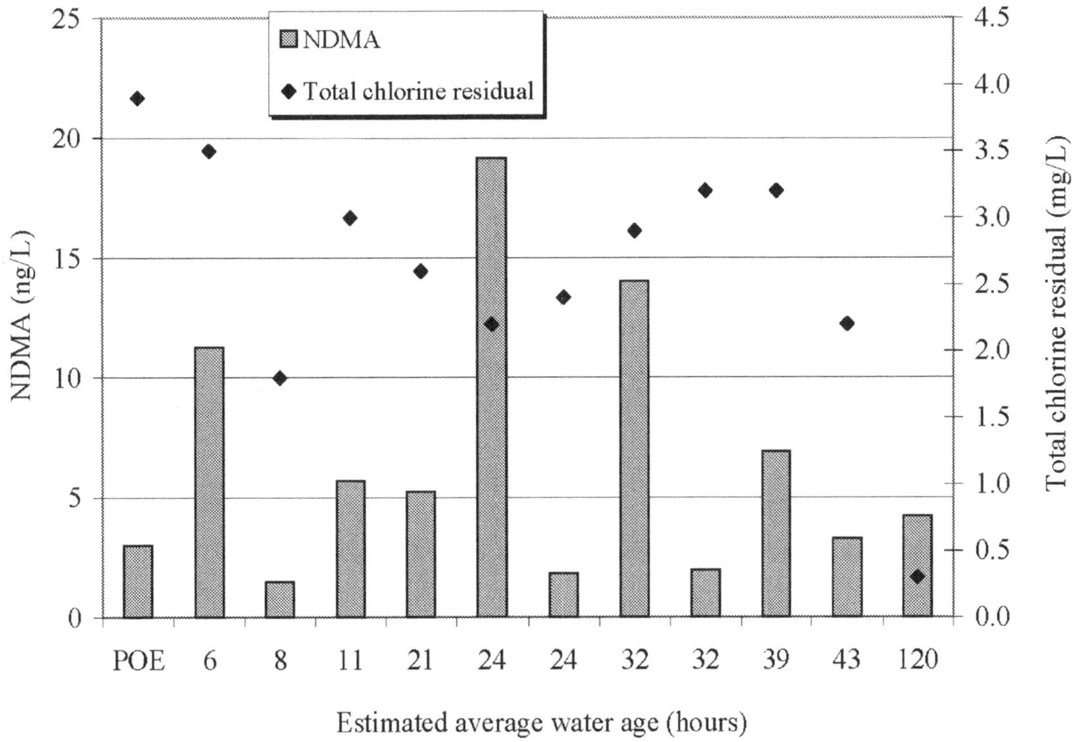

Figure 4.28 NDMA concentration and total chlorine residual as a function of water age measured during the June 11, 2003 sampling in Utility C (chloraminated system) (water temperature at the POE was 25°C, 77°F)

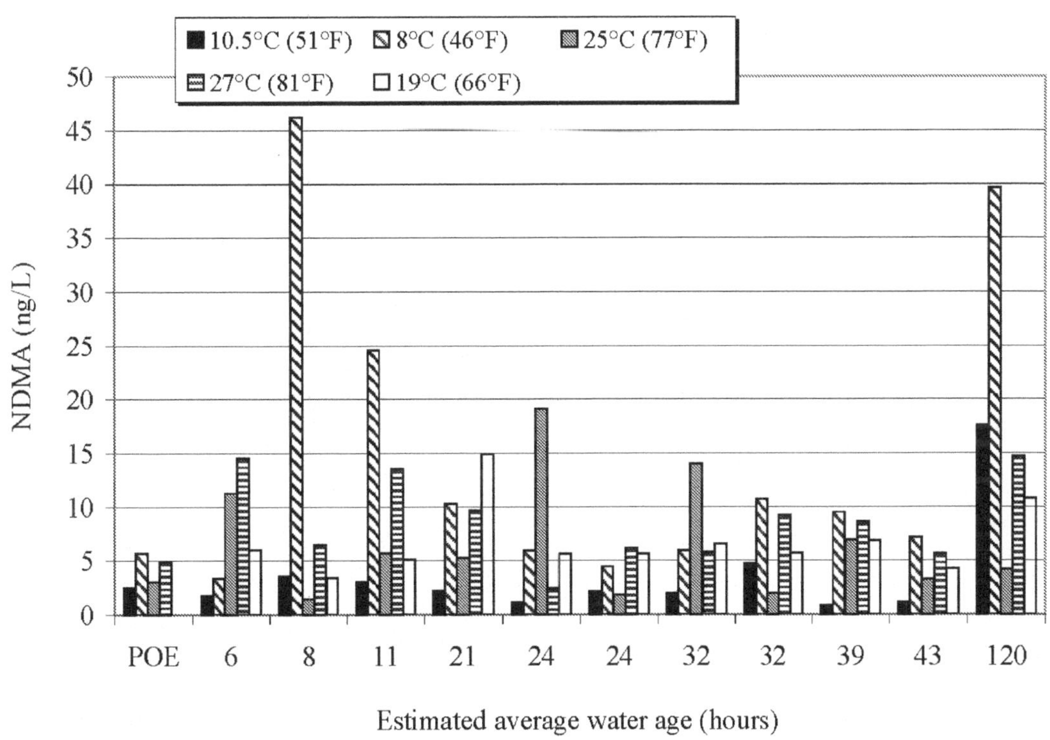

Figure 4.29 NDMA concentrations as a function of water age measured in Utility C (chloraminated system)

Free Chlorinated System

Five NDMA samplings were conducted in Utility A. A representative graph of NDMA concentration and chlorine residual plotted as a function of water age is presented on Figure 4.30. As observed for the chloraminated system (Utility C), no trend could be drawn between NDMA concentration and water age from the data obtained in Utility A, which suggests that NDMA was not forming in the distribution system. At times, NDMA concentrations seemed to increase with water age, while they seemed to decrease with water age at other times. Considering the maximum acceptable concentrations set by some regulatory agencies (9-10 ng/L), concentrations measured in this free chlorinated distribution system were high: over 20 ng/L were measured during two of the five samplings conducted at Utility A.

Figure 4.31 summarizes data obtained from the samplings conducted under the coldest and warmest conditions. It appears that NDMA concentrations were slightly greater in warm water than in colder water, as observed in the chloraminated system, Utility C.

Chloraminated System Periodically Switching to Free Chlorine

NDMA concentrations were measured in Utility E during the six samplings conducted before, during and after the switch to free chlorine as the final disinfectant. Figure 4.32 shows NDMA concentrations for all water ages categorized by POE water. Although important variations in NDMA concentration were observed at the sampling locations originating from both POEs, results suggest that the POE water did not have any discernable effect on NDMA concentrations.

NDMA concentrations and chlorine residuals measured in Utility E were plotted as a function of water age (Figure 4.33). As observed in the other systems studied, no trends with increasing water age could be isolated from the data obtained, which suggests that NDMA was not forming in the distribution system. In general, NDMA concentrations were high in Utility E's distribution system with either final disinfectant used; 8 samples showed NDMA concentrations over 40 ng/L, one of which was the POE.

Results also suggest that NDMA concentrations increased throughout the sampling period. Concentrations seemed lowest during the first two samplings conducted before the switch to free chlorine, greater during the free chlorination period, and greatest following the return to chloramination. These results are surprising as greater NDMA concentrations were expected in chloraminated water than in free chlorinated water. Increases in water temperature from 6 to 18°C (43 to 64°F) at the POEs of Utility E may explain the corresponding increases in NDMA concentrations. This observation is consistent with the greater NDMA concentrations observed in warm water than in colder water in the other systems examined.

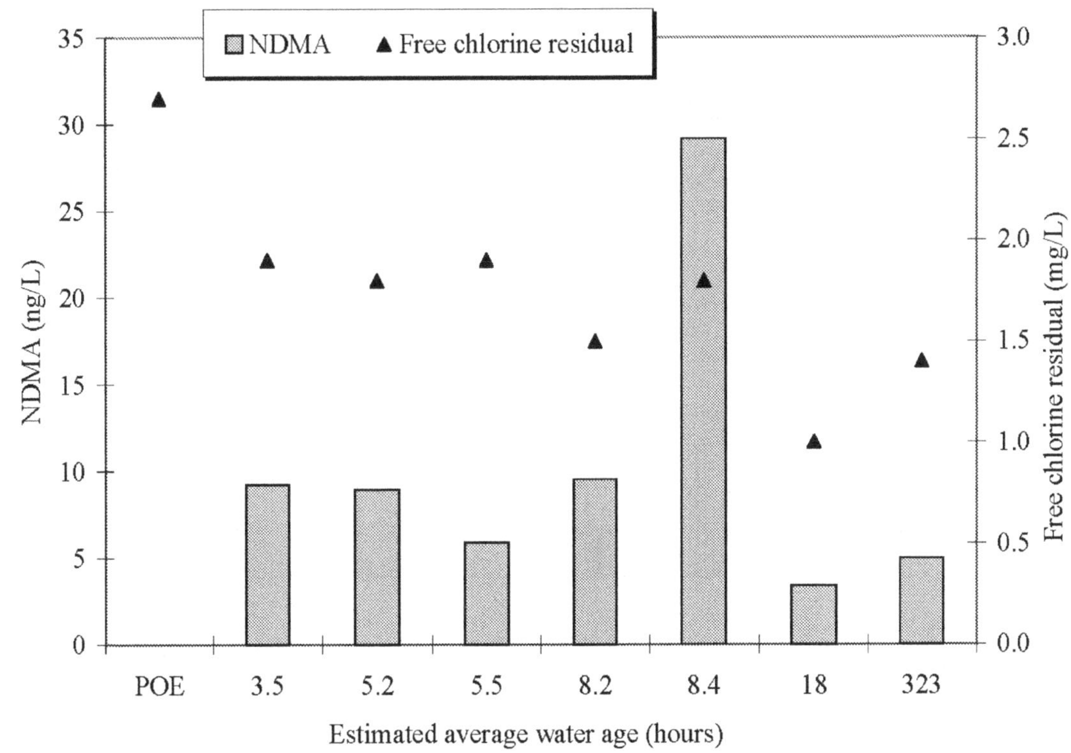

Figure 4.30 NDMA concentration and free chlorine residual as a function of water age measured during the February 25, 2003 sampling in Utility A (free chlorinated system) (water temperature at the POE was 4.0°C, 39°F)

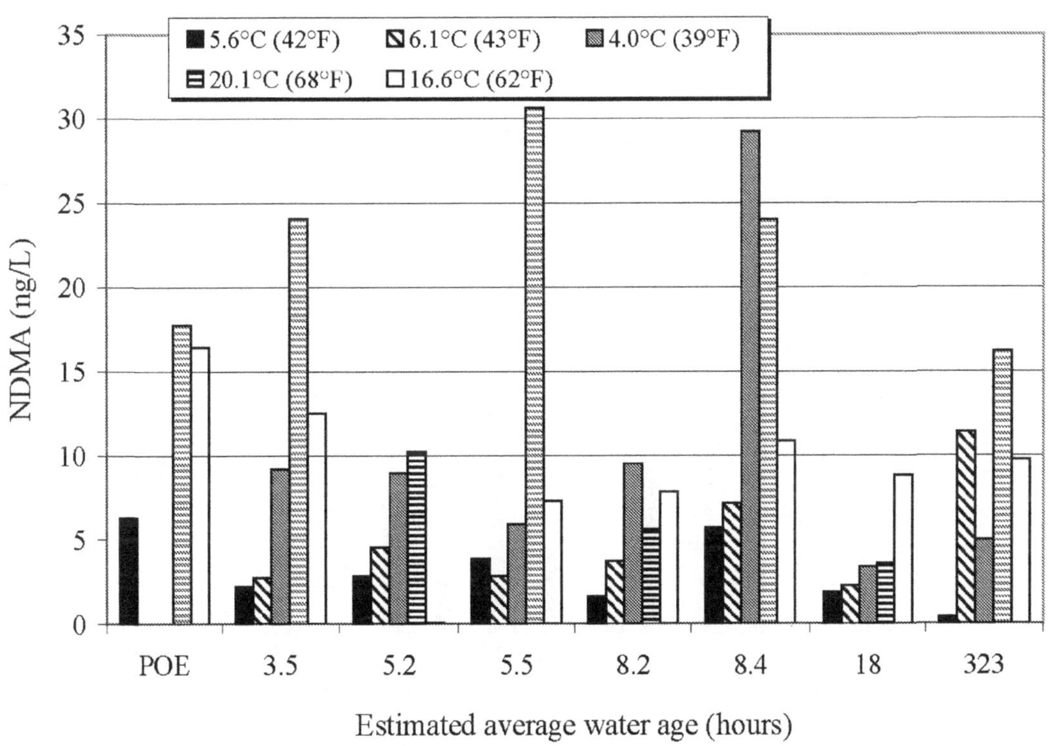

Figure 4.31 NDMA concentrations as a function of water age measured in Utility A (free chlorinated system)

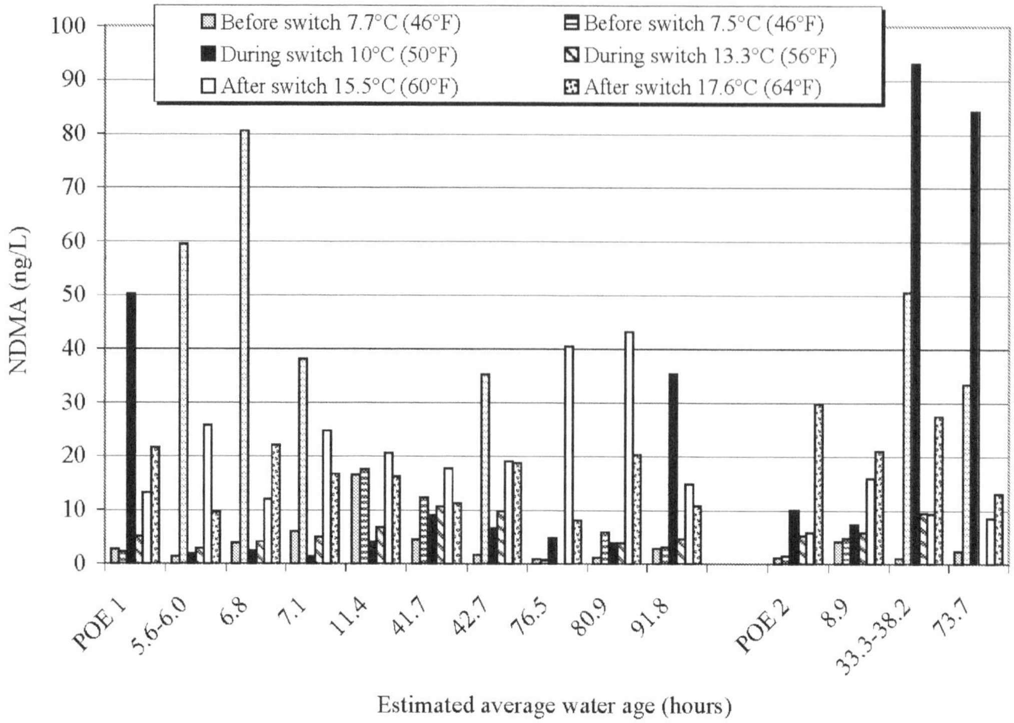

Figure 4.32 NDMA concentrations as a function of water age for all samplings sorted by POE water in Utility E

a) During chloramination, before the switch to free chlorine (water temperature at the POEs was 6.4 to 9.0°C [44 to 48°F] on February 11, and 7.0 to 8.0°C [45 to 46°F] on February 28, 2003)

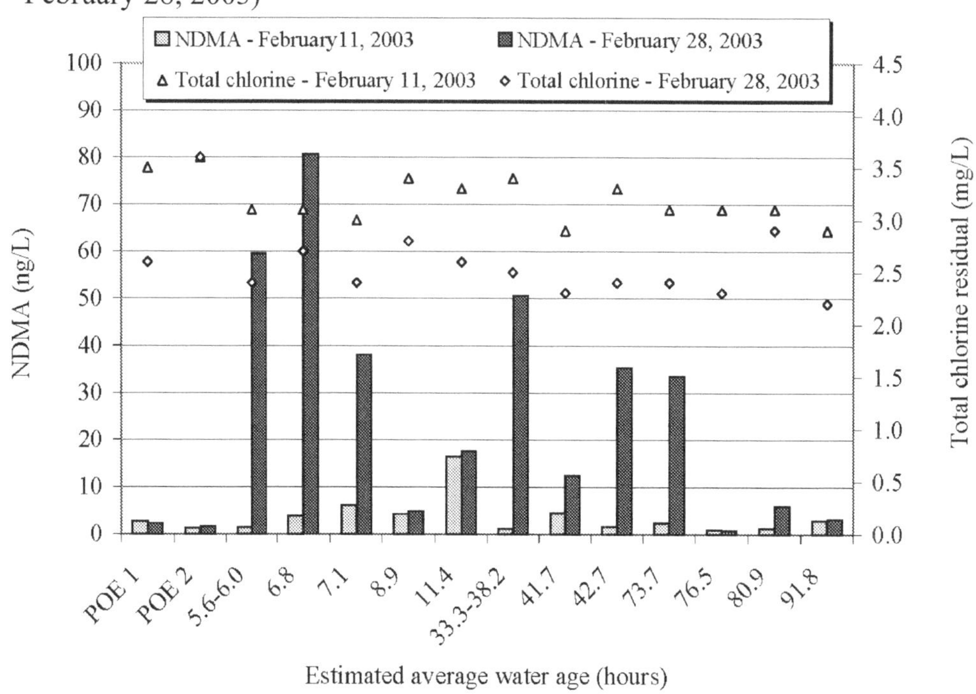

Figure 4.33 NDMA concentration and chlorine residual as a function of water age measured in Utility E (continued)

Figure 4.33 (continued)

b) During the free chlorination period (water temperature at the POEs was 9.9 to 10.2°C [50°F] on March 11, and 13.3°C [56°F] on March 31, 2003)

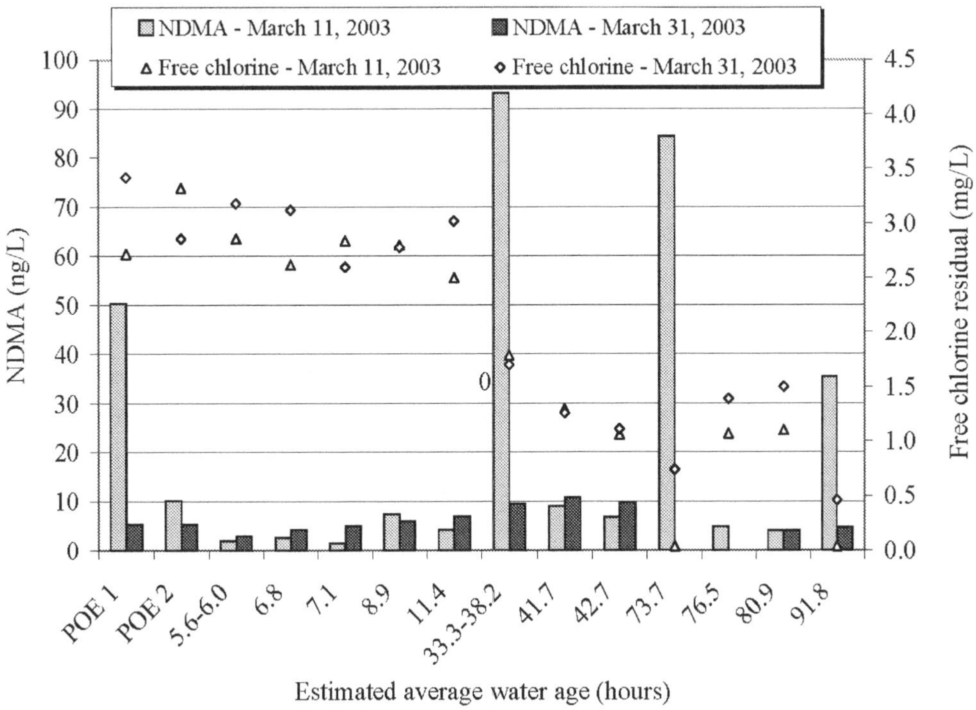

c) Following the return to chloramines (water temperature at the POEs was 15.5 to 15.6°C [60°F] on April 29, and 17.5 to 17.7°C [64°F] on May 20, 2003)

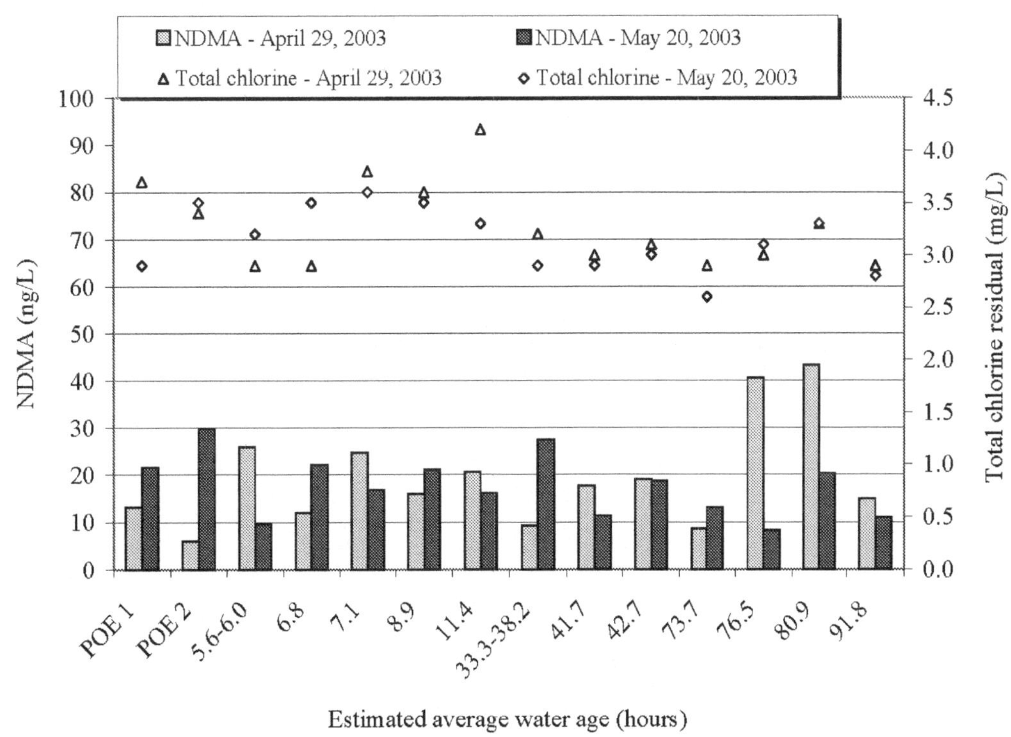

System Comparison

Because all NDMA samples were analyzed by the same laboratory, a brief comparison across the systems studied was done. Results are presented in Table 4.19 and suggest that there was little differences between the free chlorinated and the chloraminated conditions. Although the average value was higher in the chloraminated system (Utility C) than in the free chlorinated system (Utility A), the median values were similar. Trends are more difficult to see in Utility E, where important differences were observed between both samplings conducted during the original chloramination period, before the switch to free chlorine. Important differences were also observed between the average values obtained during the free chlorination period. The median values however suggest greater NDMA concentrations following the return to chloramination than during the free chlorination period, although it is possible that the higher water temperature measured following the return to chloramination may have lead to higher NDMA concentrations. The absence of significant differences between the free chlorinated and the chloraminated systems is not entirely surprising, as it is likely that there were different levels of NDMA precursors in each system, regardless of their sources (i.e., present in the raw water, or introduced during treatment, such as through the use of polymers).

To complement this analysis, the occurrence of NDMA concentrations greater than the MCL of 4 ng/L was also examined (Table 4.20). Results obtained confirmed the above observations, in that there was no significant difference between the chloraminated system (Utility C) and the free chlorinated system (Utility A), and that greater NDMA concentrations were observed following the return to chloramination, which also coincided with a warmer water temperature.

Table 4.19
Comparison of NDMA results across utilities

	Disinfectant residual	Sampling date	Average (ng/L)	Median (ng/L)
Utility A	Free chlorine	Combined samplings	9	7
Utility C	Chloramines	Combined samplings	12	6
Utility E	Chloramines	February 11, 2003	2	0
	Chloramines	February 28, 2003	24	15
	Free chlorine	March 11, 2003	22	7
	Free chlorine	March 31, 2003	5	5
	Chloramines	April 29, 2003	19	17
	Chloramines	May 20, 2003	18	18

Concentrations below the MDL were considered 0 for calculation purposes.

Table 4.20
Occurrence of NDMA concentrations greater than the MCL (percent of samples)

	Disinfectant residual	Sampling date	Occurrence (percent)
Utility A	Free chlorine	Combined samplings	73
Utility C	Chloramines	Combined samplings	71
Utility E	Chloramines	February 11, 2003	29
	Chloramines	February 28, 2003	71
	Free chlorine	March 11, 2003	79
	Free chlorine	March 31, 2003	71
	Chloramines	April 29, 2003	100
	Chloramines	May 20, 2003	100

OTHER WATER QUALITY PARAMETERS

Chlorine Residuals and Heterotrophic Plate Counts

As for the previous sections, the trends in chlorine residuals and HPCs observed in the free chlorinated systems were analyzed separately from the trends observed in the chloraminated systems.

Free Chlorinated Systems

Figures 4.34 and 4.35 present representative graphs of free chlorine residuals and HPCs as water age increased in Utility A and Utility B, respectively. As expected, HPCs generally increased with increasing water age in the distribution systems, with concentrations reaching 2.0×10^4 CFU/mL at the MRTL. Surprisingly HPCs were generally high in Utility A's distribution system, despite the presence of a free chlorine residual of 0.5 mg/L or greater. Free chlorine residual varied from 2.4 to 3.6 mg/L (average of 2.8 mg/L) at the POE, and decreased gradually with increasing water age in the distribution system. At the MRTL, free chlorine residual varied from <0.1 to 1.4 mg/L (average of 0.5 mg/L), with highest residuals observed during the colder months, as expected.

In Utility B, elevated HPCs were measured in the distribution system (up to 2.2×10^4 CFU/mL) even at sampling locations with free chlorine residuals >1.1 mg/L (Figure 4.35). Again, the simultaneous presence of heterotrophic bacteria and chlorine residual was surprising as it is opposed to the general understanding and the fundamentals of the Surface Water Treatment Rule and the Total Coliform Rule. As opposed to the trends observed in Utility A, there was no apparent pattern in HPCs with increasing water age in Utility B; HPCs sometimes increased with water age, sometimes decreased, or remained relatively constant at other times. As observed in Utility A, free chlorine residual decreased with increasing water age in Utility B, although the decrease was less pronounced than in Utility A, most likely as a result of the lower water ages throughout Utility B's distribution system. Also, free chlorine residuals did not decrease as much in cold water as in warmer water. As a result, free chlorine residual ranged from 1.7 to 2.0 mg/L (average of 1.8 mg/L) at the POE, and decreased to 1.2-1.5 mg/L during the colder months, and 0.9-1.5 mg/L (average of 1.2 mg/L) during the warmer months at the MRTLs.

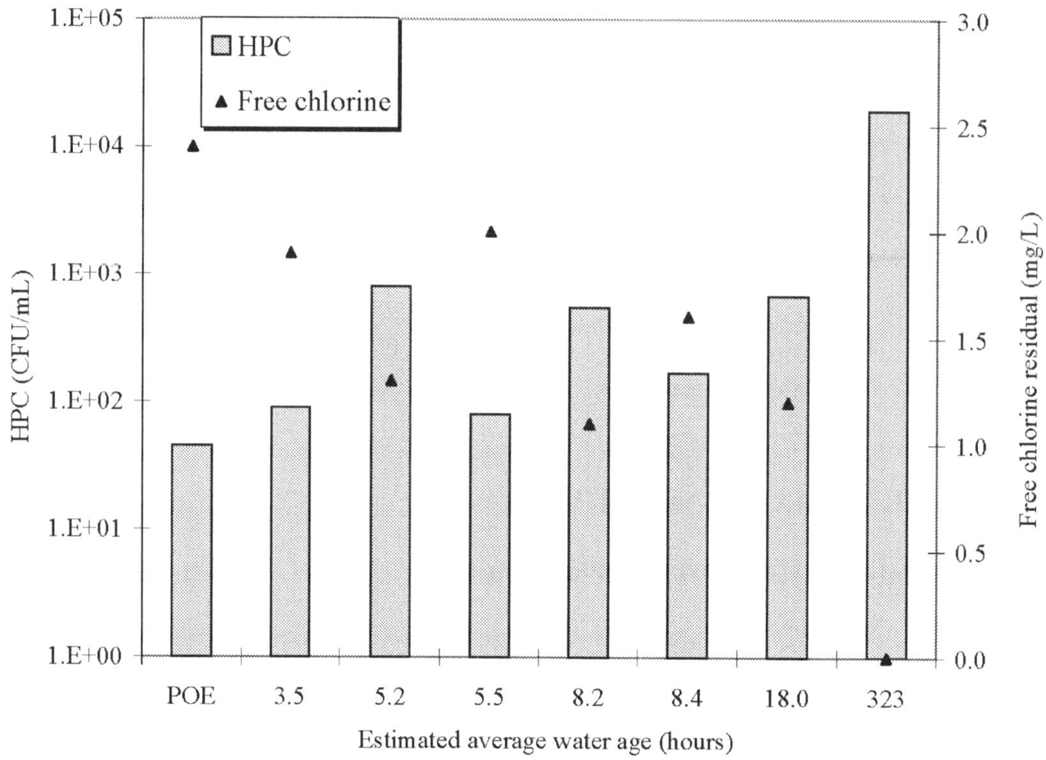

Figure 4.34 Free chlorine residual and HPC as a function of water age measured during the December 3, 2002 sampling in Utility A (free chlorinated system) (water temperature at the POE was 5.6°C, 42°F)

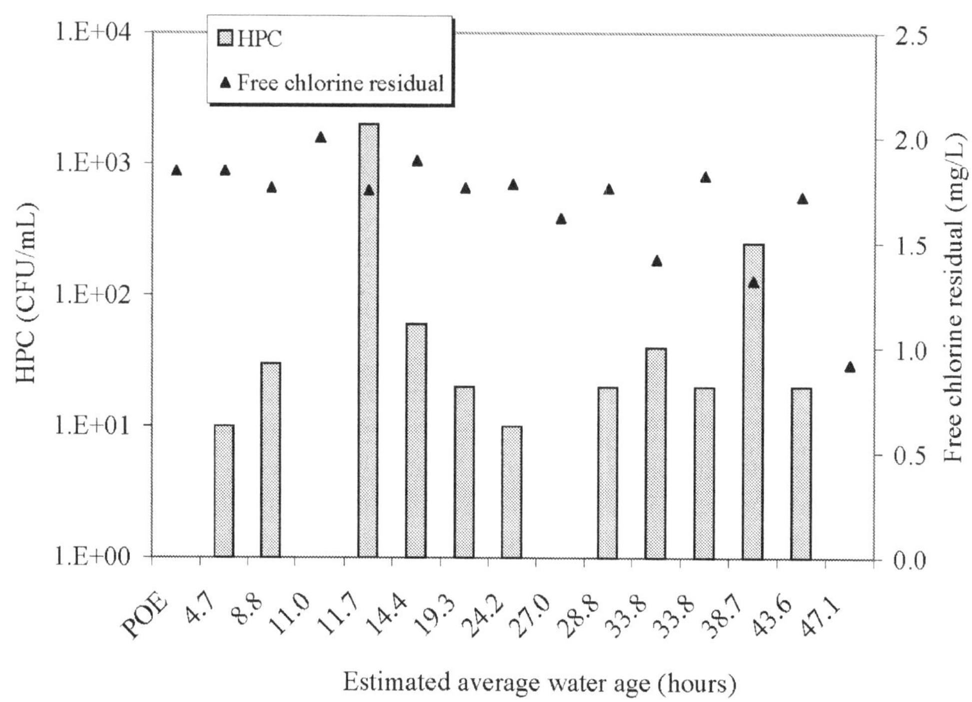

Figure 4.35 Free chlorine residual and HPC as a function of water age measured during the August 12, 2003 sampling in Utility B (free chlorinated system) (water temperature at the POE was 27°C, 81°F)

Chloraminated Systems

Figure 4.36 presents representative results of total chlorine residuals and HPCs as water age increased in Utility C. HPCs generally increased with increasing water age, and peaks in HPCs (up to 1.2×10^4 CFU/mL) usually occurred at sampling locations where low total chlorine residuals were measured, such a the MRTL. In general, total chlorine residuals decreased with increasing water age in Utility C. These trends are consistent with the general understanding of the fate of heterotrophic bacteria and chlorine residual in distribution systems. Total chlorine residuals ranged from 3.3 to 4.0 mg/L at POE (average of 3.7 mg/L), and decreased to <0.1 to 2.0 mg/L at the MRTL (average of 0.5 mg/L). Decreases in total chlorine residual and increases in HPCs were also seen at an intermediate location with a relatively low estimated average water age (8 hours). It is possible that the water age was not really 8 hours at this location due to possible modeling errors. No seasonal trend was apparent in Utility C.

In Utility D, trends observed can be grouped in two categories based on the hydraulic scenario observed at the time of sampling. When both POEs were used, water age remained low throughout the distribution system, which resulted in lower HPCs and more stable total chlorine residual (Figure 4.37a). HPCs remained below 10^2 CFU/mL, with a concentration of 7.9×10^2 CFU/mL measured at one location during one of the samplings. Total chlorine residual ranged from 2.9 to 3.7 mg/L at the POEs, and decreased to 0.6 to 2.8 mg/L at MRTLs. When only one POE was used, water age was allowed to increased in Utility D, which resulted in significant fluctuations in HPCs and decreases in total chlorine residuals (Figure 4.37b). HPCs were highest (up to 2.6×10^3 and 2.0×10^4 CFU/mL) at the locations with the lowest total chlorine residuals, as expected. Total chlorine residual ranged from 2.2 to 3.0 mg/L at the POE, and decreased to 0.1 mg/L at the MRTLs in Utility D.

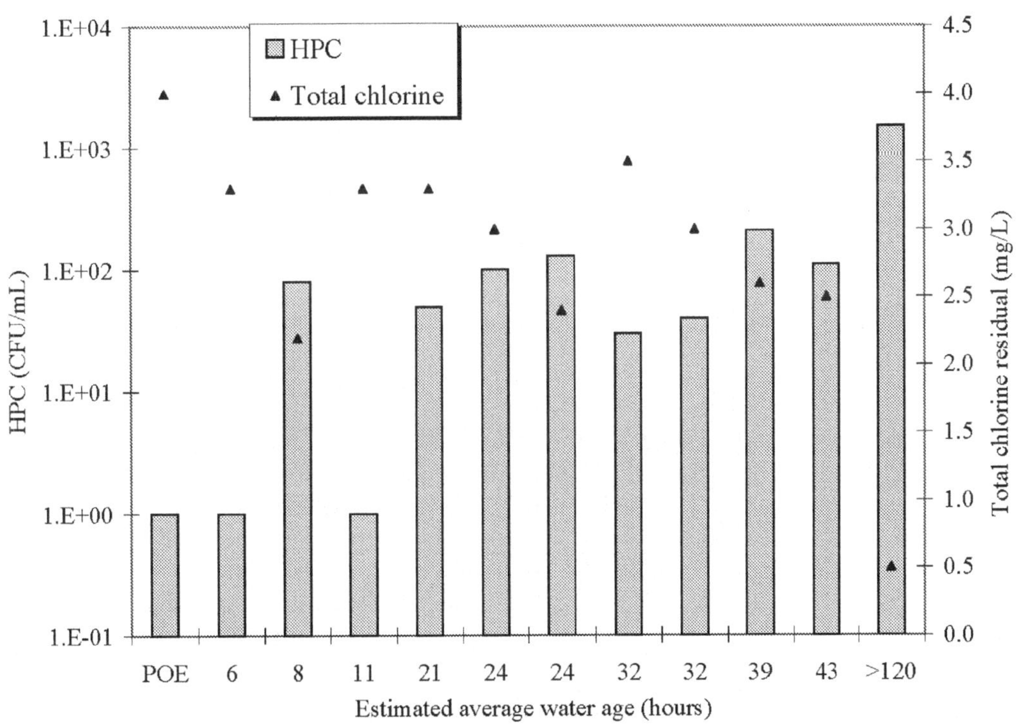

Figure 4.36 Total chlorine residual and HPC as a function of water age measured during the October 15, 2002 sampling in Utility C (chloraminated system) (water temperature at the POE was 22°C, 72°F)

a) Both POEs were used (April 8, 2003) (water temperature at the POEs was 18°C, 64°F)

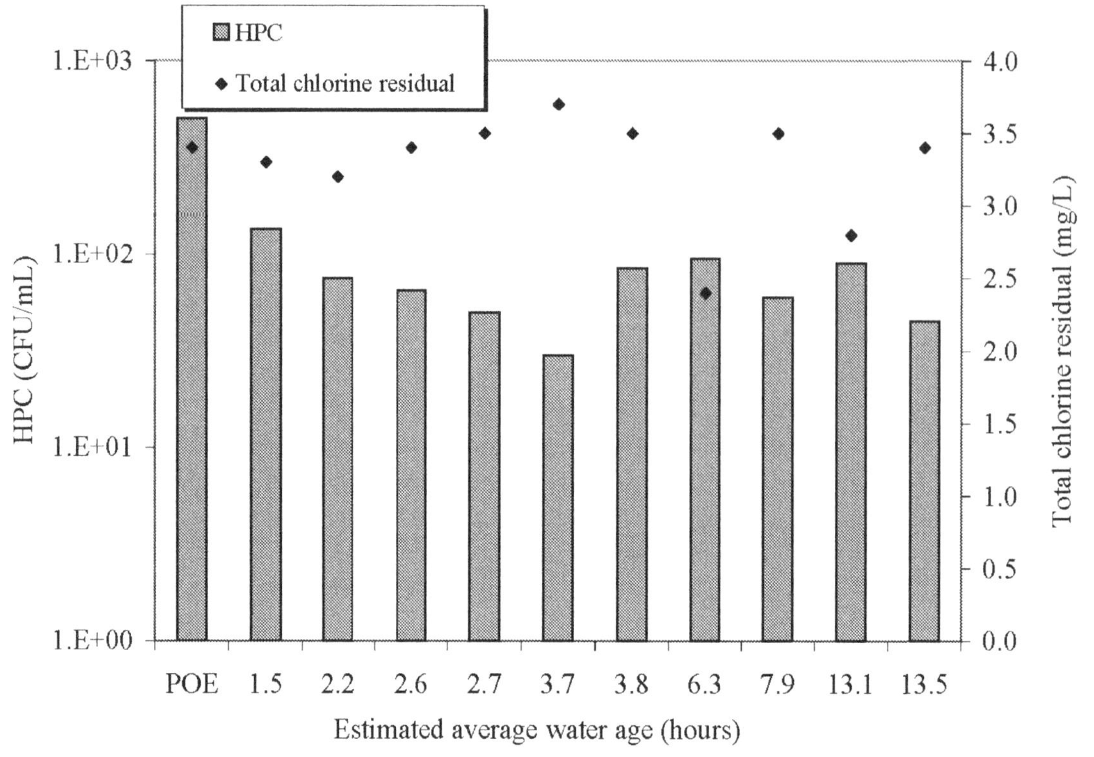

b) One POE was used (January 7, 2003) (water temperature at the POE was 15.6°C, 60°F)

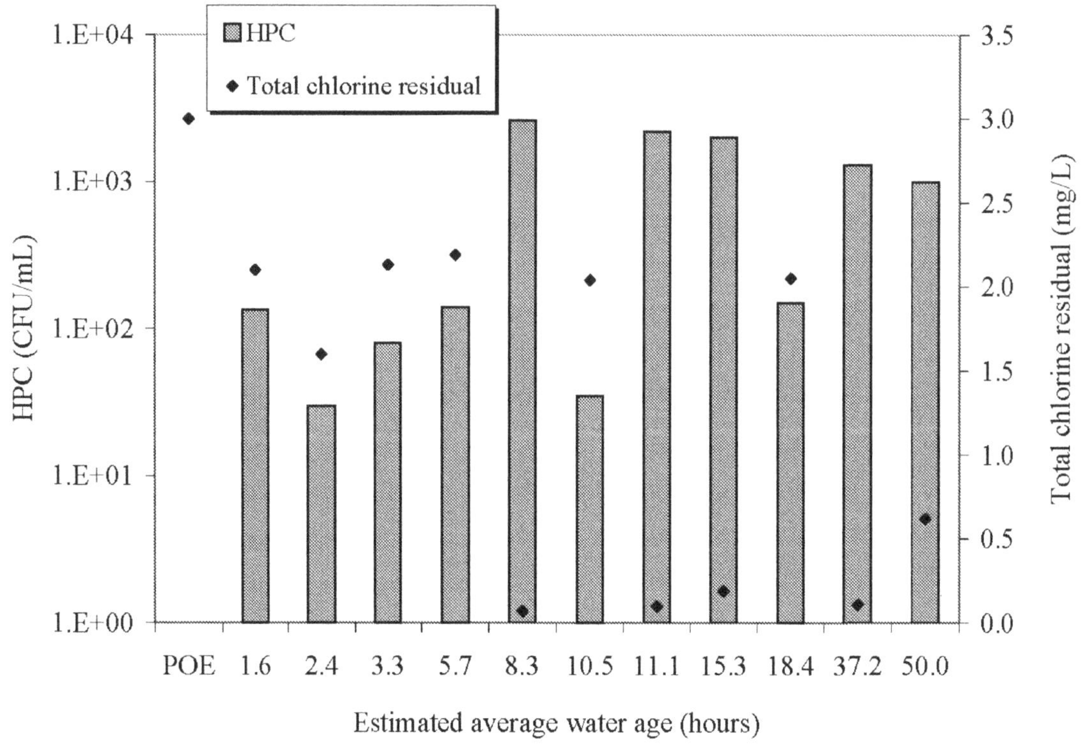

Figure 4.37 Total chlorine residual and HPC as a function of water age measured during the sampling in Utility D (chloraminated system)

Chloraminated System Periodically Switching to Free Chlorine

Figure 4.38 shows HPCs for all water ages categorized by POE water. Results suggest that the POE water influenced HPCs at the sampling sites in Utility E; sampling sites under the influence of Site 2 POE water showed higher HPCs when compared to sampling sites receiving Site 1 POE water.

Figure 4.39 shows chlorine residuals for all water ages categorized by POE water. Unlike the HPC data, results suggest that the POE water did not influence chlorine residual. This figure also indicates that similar total chlorine residuals were measured during both chloramination periods.

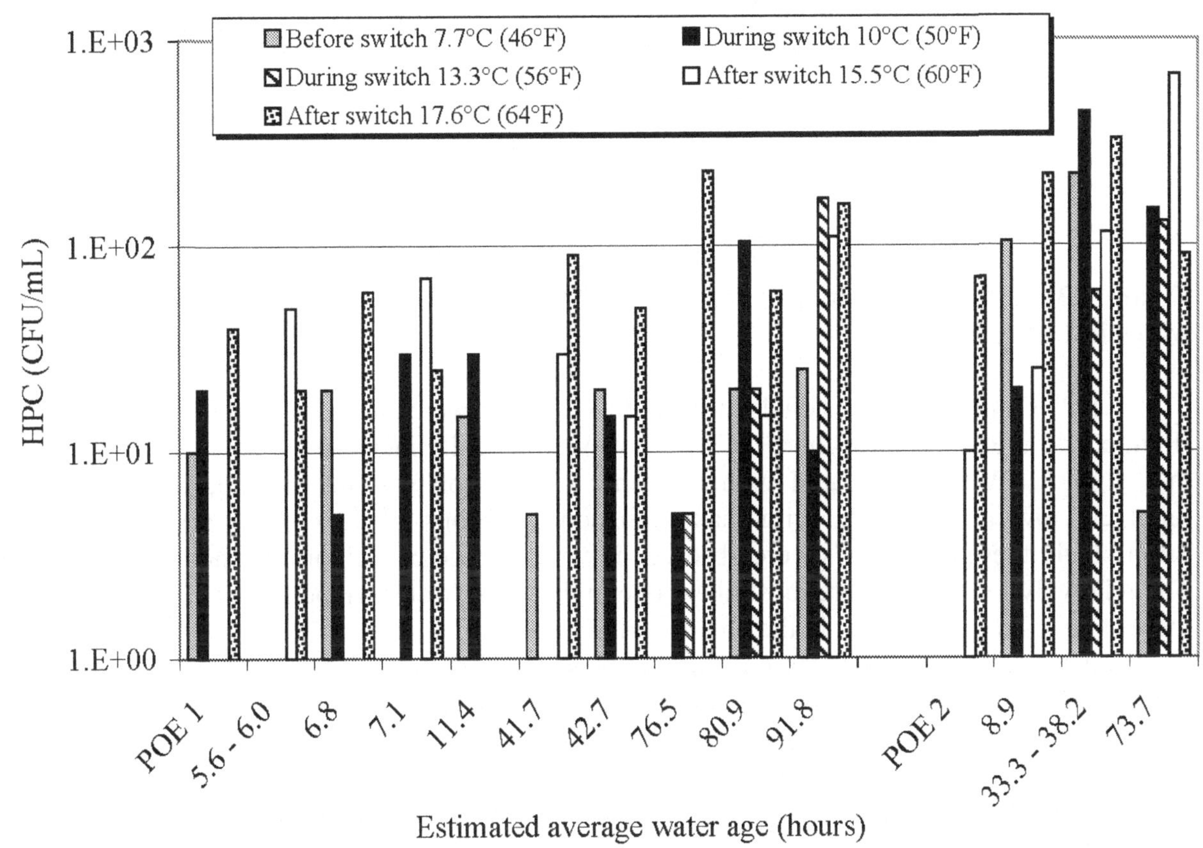

Figure 4.38 HPC as a function of water age for all samplings sorted by POE water in Utility E

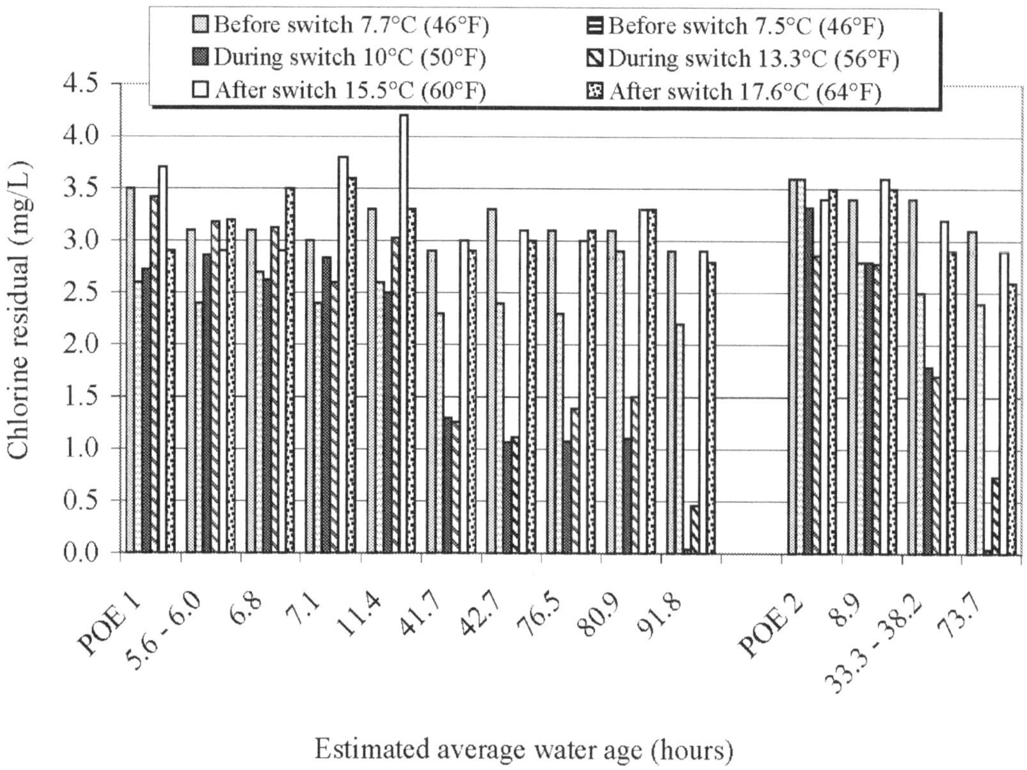

Figure 4.39 Chlorine residual as a function of water age for all samplings sorted by POE water in Utility E

HPCs and chlorine residuals measured in Utility E were plotted as a function of water age (Figure 4.40). Results show that HPCs did not systematically increase with increasing water age in Utility E. When compared with the other distribution systems studied, HPCs remained relatively low in Utility E, with maximum concentrations of $1.7x10^2$ to $6.8x10^2$ CFU/mL. Surprisingly these high HPCs were observed in the presence of a chlorine residual (>2.0 mg/L). The relatively low HPCs were most likely due to the lower water temperature (<18°C, <64°F) as these samples were collected in the winter/spring.

Although the water temperature increased throughout the samplings conducted in Utility E, HPCs did not increase significantly from one sampling to the next. Also, the type of distribution system disinfectant residual used did not significantly affect HPCs. These results are surprising as previous studies conducted in Utility E's distribution system (DiGiano 2003) showed that chloramine disinfection usually results in lower HPCs than free chlorine.

As shown on Figure 4.40, the total chlorine residuals remained constant or decreased slightly with increasing water age during the chloramination periods, whereas the free chlorine residuals decreased significantly during the free chlorination period. Such results were expected considering the greater stability (lower reactivity) of chloramines compared to free chlorine. During the chloramination periods, the total chlorine residual ranged from 2.6 to 3.7 mg/L (average of 3.2 mg/L) in Site 1 POE, and ranged from 3.4 to 3.6 mg/L (average of 3.5 mg/L) in Site 2 POE. Minimum total chlorine residuals reached at the MRTLs of the distribution system ranged from 2.2 to 2.9 mg/L, showing very little decrease with increasing water age. The relatively high chlorine residuals maintained in Utility E's distribution system during the chloramination periods may have limited microbial regrowth, as shown by the relatively low HPCs and absence of HAA biodegradation, as presented above.

a) During chloramination, before the switch to free chlorine (February 11, 2003) (water temperature at the POEs was 6.4 and 9.0°C, 44 and 48°F)

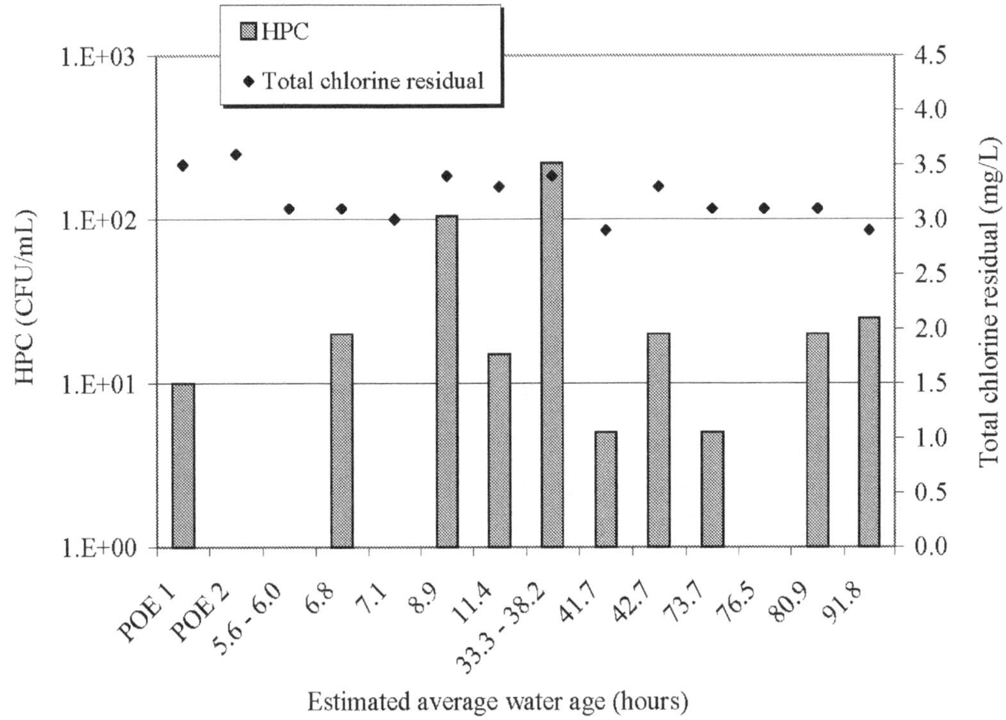

b) During the free chlorination period (water temperature at the POEs was 9.9 to 10.2°C [50°F] on March 11, and 13.3°C [56°F] on March 31, 2003)

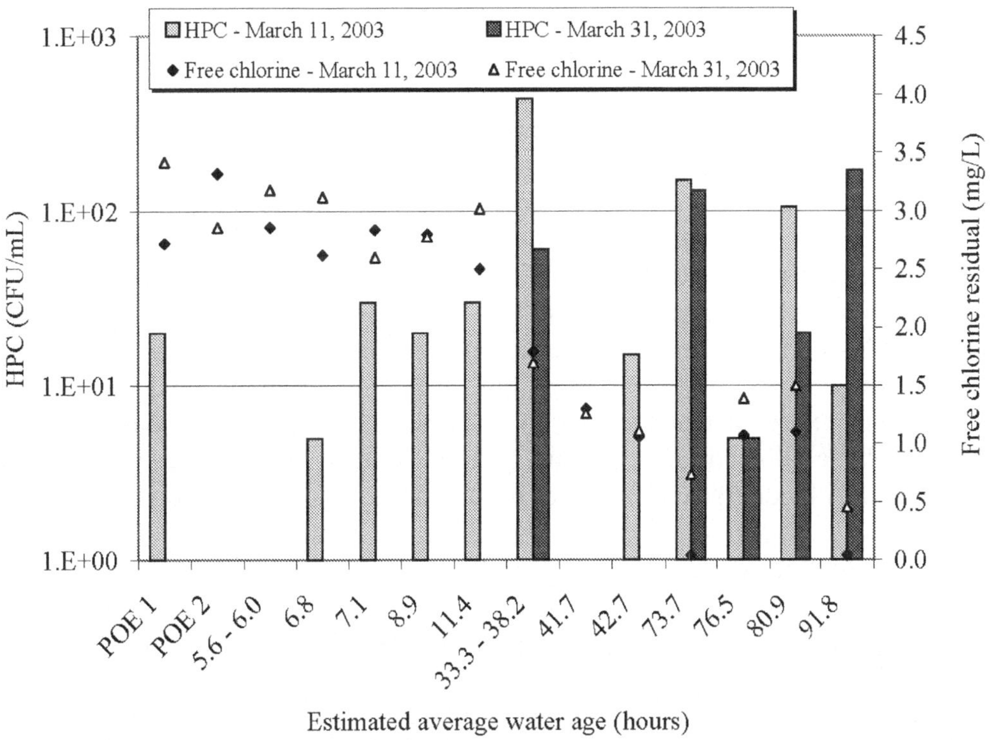

Figure 4.40 HPC and chlorine residual as a function of water age measured in Utility E (continued)

Figure 4.40 (continued)

c) Following the return to chloramines (water temperature at the POEs was 15.5 to 15.6°C [60°F] on April 29, and 17.5 to 17.7°C [64°F] on May 20, 2003)

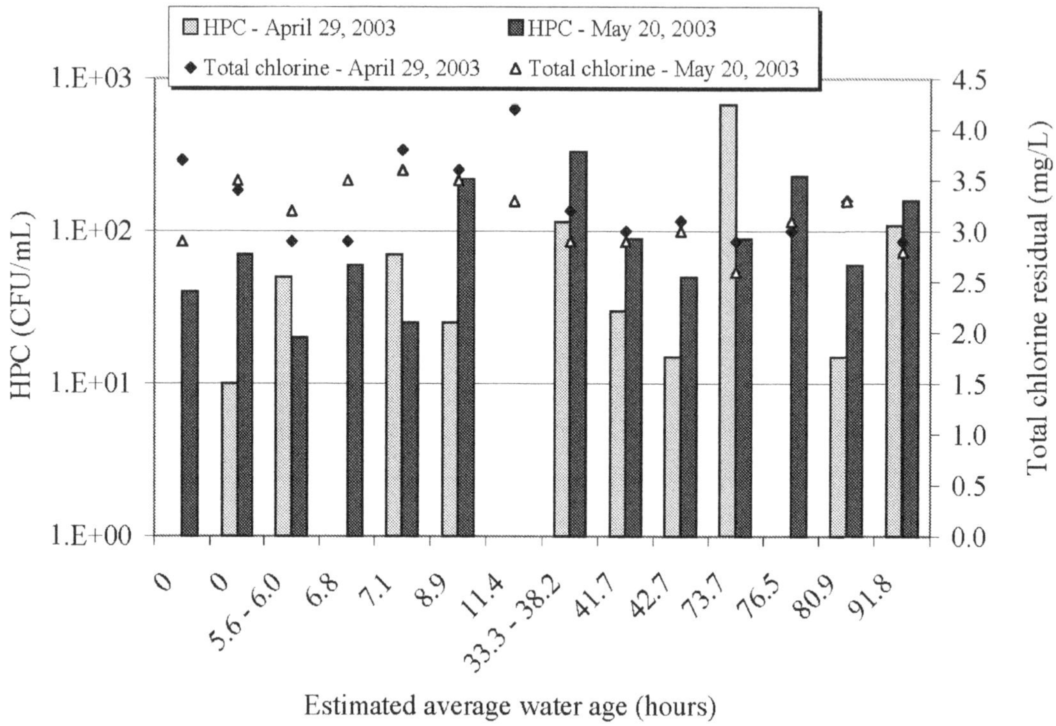

During the free chlorination period, the free chlorine residual ranged from 2.7 to 3.4 mg/L at the POEs, and decreased to <0.1 to 0.5 mg/L at the MRTLs. During the first sampling conducted under free chlorination (March 11, 2003), the free chlorine residual decreased to zero at two of the sampling sites with high water ages. However, both water treatment plants were in the process of increasing chlorine doses in an attempt to maintain a residual at all points within the distribution system. Therefore, the chlorine doses fed at the water treatment plants on the day of sampling were higher than the corresponding chlorine doses associated with the water sampled at MRTLs, due to the lag time (water age) between the POEs and these MRTLs. By the time of the second free chlorine sampling (March 31, 2003), all locations within the distribution system had a free chlorine residual of at least 0.5 mg/L. Although it was expected that the loss in chlorine residual observed during the free chlorine period would have allowed heterotrophic bacteria to grow, such effect was not observed, and consistently, HAA biodegradation was not observed during the free chlorination period.

Natural Organic Matter

Three parameters were used to characterize NOM: TOC concentration, UV absorbance measured at 254 nm, and AOC concentration. TOC concentration and UV absorbance were measured at each sampling location in some distribution systems, and only at the POE of the other distribution systems. During each sampling and for each system studied, AOC concentration was measured at the POE only. The rational for measuring these parameters at the

POE only is that POE is the only location where these parameters are measured in most systems. Unless a special project is conducted, utilities do not measure TOC concentration or UV absorbance throughout the distribution system. As such, a discussion of the effect of NOM characteristics measured in the distribution system to predict DBP formation would not be practical to utilities.

Free Chlorinated Systems

Table 4.21 summarizes TOC concentration, UV absorbance and AOC concentration measured at the POE of Utilities A and B during each sampling. Over the course of the study in Utility A, these parameters tended to follow an expected seasonal pattern with lower concentrations in cold water and higher concentrations in warmer water, although the trend was not as clear with AOC. Lower TOC concentrations and UV absorbance were also observed in cold water than in warmer water in Utility B, although the trend was not as pronounced, possibly because the seasonal change in water temperature was not as pronounced in this system (11-27°C, 52-81°F, in Utility B, compared to 4-20°C, 39-68°F, in Utility A).

Table 4.21
TOC concentration, UV absorbance at 254 nm and AOC concentration measured at the POE of Utilities A and B (free chlorinated systems) during each sampling

Sampling date	Water temperature at the POEs (°C, °F)	TOC concentration (mg/L C)	UV absorbance (cm^{-1})	AOC concentration (μg/L acetate C)
Utility A:				
October 2002	13.3°C (56°F)	2.40	0.026	740
December 2002	5.6°C (42°F)	1.24	0.029	453
January 2003	6.1°C (43°F)	1.19	0.029	270
February 2003	4.0°C (39°F)	2.80	0.041	711
June 2003	15.0°C (59°F)	2.13	0.045	550
August 2003	20.1°C (68°F)	2.46	0.033	513
September 2003	16.6°C (62°F)	2.85	0.034	203
Utility B:				
October 2002	26°C (79°F)	1.68	0.022	732
November 2002	17°C (63°F)	1.70	0.029	274
January 2003	11°C (52°F)	1.47	0.020	127
February 2003	12°C (54°F)	1.56	0.023	278
April 2003	22°C (72°F)	1.54	0.024	139
June 2003	26°C (79°F)	1.66	0.027	387
August 2003	27°C (81°F)	1.48	0.029	399
October 2003	22°C (72°F)	1.40	0.020	71

Chloraminated Systems

Table 4.22 summarizes TOC concentration, UV absorbance and AOC concentration measured at the POE of Utilities C and D during each sampling. Over the course of the study, TOC concentrations and UV absorbance remained relatively constant at the POE to Utility C's distribution system. Corresponding AOC concentrations showed some fluctuations, although no seasonal pattern could be distinguished. Likewise, TOC concentrations, UV absorbance and AOC concentrations at the POE to Utility D's distribution system showed some fluctuations during the study, although no seasonal pattern could be distinguished.

Chloraminated System Periodically Switching to Free Chlorine

At both POEs to Utility E's distribution system, samples were collected and analyzed for TOC concentration, UV absorbance at 254 nm and AOC concentration. Table 4.23 summarizes the data obtained. Results suggest that NOM concentration was higher for Site 2 POE than Site 1 POE based on the analyses conducted during this study, which is consistent with historical organic carbon measurements at these locations. Slightly greater TOC concentrations were observed during the free chlorination period, but UV absorbance did not follow this trend. The AOC concentrations from both POEs were highly variable from one sampling to the next, which is consistent with historical data observed at Utility E.

Table 4.22
TOC concentration, UV absorbance at 254 nm and AOC concentration measured at the POE of Utilities C and D (chloraminated systems) during each sampling

Sampling date	Water temperature at the POEs (°C, °F)	TOC concentration (mg/L C)	UV absorbance (cm^{-1})	AOC concentration (µg/L acetate C)
Utility C:				
October 2002	22.0°C (72°F)	3.10	0.048	Not analyzed
November 2002	16.0°C (61°F)	2.53	0.049	262
January 2003	10.5°C (51°F)	3.17	0.053	293
February 2003	8.0°C (46°F)	3.10	0.059	421
April 2003	20.0°C (68°F)	2.97	0.050	224
June 2003	25.0°C (77°F)	3.07	0.062	307
August 2003	27.0°C (81°F)	3.31	0.064	211
October 2003	19.0°C (66°F)	3.19	0.030	211
Utility D:				
November 2002	21.1°C (70°F)	1.57	0.053	138
January 2003	15.6°C (60°F)	1.70	0.056	188
April 2003	18.3°C (65°F)	1.92	0.047	253
June 2003	17.8°C (64°F)	1.71	0.050	107
July 2003	21.1°C (70°F)	2.58 at POE 1 2.55 at POE 2	0.054 at both POEs	126 at POE 1 89 at POE 2
September 2003	21.1°C (70°F)	2.28 at POE 1 2.29 at POE 2	0.049 at POE 1 0.050 at POE 2	108 at POE 1 81 at POE 2

Table 4.23
TOC concentration, UV absorbance at 254 nm and AOC concentration measured at the POEs of Utility E during each sampling

Sampling date	Sampling location	Water temperature at POEs (°C, °F)	TOC concentration (mg/L C)	UV absorbance (cm^{-1})	AOC concentration (µg/L acetate C)
February 11, 2003	Site 1 POE	6.4°C (44°F)	2.20	0.046	193
	Site 2 POE	9.0°C (48°F)	2.22	0.050	Not sampled
February 28, 2003	Site 1 POE	7.0°C (45°F)	1.81	0.046	519
	Site 2 POE	8.0°C (46°F)	2.12	0.049	361
March 11, 2003	Site 1 POE	9.9°C (50°F)	2.01	0.026	Not sampled
	Site 2 POE	10.2°C (50°F)	2.55	0.031	Not sampled
March 31, 2003	Site 1 POE	13.3°C (56°F)	2.01	0.040	Not analyzed
	Site 2 POE	13.3°C (56°F)	2.55	0.040	347
April 29, 2003	Site 1 POE	15.6°C (60°F)	1.57	0.044	204
	Site 2 POE	15.5°C (60°F)	1.95	0.049	198
May 20, 2003	Site 1 POE	17.7°C (64°F)	1.57	0.037	Not sampled
	Site 2 POE	17.5°C (64°F)	1.82	0.043	Not sampled

Nitrification in the Chloraminated Systems

Nitrification in drinking water distribution systems is a common challenge encountered by most utilities using chloramines as the final disinfectant. Nitrification is characterized by increases in nitrite concentration and possibly nitrate concentration, decreases in chlorine residual, decreases in ammonia concentration, and possibly increases in microbial counts.

Table 4.24 summarizes results of total ammonia, nitrite and nitrate concentrations measured in the absence of nitrification in the distribution systems of Utilities C and D over the course of the study. In the absence of nitrification, the concentrations of nitrogen species remained relatively constant with increasing water age in Utility C's distribution system (Figure 4.41). However nitrification was obvious at the MRTL (water age of >120 hours), where nitrite concentration increased to reach levels ranging from 391 to 818 µg/L NO_2^- -N, nitrate concentration increased to levels of 0.341 to 0.928 mg/L NO_3^- -N, and total ammonia concentration decreased to 44.1 to 97.5 percent. The only sampling that did not show such strong signs of nitrification occurred in February 2003, during which nitrite concentration increased to 14 µg/L NO_2^- -N, nitrate concentration increased to 0.091 mg/L NO_3^- -N, and total ammonia concentration did not decrease. The MRTL also showed low total chlorine residuals and high HPCs during each sampling, which confirmed the occurrence of nitrification at this location, except during the February 2003 sampling, consistent with the above observations. In Utility C, modest signs of nitrification were also observed at times at the sampling locations with water ages of 8 and 24 hours, although the changes in nitrite, nitrate and ammonia concentrations were much smaller than at the MRTL.

Table 4.24
Concentrations of total ammonia, nitrite and nitrate measured in absence of nitrification in the distribution systems of Utilities C and D (chloraminated systems)

Distribution system	Minimum	Average	Maximum
Utility C:			
Total ammonia (mg/L NH_3-N)	0.422	0.575	0.793
Nitrite (µg/L NO_2^- -N)	ND	2.43	9.47
Nitrate (mg/L NO_3^- -N)	ND	0.042	0.090
Utility D:			
Total ammonia (mg/L NH_3-N)	0.07	0.40	0.53
Nitrite (mg/L NO_2^- -N)	ND	0.017	0.101
Nitrate (mg/L NO_3^- -N)	2.0	3.2	4.0

ND: Not detected

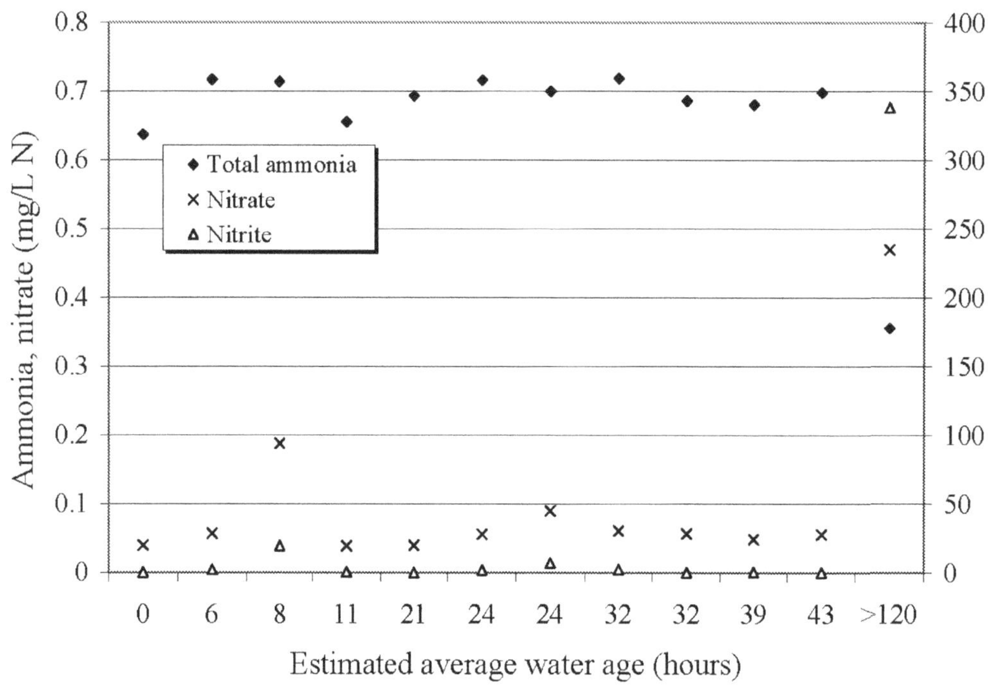

Figure 4.41 Nitrogen species as a function of water age measured during the August 12, 2003 sampling in Utility C (chloraminated system) (water temperature at the POE was 27.0°C, 81°F)

In Utility D, trends observed differed according to the hydraulic scenario present in the distribution system. When both POEs were used (resulting in lower water ages throughout the distribution system), Site 1 and Site 2 POEs had similar water quality in terms of nitrogen species concentrations, and nitrogen species remained relatively constant with increasing water age in the distribution system (Figure 4.42a), with the following exceptions. During the April 2003 sampling, a very high nitrite concentration (0.021 mg/L NO_2^- -N) was measured at the sampling location with a water age of 6.3 hours. This location also showed a lower total chlorine residual (2.4 mg/L) and a high HPC (95 CFU/mL) during this particular sampling, consistent with the occurrence of nitrification. None of the other sampling locations or parameters showed signs of nitrification. The other exception was observed in September 2003, when the MRTL (water age of 12.7 hours) showed a high nitrite concentration (0.027 mg/L NO_2^- -N) and a slight decrease in total ammonia concentration (0.37 mg/L NH_3-N). The HPC was also higher at this location (7.9×10^2 CFU/mL), although the total chlorine residual was not significantly lower.

When only one POE was used (resulting in greater water ages throughout the distribution system), the concentrations of all nitrogen species fluctuated in the distribution system and nitrification was observed at several locations during all of the samplings (Figure 4.42b). Nitrite concentrations of up to 0.332 mg/L NO_2^- -N, nitrate concentrations of up to 4.9 mg/L NO_3^- -N and decreases in total ammonia concentrations were observed throughout the distribution system. These locations also showed lower total chlorine residuals and higher HPCs, confirming that nitrification was occurring during these samplings.

Chloraminated System Periodically Switching to Free Chlorine

In Utility E's distribution system, nitrite concentrations were consistently lower than 0.1 µg/L NO_2^- -N during all samplings conducted. Nitrate concentrations ranged from 0.10 to 0.40 mg/L NO_3^- -N, and concentrations remained constant with increasing water age in the distribution system. Figure 4.43 shows free ammonia concentrations for all water ages categorized by POE water. Results suggest that the POE water did not influence the free ammonia concentration. Free ammonia concentrations usually ranged from 0.00 to 0.42 mg/L NH_3-N (average of 0.19 mg/L N) during the chloramination periods, however abnormally high free ammonia concentrations were observed during the February 28, 2003 sampling, ranging from 0.46 to 0.72 mg/L NH_3-N. Concentrations fluctuated in the distribution system during all samplings (Figure 4.43). As expected, ammonia concentrations were lower than 0.05 mg/L NH_3-N during the free chlorination period.

These data suggest that nitrification was not observed in Utility E's distribution system during this study. Although the ammonia concentration decreased slightly on several occasions, these decreases did not correspond to increases in nitrite or nitrate concentrations. In addition, total chlorine residuals above 2.5 mg/L Cl_2 were maintained throughout the distribution, which most likely limited microbial regrowth in Utility E's distribution system. The absence of nitrification is this distribution system was most likely due to the lower water temperature when these samples were collected.

a) Both POEs were used (July 22, 2003) (water temperatures at the POEs were 21.1 and 23.3°C, 70 and 73°F)

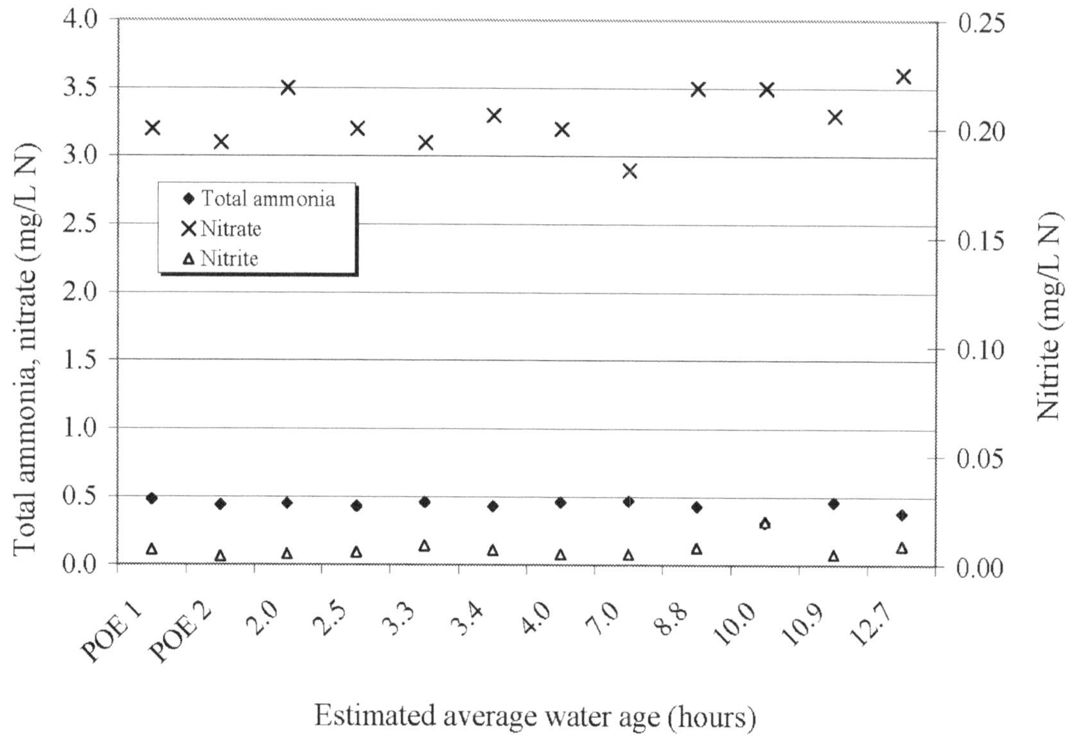

b) One POE was used (November 18, 2002) (temperature at the POE was 21.1°C, 70°F)

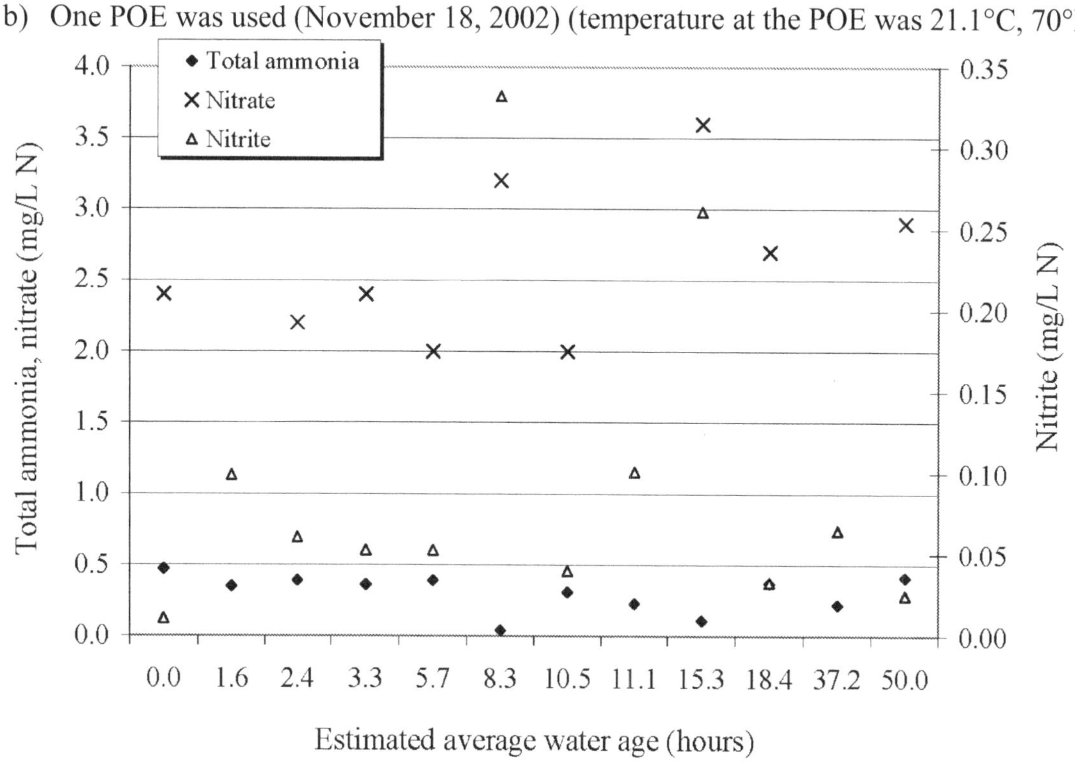

Figure 4.42 Nitrogen species as a function of water age measured in Utility D (chloraminated system)

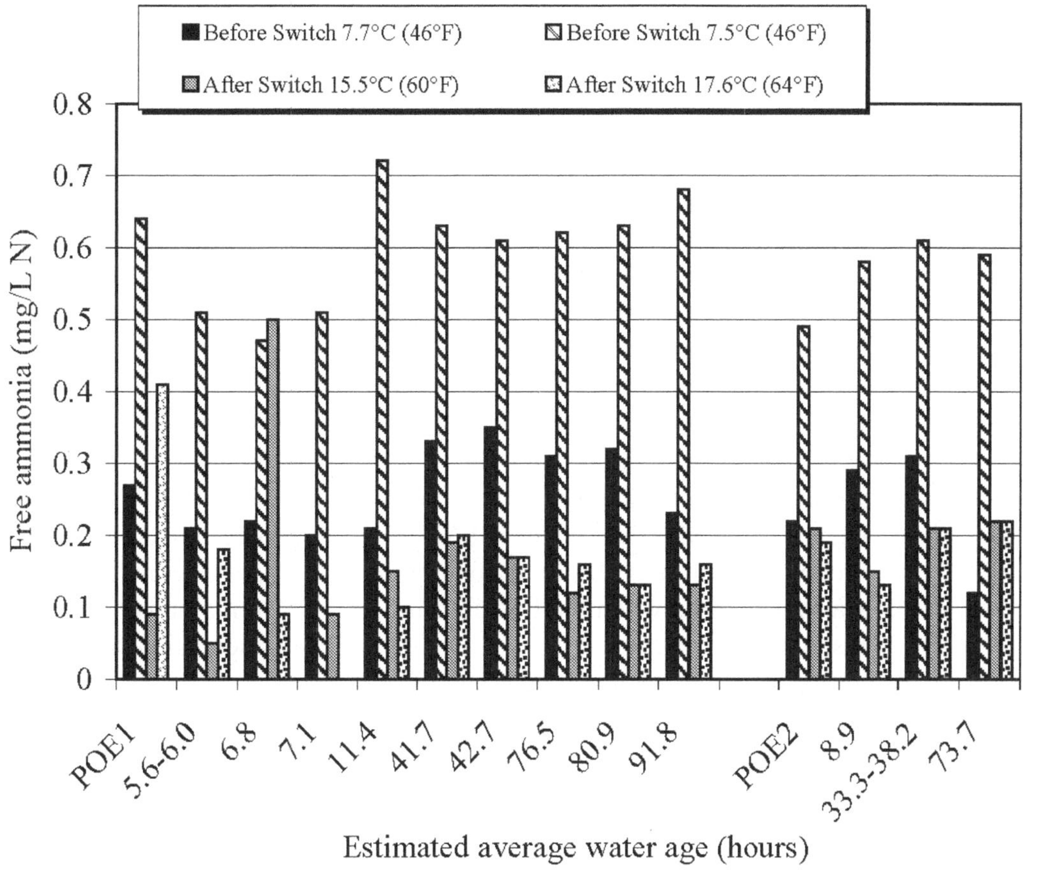

Figure 4.43 Free ammonia concentration as a function of water age for all samplings sorted by POE water in Utility E

SUMMARY AND CONCLUSIONS

This study led to interesting conclusions, some expected and some not. Below is a summary of these observations.

Trihalomethanes and Haloacetic Acids

Table 4.25 summarizes observations made throughout the project regarding the fate of THM4 and HAA9 in the five full-scale distribution systems studied.

Table 4.25
Fate of THM4 and HAA9 in full-scale distribution systems

Utility	Expected trends	Unexpected trends
Utility A - free chlorine	• THM4 concentrations increased as water age increased and chlorine residual decreased. • The chlorinated fraction (i.e., $CHCl_3$ and $CHBrCl_2$) comprised most of the THMs because of the lower source water bromide concentrations. • HAA9 concentrations generally increased or remained constant as water age increased and chlorine residual decreased, except at the MRTLs, where HAA biodegradation was observed, mainly in warm water and when the free chlorine residual was low. • The HAA5 represented 72 percent of the HAA9, and the trends observed with HAA5 and HAA9 were similar. • The dominant HAA species were X_2AA and X_3AA. Decreases in HAA concentrations were generally associated with losses in XAAs and X_2AAs; X_3AAs remained relatively conserved with increasing water age.	• The fraction of chlorine-containing THM4 decreased slightly with increasing water age during some sampling events or remained stable. • The fraction of chlorine-containing HAAs (i.e., ClAA, Cl_2AA, Cl_3AA and $BrCl_2AA$) decreased slightly during some sampling events or remained constant with increasing water age.
Utility B - free chlorine	• THM4 concentrations increased as water age increased and chlorine residual decreased. • The chlorinated fraction (i.e., $CHCl_3$ and $CHBrCl_2$) comprised most of the THMs because of the lower source water bromide concentrations. • The fraction of chlorine-containing THMs increased slightly with increasing water age. • HAA9 concentrations generally increased or remained constant as water age increased and chlorine residual decreased, except at the MRTLs, where a decrease in HAA9 concentrations was measured. In this system, HAA biodegradation was not as pronounced, consistent with more stable free chlorine residuals.	

(continued)

Table 4.25 (Continued)

Utility	Expected trends	Unexpected trends
Utility B - free chlorine	• The HAA5 represented 79 percent of the HAA9, and the trends observed with HAA5 and HAA9 were similar. • The dominant HAA species were X_2AAs and X_3AAs. Decreases in HAA concentrations were generally associated with losses in XAAs and X_2AAs; X_3AAs remained relatively conserved with increasing water age. • The fraction of chlorine-containing HAAs (i.e., ClAA, Cl_2AA, Cl_3AA and $BrCl_2AA$) increased slightly with increasing water age during most sampling events, particularly in warm water.	
Utility C - chloramines	• THM4 concentrations were relatively stable as water age increased. • Because of drought conditions when sampling began, raw water bromide concentrations were higher during the first two sampling events and the treated water contained mainly the brominated THM species (i.e., $CHBr_2Cl$ and $CHBr_3$). Subsequent samplings were dominated by $CHCl_3$ and $CHCl_2Br$.	• Occasional decreases in THM4 concentration were observed at the MRTL in warmer water, consistent with the occurrence of nitrification at these locations. • The proportion of chlorinated THMs decreased slightly with increasing water age during some sampling events or remained stable. The brominated THMs were conserved throughout the distribution system.
	• HAA9 levels were relatively constant with increasing water age, with the exception of the MRTLs, where HAA biodegradation occurred, consistent with elevated HPCs, loss in disinfectant residual, and signs of nitrification. • The HAA5 represented 77 percent of the HAA9, and the trends observed with HAA5 and HAA9 were similar. • The dominant HAA species were X_2AA and X_3AA. Decreases in HAA concentrations were generally associated with losses in XAAs and X_2AAs; X_3Aas remained relatively constant with increasing water age.	• The proportion of chlorinated HAAs followed various trends with increasing water age: sometimes they increased, they decreased, and at other times they remained stable.

(continued)

Table 4.25 (Continued)

Utility	Expected trends	Unexpected trends
Utility C - chloramines	• Because of the high raw water bromide concentration at the beginning of the study, samples collected during the first two sampling events contained significant concentrations of brominated HAAs. Subsequent sampling events were dominated by the chlorinated HAAs.	
Utility D - chloramines	• When both POEs were used (same source water), resulting in lower water ages throughout the distribution system, total chlorine residual remained stable and THM4 concentrations remained constant with increasing water age. Concurrently, HPCs remained low and nitrification did not occur, except in rare cases. • Raw water bromide concentrations were relatively high, resulting in a high proportion of brominated THMs (i.e., $CHBr_3$ and $CHBr_2Cl$).	• When only one POE was used, resulting in higher water ages, decreases in total chlorine residuals were observed and THM concentrations decreased at the sampling locations with higher water ages and low chlorine residuals. Concurrently, significant fluctuations in HPCs were observed, along with signs of nitrification. • The proportion of chlorinated THMs did not follow any trends with increasing water age.
	• When both POEs were used and low water ages wee encountered, HAA9 concentrations remained constant with increasing water age and total chlorine residual was more stable. HPCs were lower and nitrification was not observed, suggesting low biological activity. • When only one POE was used and high water ages were experienced, HAA9 concentrations remained uniform and then decreased at MRTLs. HPCs fluctuated significantly, total chlorine residuals decreased, and nitrification occurred, indicating biological activity. • The trends observed with HAA5 and HAA9 were similar.	• Because of the high source water bromide concentrations, the HAA5 represented only 39 percent of the HAA9. • The proportion of chlorinated HAAs followed various trends with increasing water age: sometimes they increased, sometimes they decreased, and at other times they remained stable.

(continued)

Table 4.25 (Continued)

Utility	Expected trends	Unexpected trends
Utility D - chloramines	• The dominant HAA species were X_2AA and X_3AA. When decreases in HAA concentrations were observed, they were generally associated with losses in XAAs and X_2AAs; X_3Aas remained relatively consistent with increasing water age. • Because of the relatively high bromide concentration present in the source water, the fraction of bromine-containing HAAs was similar to the fraction of chlorine-containing HAAs.	
Utility E - chloramines periodically switching to free chlorine	• THM4 concentrations increased with increasing water age. Greater increases and higher concentrations were observed during the free chlorination period than during the chloramination periods. • THM4 concentrations did not decrease at MRTLs, as observed in some of the other distribution systems. Concurrently, relatively low HPCs were measured and nitrification was not detected. • The $CHCl_3$ fraction comprised most of the THM4 measured because of the relatively low source water bromide concentration. The $CHCl_3$ fractions increased with increasing water age or remained stable. • The speciation of THM4 was not affected by disinfectant type. • HAA9 concentrations remained relatively stable with increasing water age during both chloramination periods; only minor increases in HAA9 concentrations were noticed. During the free chlorination period, however, HAA9 concentrations increased significantly with water age. • The HAA5 represented >84 percent of the HAA9, and the trends observed with HAA5 and HAA9 were similar.	

(continued)

Table 4.25 (Continued)

Utility	Expected trends	Unexpected trends
Utility E - chloramines periodically switching to free chlorine	• The dominant HAA species were X_2AA and X_3AA. • When chloramines were used as the final disinfectant, Cl_2AA and Cl_3AA were the predominant HAA species, and were approximately equal in concentration. A slightly higher concentration of Cl_3AA was measured while using free chlorine compared to Cl_2AA.	• No clear decrease in HAA concentration was observed at the MRTLs, even though the maximum water age was relatively high (92 hours), suggesting that HAA biodegradation did not occur over the course of this study, which was confirmed by minimum microbial regrowth in the distribution system.

N-Nitrosodimethylamine

Surprisingly high concentrations of NDMA were measured in all systems studied. Although most sampling locations in the chloraminated system studied (Utility C) had concentrations below 20 ng/L, with the majority of samples having concentrations below 10 ng/L, high concentrations (40 and 46 ng/L) were measured on two occasions. Even in the free chlorinated system (Utility A), NDMA concentrations over 20 ng/L were measured during two of the five sampling events. In the chloraminated system periodically switching to free chlorine (Utility E), NDMA concentrations were high, with 8 samples showing NDMA concentrations over 40 ng/L. There were no obvious differences between the concentrations measured in the free chlorinated and the chloraminated systems. It is likely that there were different levels of NDMA precursors in each system, regardless of their sources (i.e., present in raw water, or introduced during treatment, such as through the use of polymers).

Attempts were made to examine the NDMA results obtained as a function of the type of polymers used at the utilities studied. Unfortunately, the chemical formulae of the polymers used were proprietary in nature and were not available.

No particular trend in NDMA concentrations could be identified with increasing water age in a chloraminated system (Utility C) or a free chlorinated system (Utility A). Trends could not be found under either chloramination conditions or free chlorination conditions in Utility E. As a result, it appears that water age does not affect NDMA concentration in the distribution system. In addition, the type of final disinfectant did not affect NDMA concentrations.

CHAPTER 5
GUIDELINES TO SELECT DISTRIBUTION SYSTEM SAMPLING LOCATIONS FOR DISINFECTION BY-PRODUCT MONITORING

The aim of this chapter is to present regulatory information and guidance along with key results of this study to help utilities determine distribution system sampling locations to meet Stage 2 DBPR requirements, particularly locations of maximum THM and HAA concentrations. To reach this goal, a variety of tools including those presented in the USEPA guidance manuals combined with the findings of this study are presented. A case study of a Standard Monitoring Program (SMP) conducted with one of this study's participating utilities is also presented. The information presented in this chapter is based on the draft USEPA Stage 2 DBPR IDSE (Draft DBPR-IDSE) of July 2003. The Stage 2 DBPR and IDSE guidance manual are still under review and information presented here may need to be revised accordingly.

STAGE 2 DISINFECTANTS / DISINFECTION BY-PRODUCTS RULE

On August 18, 2003, the USEPA proposed the Stage 2 DBPR, which builds upon the Stage 1 DBPR by providing more consistent protection from DBPs across the entire distribution system and by focusing on the reduction of DBP peaks. Some of the key provisions of the Stage 2 DBPR include the following items:

- An IDSE to identify compliance monitoring locations that represent high THM4 and high HAA5 concentrations. The IDSE is the first step in complying with the Stage 2 DBPR.
- Locational running annual averages (LRAAs) calculated for each monitoring location in the distribution system for THM4 and HAA5. Unlike the RAA, which was used in the Stage 1 DBPR to calculate system-wide annual average DBP concentrations of all samples, the LRAA calculates a location-specific annual average of DBP concentrations at each monitoring location. In the RAA, high DBP concentrations in one or more locations are averaged with, and dampened by, lower concentrations elsewhere in the distribution system. Thus the LRAA calculation will help avoid the distortion of DBP data observed with the RAA calculation.

All community water systems (CWSs) that add a primary or distribution system residual disinfectant other than UV, and consecutive CWSs that deliver water that has been treated with a primary or residual disinfectant other than UV, are required to comply with the Stage 2 DBPR.

INITIAL DISTRIBUTION SYSTEM EVALUATION

Recognizing that THMs and HAAs vary over time and space in distribution systems, the USEPA will require systems to select distribution system sampling sites with high DBP concentrations for compliance monitoring by performing the IDSE. All systems that are required to comply with the Stage 2 DBPR, as stated above, need to conduct an IDSE. However, waivers are possible for the following systems: non-transient non-community water systems (NTNCWSs) serving less than 10,000 people, systems serving fewer than 500 people (Small

System Waiver), and systems that demonstrate historically low distribution system DBP concentrations (40/30 Certification). Systems need to determine their eligibility for IDSE waivers and any further waiver requirements by using the guidelines in Section 2.0 of the Draft DBPR-IDSE (USEPA 2003a) and consult with their state/primacy agency.

IDSE Schedule and System Classification

For schedule purposes, systems fall in either one of two categories: Early Schedule or Late Schedule. At this point, systems in the early schedule are required to conduct their IDSE and submit their IDSE report to their state/primacy agency within 2 years of the promulgation of the final Stage 2 DBPR, whereas those in the late schedule have 4 years after rule promulgation to conduct the IDSE and submit their IDSE report to their state/primacy agency (USEPA 2003a). These requirements may be changed as the final Stage 2 DBPR is promulgated. To determine IDSE requirements and schedule, systems are categorized as shown in Figure 5.1.

Initial Distribution System Evaluation Options

The IDSE is based on either one of the following:

1. Standard Monitoring Program (SMP), which includes a one-year monitoring program, or
2. System Specific Study (SSS), which uses historical system-specific data or a well-calibrated distribution system hydraulic or water quality model that provides equivalent or better information than the one-year monitoring.

Since the choice between SMP and SSS is system-specific, each utility is encouraged to consult with its state/primacy agency prior to proceeding with the chosen IDSE technique (SMP or SSS). The overall IDSE process can be summarized as per Figure 5.2. Further information about each element of the process can be found below, as well as in various sections of the Draft DBPR-IDSE (USEPA 2003a) as specified in Figure 5.2.

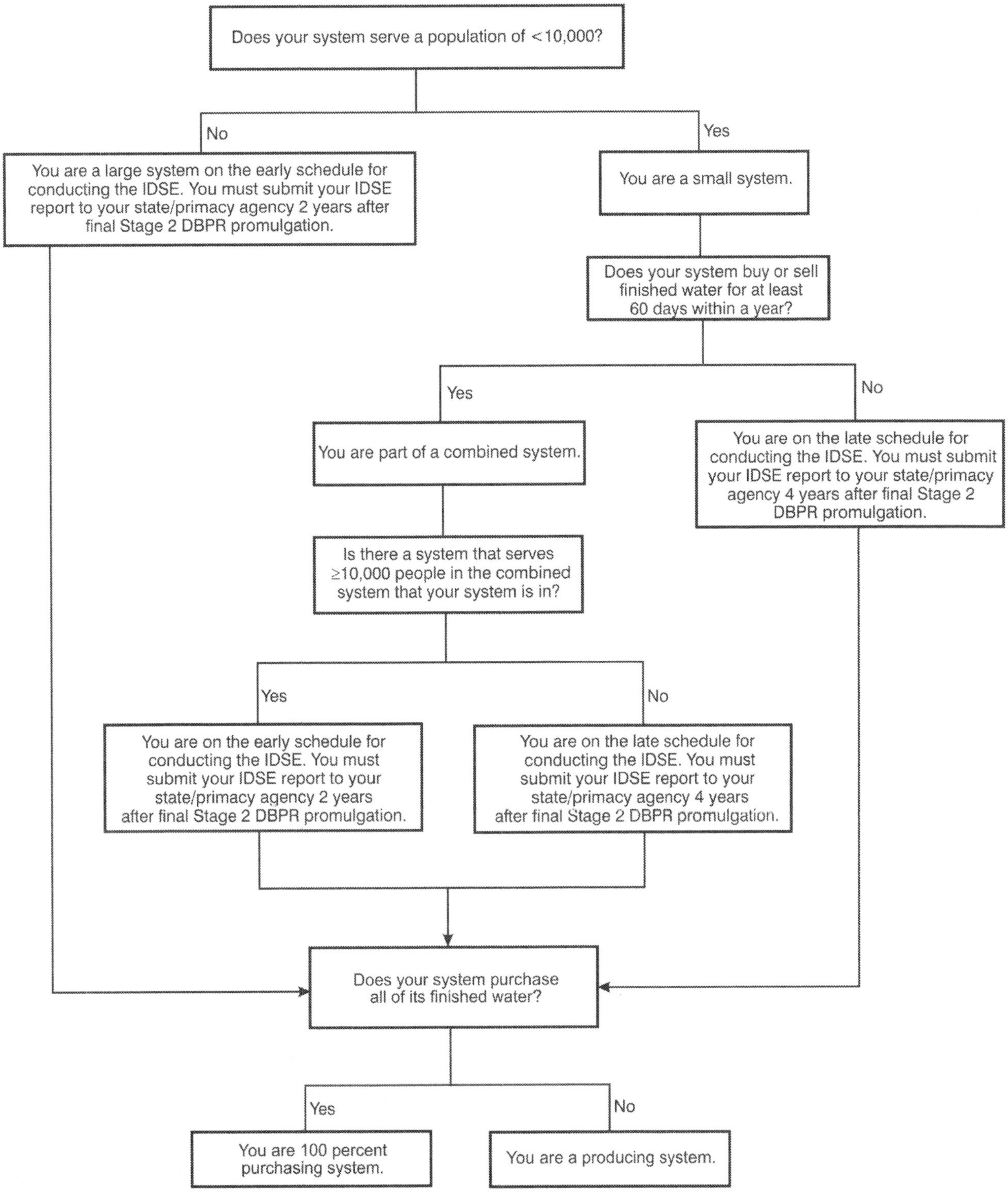

Source: Modified from the Draft DBPR-IDSE (USEPA 2003a)

Figure 5.1 System classification and IDSE schedule flow chart

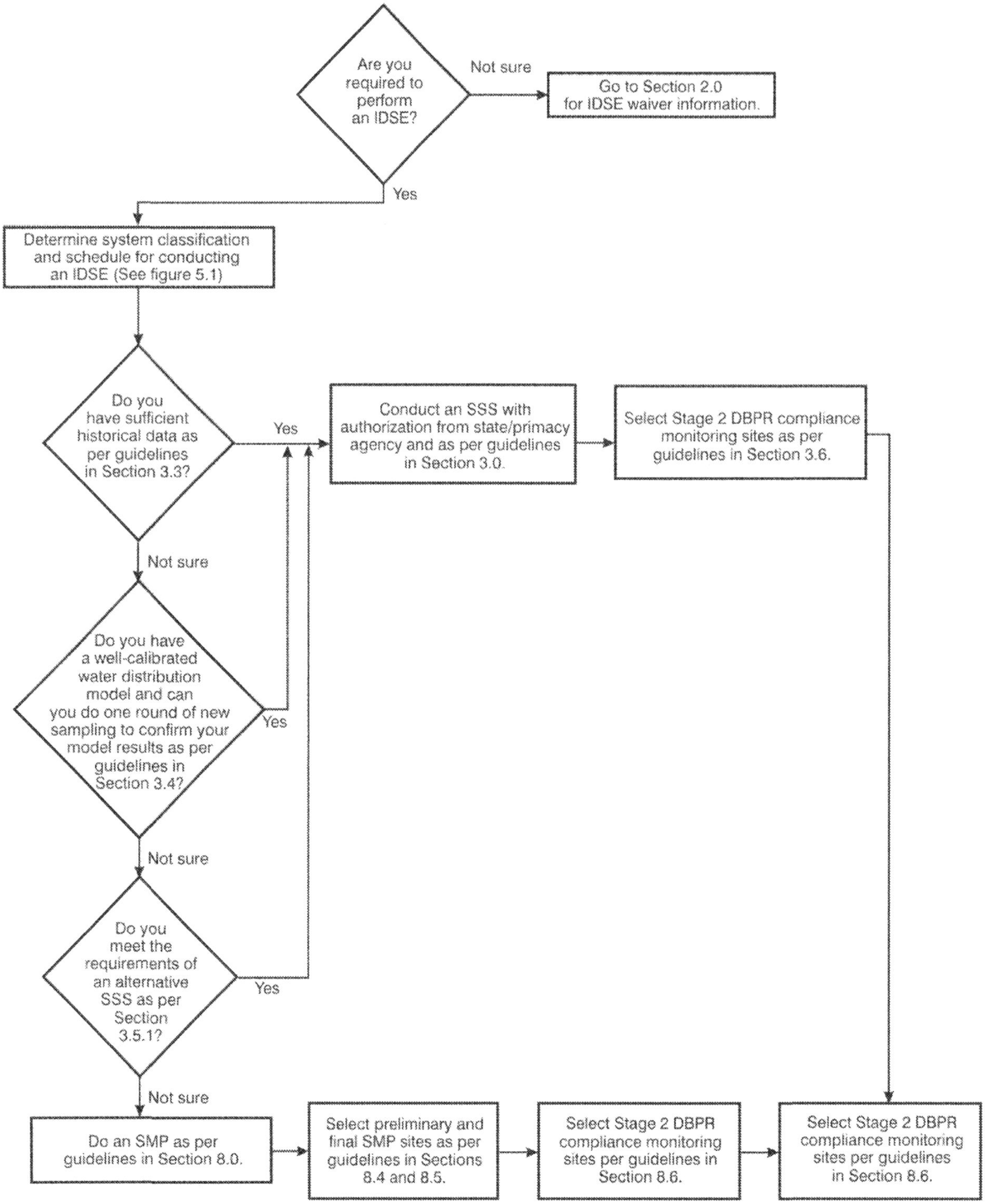

Source: Adapted from Draft DBPR-IDSE (USEPA 2003a)

Figure 5.2 Guidelines for IDSE options

Standard Monitoring Program

The SMP entails a one-year distribution system monitoring effort to determine areas/locations which routinely show higher DBP concentrations. Tables 5.1 and 5.2 summarize the SMP sampling site requirements for 100-percent purchasing systems and for producing systems, respectively. These two tables also include Stage 2 compliance monitoring site requirements, which will be discussed in a later section. The general approach for selecting SMP monitoring sites is to use available data sources and analytical tools to select a large number of potential sites (referred to as preliminary sites). From this group of sites, systems should consider geographic coverage and other distribution system factors to narrow down preliminary sites to final SMP sites (left side of Tables 5.1 and 5.2) that will be used during the SMP samplings. Requirements for selecting SMP sampling sites are different depending on system classification as shown on Tables 5.1 and 5.2. These requirements may change as the Stage 2 DBPR is still under review.

SMP Sampling Locations. As shown in Tables 5.1 and 5.2, there are four types of sampling locations defined for the SMP: near-POE, average residence time, high THM4 concentrations, and high HAA5 concentrations.

Near-POE Point SMP Sites. The near-POE SMP sites represent the minimum water age found in the distribution system and can be used as a baseline for interpreting changes in water quality as water travels through the distribution system. The near-POE SMP sites should be located between the treatment facility/consecutive system POE and before the first group of customers. More information for selecting near-POE SMP sites for 100-percent purchasing systems with specific scenarios is provided in Section 8.2.1.1 of the Draft DBPR-IDSE (USEPA 2003a).

SMP Sites with Average Residence Time. Average residence time sites should represent the average water age that is delivered to the majority of the customers in the distribution system. It is the flow-weighted or population-weighted average water age. Thus, due to the complexity of this value, most systems can only make rough estimates using various data tools or hydraulic models as presented below.

SMP Sites with High THM4 and High HAA5 Concentrations. Locations with high THM4 and high HAA5 concentrations represent areas in the distribution system with the highest annual average DBP concentrations. These locations are more difficult to identify as DBPs form and decay in distribution systems, and THM4 and HAA5 concentrations may peak at different sites within the same distribution system, as shown in this study.

Use of Data Sources and Tools for Selecting SMP Sites with High THM4 Concentrations, High HAA5 Concentrations, and Average Residence Time. SMP site selection is a multi-step process in which a combination of the following data sources and tools is used:

- Water distribution system model outputs
- Tracer study results
- Disinfectant residual data
- DBP data.

Table 5.1
SMP sampling and Stage 2 DBPR compliance monitoring requirements for 100-percent purchasing systems

100-percent purchasing systems											
	Final SMP site sampling requirements						Stage 2 DBPR compliance monitoring requirements				
	Number of distribution system sites per system				Total number of sites per system	Monitoring frequency*	Number of distribution system sites per system			Total number of sites per system	Monitoring frequency*
System size (population served)	Near-POE	Average residence time	High THM4	High HAA5			Stage 1 average residence time site	Highest THM4	Highest HAA5		
Surface water systems:											
<500	-	-	1	1	2	Every 180 days	-	1	1	2†	Every 365 days
500 - 4,999	-	-	1	1	2	Every 90 days	-	1	1	2†	Every 90 days
5,000 - 9,999	-	1	2	1	4	Every 90 days	-	1	1	2	Every 90 days
10,000 - 24,999	1	2	3	2	8	Every 60 days	1	2	1	4	Every 90 days
25,000 - 49,000	2	3	4	3	12	Every 60 days	1	3	2	6	Every 90 days
50,000 - 99,999	3	4	5	4	16	Every 60 days	2	4	2	8	Every 90 days
100,000 - 499,999	4	6	8	6	24	Every 60 days	3	6	3	12	Every 90 days
500,000 - < 1.5 million	6	8	10	8	32	Every 60 days	4	8	4	16	Every 90 days
1.5 million - < 5 million	8	10	12	10	40	Every 60 days	5	10	5	20	Every 90 days
≥ 5 million	10	12	14	12	48	Every 60 days	6	12	6	24	Every 90 days
Ground water systems:											
< 500	-	-	1	1	2	Every 180 days	-	1	1	2†	Every 365 days
500 - 9,999	-	-	1	1	2	Every 90 days	-	1	1	2	Every 365 days
10,000 - 99,999	1	1	2	2	6	Every 90 days	1	2	1	4	Every 90 days
100,000 - 499,999	1	1	3	3	8	Every 90 days	1	3	2	6	Every 90 days
≥ 500,000	2	2	4	4	12	Every 90 days	2	4	2	8	Every 90 days

Source: Adapted from the Draft DBPR-IDSE (USEPA 2003a)

* During each sampling, a dual sample set must be collected at each location, unless otherwise noted. A dual sample set is one THM4 and one HAA5 sample collected at the same time and location.

† If the highest THM4 and the highest HAA5 LRAA values occur at the same location, a dual sample set is collected at that location. If the highest THM4 LRAA and HAA5 LRAA values occur at different locations, then one THM4 and one HAA5 sample are collected at the highest THM4 and the highest HAA5 sites, respectively.

Table 5.2
SMP sampling and Stage 2 DBPR compliance monitoring requirements for producing systems

		Producing systems										
	Final SMP site sampling requirements							Stage 2 DBPR compliance monitoring requirements				
		Number of distribution system sites per plant						Number of distribution system sites per plant				
System size (population served)	Residual disinfectant	Near-POE	Average residence time	High THM4	High HAA5	Total number of sites per plant	Monitoring frequency*	Stage 1 average residence time site	Highest THM4	Highest HAA5	Total number of sites per plant	Monitoring frequency*
Surface water systems:												
<500	Chlorine or chloramine	-	-	1	1	2	Every 180 days	-	1	1	2†	Every 365 days
500 - 9,999	Chlorine or chloramine	-	-	1	1	2	Every 90 days	-	1	1	2	Every 90 days
≥ 10,000	Chlorine	1	2	3	2	8	Every 60 days	1	2	1	4	Every 90 days
	Chloramine	2	2	2	2	8						
Ground water systems:												
< 10,000	Chlorine or chloramine	-	-	1	1	2	Every 180 days	-	1	1	2†	Every 365 days
≥ 10,000	Chlorine or chloramine	-	-	1	1	2	Every 90 days	-	1	1	2	Every 90 days

Source: Adapted from the Draft DBPR-IDSE (USEPA 2003a)

* During each monitoring a dual sample set must be collected at each location, unless otherwise noted.

† If the highest THM4 and the highest HAA5 LRAA values occur at the same location, a dual sample set is collected at that location. If highest THM4 LRAA and HAA5 LRAA values occur at different locations, then one THM4 and one HAA5 sample are collected at the highest THM4 and highest HAA5 sites respectively.

These data tools should be used in combination with maps to identify preliminary SMP sites with average residence time, high THM4 concentration, and high HAA5 concentration. The areas with high THM4 concentration, high HAA5 concentration, and average residence time identified by each data source or tool should be plotted on a detailed distribution system map. Of these plotted sites, preliminary SMP sites should be selected from areas that are repeatedly identified by the different data sources and tools. Figure 5.3 presents a decision tree that helps select preliminary SMP sites to capture average residence time, high THM4 and high HAA5 sites. The findings of this study can also serve this purpose and are discussed below.

After selecting preliminary SMP sites, systems should identify their final SMP sites that will be used during the SMP samplings. The Draft DBPR-IDSE (USEPA 2003a) provides guidelines to narrow down the preliminary SMP sites into final SMP sites, as summarized in Figure 5.4. Note that the final SMP sites selected cannot be the same as any of the Stage 1 DBPR compliance monitoring sites.

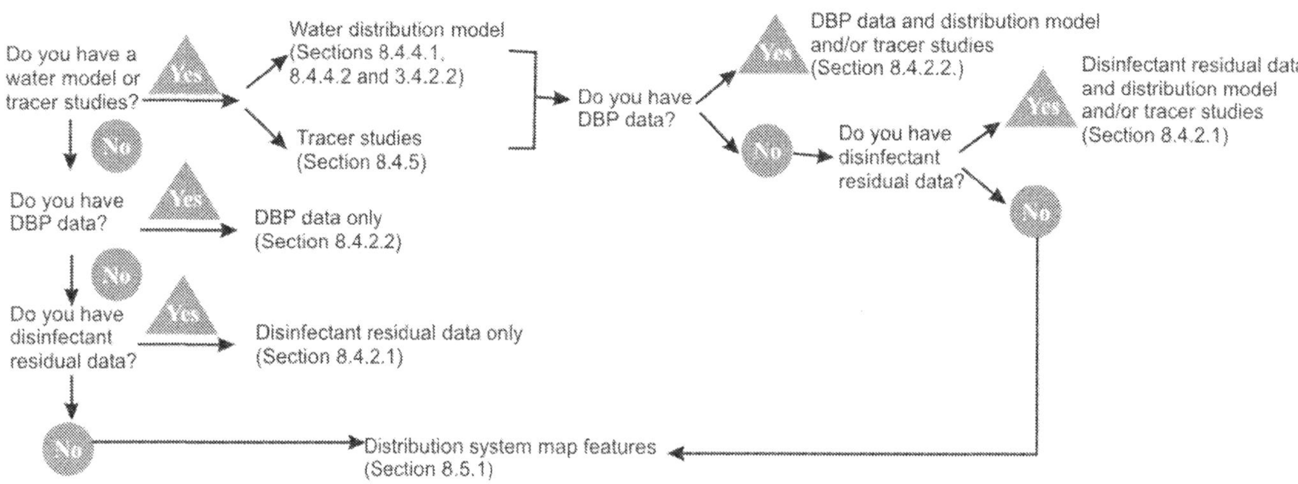

Source: Adapted from Draft DBPR-IDSE (USEPA 2003b)

Figure 5.3 Selection of preliminary SMP sites

1. Selecting near-POE final SMP sites →

- Locate sites between the treatment facility or consecutive system POE and before or near the first customer.
- Multiple consecutive POEs considered as one plant
  - A location near any POE or prior to the first group of customers for any one POE should be acceptable. However, if the average flow differs significantly between the POEs, the POE with the greatest flow should be used to locate the near POE SMP site.
- For 100-percent purchasing systems:
  - If the required number of near-POE SMP sites is *less than* the actual number of consecutive system POEs:
    1. Select sites at the POEs delivering surface water in order from the highest to lowest flow.
    2. Select the remaining sites at the POEs delivering ground water, in order from the highest to lowest flow, until the required number of SMP sites have been identified.
  - If the required number of near-POE SMP sites is *more than* the actual number of consecutive system POEs:
    1. Distribute "excess" near-POE sites among high THM4 sites
    2. Distribute the remaining "excess" near-POE sites among high HAA5 sites.
- For producing systems:
  - Multiple ground water wells drawing from one aquifer and delivering water directly to the distribution system may be considered a single plant with approval from state. Locations prior to the first group of customers of any well are acceptable near-POE SMP sites.

2. Selecting final SMP sites with high THM4 and HAA5 concentrations →

- Select SMP sites in hydraulically and geographically different areas, and separate from existing Stage 1 DBPR sites.
- Prioritize sites that meet the most siting criteria and those identified based on more than one data source.
- Select high THM4 sites located after storage facilities (Section 8.5.2.1, of the IDSE Guidance Manual, USEPA 2003a) and booster disinfection stations.
- Select high THM4 sites in areas with the lowest or no residual disinfectant (unless your system uses booster disinfection).
- Generally, select high HAA5 sites with a minimum of 0.2 mg/L free chlorine residual or 0.5 mg/L chloramine residual for all observations.
- Locate at least one of your high THM4 sites in a remote area of your distribution system. If you are only required to select one high THM4 site, it is strongly recommended that you locate this site away from the treatment plant, near the last group of customers (but prior to the last fire hydrant).
- Historical DBP data are important to help prioritize one site over another, as long as they represent your current system configuration.

3. Selecting final SMP site with average residence time → Do you serve at least 10,000 people?

- **Yes** → Select final average residence time sites that are geographically diverse from other SMP sites and from existing Stage 1 DBPR average residence time sites.
- **No** → No average residence time site needs to be selected.

Source: Modified from Draft DBPR-IDSE (USEPA 2003a)

Figure 5.4 Selection of final SMP sites

Selection of Stage 2 DBPR Compliance Monitoring Sites Based on SMP Results. After selecting final SMP sites, systems need to monitor at the final SMP sites for one year according to their respective sampling schedule. The average THM4 and HAA5 concentrations measured over the SMP monitoring period are then calculated to give THM4 and HAA5 LRAA values at each monitoring site. Stage 2 DBPR sites are selected among the final SMP sites and existing Stage 1 DBPR sites using the THM4 and HAA5 LRAA data at these locations. The Stage 2 DBPR monitoring sites are generally selected as those sites with the highest THM4 and HAA5 LRAA values. Figure 5.5 outlines the protocols to select Stage 2 DBPR compliance monitoring sites, as per the guidelines of the Draft DBPR-IDSE (USEPA 2003a). Systems also need to determine their required number of sites and their monitoring frequency by referring to Tables 5.1 and 5.2.

System-Specific Study

In order for a system to use the SSS approach to select the monitoring sites for Stage 2 DBPR, the systems must have one of the following:

- Sufficient historical THM4 and HAA5 data that are equivalent or superior to data that would be obtained under the IDSE SMP.
- A well-calibrated water distribution system model and performance of at least one round of new sampling.
- Systems can use alternative options to conduct an SSS (e.g., historical data combined with new data), but they need to evaluate their methodology very carefully.

Before deciding to conduct an SSS, systems should review Section 3.0 of the Draft DBPR-IDSE (USEPA 2003a). They should also carefully check the adequacy of their data and information available for completing an SSS. Systems are encouraged to contact their State/primacy agency to verify the adequacy of their SSS plan.

SSS Using Historical DBP Data. Before using historical DBP data in an SSS, it is important to determine whether these data are sufficient and appropriate for use in an SSS:

- Historical data should meet overall SMP requirements with respect to number of sites, site locations, number of dual sample sets per site (a dual sample set is one THM4 and one HAA5 sets of samples collected at the same time and location), and sampling frequency. Some of these requirements are presented in Tables 5.1 and 5.2.
- Historical DBP data should have been obtained using approved methods by a laboratory certified to perform these measurements under the Drinking Water Certification Program (USEPA 1997c).
- Only data collected within 10 years preceding the due date of the IDSE report should be used for an SSS, and at least 50 percent of the historical samples should have been collected within 5 years prior to the system's IDSE report due date (Figure 5.6).
- Historical data should reflect source water(s), treatment configuration, and distribution system conditions in place at the time the IDSE report is due.

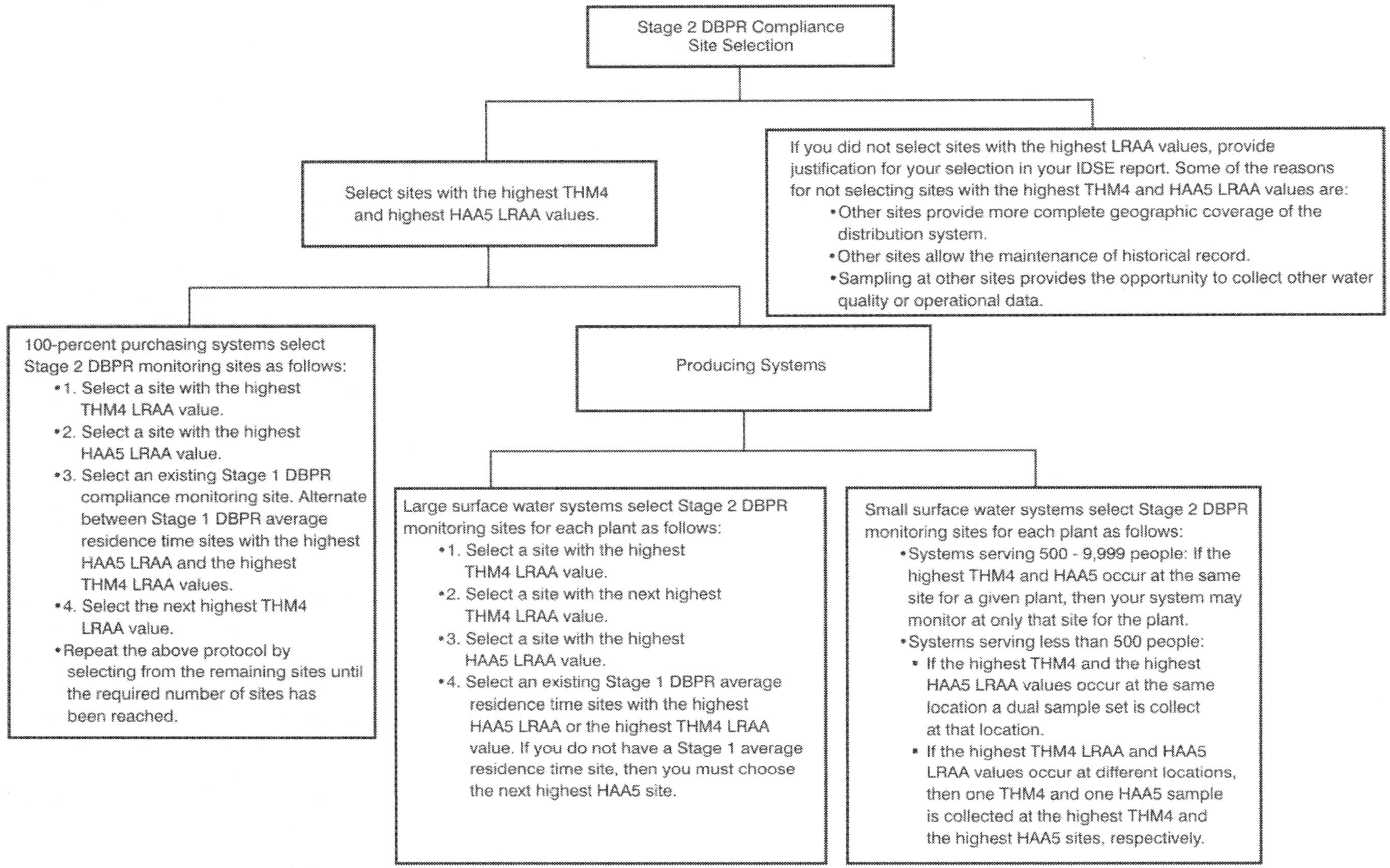

Source: Modified from Draft DBPR-IDSE (USEPA 2003a)

Figure 5.5 Stage 2 DBPR compliance monitoring site selection

a) Systems on the early schedule:

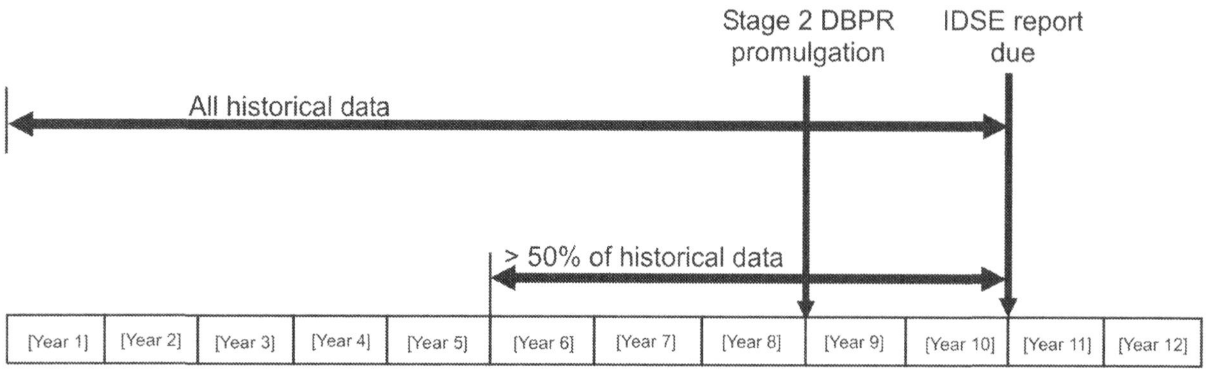

b) Systems on the late schedule:

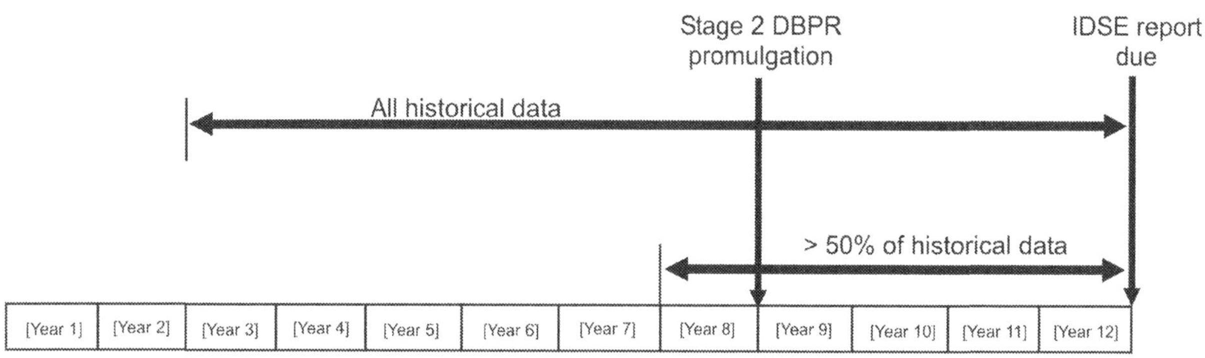

Source: Adapted from the Draft DBPR-IDSE (USEPA 2003a)

Figure 5.6 Recommended limits of sampling period for historical data for systems on the early and late schedules

SSS Using Water Distribution System Model. Two types of water distribution models can be used for the SSS: hydraulic models and water quality models. However, minimum requirements for an SSS are focused only on the hydraulic component of distribution system models. Minimum model requirements in terms of skeletonization, accurate simulation of water consumption, and model calibration must first be met, and are presented in Section 3.4.1, Draft DBPR-IDSE (USEPA 2003a).

Using a model, operation of a water distribution system including storage facilities and booster chlorination stations should be simulated over extended periods to reflect maximum residence time in the system under conditions of high THM4 and HAA5 formation potential. Model results should be combined with at least one round of THM4 and HAA5 sampling at sites that at a minimum, meet SMP requirements. Sampling should be conducted during the month of peak historical THM4 levels (or the month of peak distribution system water temperature if peak THM4 data are not available) to verify model results. These results will be used in selecting final Stage 2 DBPR compliance monitoring sites.

Alternative SSS Options. Systems may choose to use other combinations of data and analyses to conduct an SSS. Alternative SSS should generally be representative of the majority of the distribution system, including the extremities, and provide equivalent or superior data for the selection of Stage 2 DBPR monitoring sites compared to an IDSE SMP. Potential combinations include, but are not limited to:

- Historical data supplemented with new data
- New DBP data, water model, and tracer results
- Historical data and tracer study results.

Stage 2 DBPR Compliance Monitoring Site Selection Based on SSS Results. Tables 5.1 and 5.2 above present Stage 2 DBPR compliance monitoring site requirements. The following sections give guidance for the selection of Stage 2 DBPR compliance monitoring sites using SSS with historical DBP data or a water distribution system model.

SSS using Historical DBP Data: The following steps should be used in selecting Stage 2 DBPR compliance monitoring sites, and further information is available in the Draft DBPR-IDSE (USEPA 2003a):

- Step 1: Calculate the LRAA for THM4 and HAA5 concentrations at each historical data site.
- Step 2: Calculate the LRAA for THM4 and HAA5 concentrations at the Stage 1 DBPR maximum residence time site(s). Sometimes a site with slightly lower LRAA values over another site may be selected for the following reasons, although justification for the selection must be provided in the IDSE report (USEPA 2003b):
  - The lower LRAA value site provides for more complete geographic coverage of the entire distribution system
  - The lower LRAA value site allows the maintenance of historical record.
  - Sampling at the lower LRAA value site provides opportunity to collect other water quality or operational data.
- Step 3: Select Stage 2 DBPR high THM4 and high HAA5 sites starting with the highest THM4 and HAA5 LRAAs from both the Stage 1 DBPR compliance monitoring sites and historical DBP data sites.
- Step 4: Select Stage 2 DBPR average residence time site(s). Procedure for this purpose vary based on system type (producing vs. purchasing system).

SSS Using a Water Distribution System Model: If a system conducts an SSS using a water distribution system model, it must use the following steps to select its Stage 2 DBPR compliance monitoring sites:

- Step 1: Systems should confirm that the model-predicted high THM4 sites (high residence time sites), and average residence time sites are consistent with the THM4 and HAA5 concentrations from the one round of new sampling, Stage 1 DBPR compliance monitoring results, and any other historical THM4 and HAA5 data.

- Step 2: If modeled data and sampling results do not agree and Stage 2 DBPR sites are selected based on modeled results, the cause of the discrepancy between the modeled data and sampling results should be thoroughly explained in the IDSE report.
- Step 3: Sometimes a site with slightly lower LRAA values over another site may be selected as described above. In this case, justification for the selection must be provided in the IDSE report.

In addition to the above guidelines for the use of an SSS to select Stage 2 DBPR compliance monitoring sites, systems should consider other factors such as geographic coverage and population density when selecting sites. In general, best professional judgment should be used when selecting the final Stage 2 DBPR compliance monitoring sites.

FINDINGS FROM THIS STUDY

The following findings were drawn from this study and could be used as guidelines when selecting distribution system monitoring locations

Seasonal variations in THM and HAA concentrations are expected. In this study, THM4 and HAA5 concentrations were lowest in colder months and highest in warmer months. As such, the time of sampling should be carefully considered when planning monitoring programs or when using historical data. On a larger scale, climatic changes including months with high precipitation and runoff are likely to distort DBP data as well. If an SMP was conducted or historical data were obtained in a non-typical year, the collected data need to be carefully assessed before selecting DBP monitoring sites. Systems conducting an SSS are highly encouraged to use data from years with normal climatic conditions.

Some of the findings were specific to systems using free chlorine as a final disinfectant:

- THM4 concentrations increased as water age increased and free chlorine residual decreased in distribution systems. Correlation analyses with the data obtained indicated that, at times, correlations were stronger with water age, while at other times, correlations were stronger with chlorine residual, but overall, similar correlations were obtained with both parameters. In general, correlations were relatively weak, with an R^2 of 0.44 to 0.65 between THM4 concentrations and water age, and 0.33 to 0.58 between THM4 concentrations and free chlorine residual (Appendix C).
- HAA5 concentrations generally increased or remained constant as water age increased and free chlorine residual decreased, except at sampling locations with the greatest water age, where a decrease in HAA concentrations was measured. Lower HAA concentrations were observed at locations with lower free chlorine residuals (<0.5 mg/L Cl_2), which suggest that biodegradation was responsible for the decreases in HAA9 concentrations observed. Thus, a thorough examination of the data is warranted since the low disinfectant residual limit presented in the USEPA guidelines (at least 0.2 mg/L free chlorine) may allow biodegradation to occur and may not represent the location with the highest HAA concentration. In fact, when correlation analyses were performed with the data obtained, no correlations were found between HAA concentrations and water age (R^2 of 0.13 to 0.36) or between HAA concentrations and free chlorine residuals (R^2 of 0.20 to 0.34) (Appendix C).

Some of the findings were specific to systems using chloramines as the final disinfectant:

- In chloraminated systems, THM4 concentrations remained relatively constant with increasing water age, with occasional decreases in concentrations observed at MRTLs. These decreases occurred when total chlorine residuals were lower and signs of nitrification were present. As a result, high THM4 concentrations may not necessarily be found at locations with high water age. Thus utilities need to examine their system and data thoroughly when selecting their monitoring sites and relying on water age or chlorine residual data may not be appropriate, particularly when chloramine residuals are lower.
- HAA5 concentrations were relatively constant throughout the chloraminated distribution systems, with the exception of the MRTLs where biodegradation of some of the HAA species was observed. Such phenomena need to be considered when selecting distribution system sampling locations for Stage 2 DBPR compliance purposes.

Some of the findings were specific to chloraminated systems switching periodically to free chlorine as a final disinfectant:

- When switching from one disinfectant type to another, an adjustment period is often required before the distribution system responds completely to the change in disinfectants. This duration of this adjustment period is dependant on the volume of water storage in the distribution system. Such a scenario may compound attempts to capture the locations where peak DBP concentrations can be observed.
- In the system studied, increases in THM4 concentrations from the POE to the location with the maximum THM4 concentration varied from 20 to 50 percent during the chloramination periods. During the free chlorination period, increases in THM4 concentrations were much greater: 135 and 161 percent, with peak THM4 concentrations as high as 175 µg/L. As a result, chloraminated systems that periodically switch disinfectant can expect peak DBP concentrations during the free chlorination period that may result in non-compliance. In addition, results indicated that the location of peak THM4 concentrations may change when switching disinfectant.
- In the case of HAA5, concentrations remained relatively constant with increasing water age during the chloramination periods; they were much more uniform than the THM4 concentrations. However during the free chlorination period, HAA5 concentrations increased by 46 to 53 percent from the POE to the location with the maximum HAA5 concentration, with peak HAA5 concentrations of 95 to 138 µg/L. As a result, utilities that switch disinfectant can also expect to be out of compliance for HAA5 during the free chlorination period. These utilities also need to consider that the location of peak HAA5 concentration may differ when switching disinfectant.

Important observations were also made during this study regarding the use of water distribution system models for DBP monitoring site selection:

- Variations in treatment plant finished water flows, storage within the distribution system, and demand values and patterns may create possible sources of error when estimating water ages using a hydraulic model. Using Utility E, which had two POEs as an example, the modeled treated water flows from the Site 1 POE were 21, 17, and 13.5 mgd (79,490, 64,350 and 51,100 m^3/d) at maximum, average, and low flow conditions, respectively. The modeled treated water flows from the Site 2 POE were 13, 10.5, and 9.3 mgd (49,210, 39,750 and 35,200 m^3/d) at maximum, average, and low flow conditions, respectively. The POE water flows at Sites 1 and 2 during each sampling are presented in Table 5.3. The POE water flows at Site 1 were within the range of water flows examined during modeling; however, the POE water flows from Site 2 were more than 2 mgd (7,570 m^3/d) lower than the values used during modeling. In this particular case, although the POE water flows at Site 2 were observed to be lower than the modeled conditions, very little change in water age would be expected at sampling sites throughout the distribution system due to decreased water production at the Site 2 POE. However this may not be the case in other systems, and variations in POE water flows need to be considered when modeling distribution systems to obtain water ages for the determination of distribution system sampling locations.
- Models are, of course, simplifications of the actual distribution system, and numerous pipes (mostly smaller ones) are often not included in the model. Such simplifications may lead to inaccuracies when calculating water age for the purpose of the IDSE, particularly since the small-diameter pipes tend to be found at MRTLs.
- In this study, water ages throughout the distribution system were calculated using three conditions: 1) average daily flow, minimum flow hour, 2) average daily flow, average flow hour, and 3) average daily flow, maximum flow hour. In one of the systems studied, two other conditions were used: minimum daily flow, minimum flow hour; and maximum daily flow, maximum flow hour. In another system, only the average flows and demands were considered. Modeling a distribution system using various flow conditions may result in important differences in hydraulic scenarios in the distribution system, leading to significant differences in water ages.

Table 5.3

Water treatment plant flow rates (mgd, m^3/d) for both POE locations in Utility E (chloraminated system periodically switching to free chlorine)

Sampling date	Site 1 POE	Site 2 POE
February 11, 2003	16.0 mgd (60,570 m^3/d)	7.0 mgd (26,500 m^3/d)
February 28, 2003	17.2 mgd (65,110 m^3/d)	6.8 mgd (25,740 m^3/d)
March 11, 2003	13.0 mgd (49,210 m^3/d)	6.9 mgd (26,120 m^3/d)
March 31, 2003	22.5 mgd (85,170 m^3/d)	7.0 mgd (26,500 m^3/d)
April 29, 2003	22.3 mgd (84,420 m^3/d)	6.8 mgd (25,740 m^3/d)
May 20, 2003	19.2 mgd (72,680 m^3/d)	7.2 mgd (27,260 m^3/d)

- Similarly, variations in distribution system hydraulic scenario due to the use of various water sources (e.g., different POEs, introduction of well water at different locations throughout the distribution system) may change water age throughout the distribution system, which may in turn affect DBP concentrations even if the different source waters have similar water quality.
- The presence of specific devices such as autoflushers or distribution system operations such as flushing, also affect water age and need to be considered, both during modeling and sampling.
- Because of the complexity of distribution system models and the number of parameters involved, errors are often encountered when examining a model or its results. For example, the model used at Utility E considered a 4-foot (1.22-m) possible change in elevation for one of the storage reservoirs, but conversations with staff personnel indicated that the elevation change was really 6 feet (1.83 m). The model was then updated to reflect this difference. At Utility D, modeling results suggested that a valve may not have been placed at the correct location in the model, and a field investigation confirmed this suggestion. In addition to being properly located, the status of pressure-reducing valves (PRVs) also need to be correctly entered in the model in order to obtain accurate water age estimates. Another source of error may come from long sections of transmission mains. These long sections often lead to erroneous results, and as such, need to be divided into smaller sections to improve modeling result accuracy.
- Reservoirs have an important effect on water age in distribution systems and need to be considered when determining DBP sampling locations. The status of the reservoirs (filling or emptying) also need to be considered as depending on the reservoir configuration and type of inlet and outlet, aged water from the reservoir will be found at both upstream and downstream sampling locations when the reservoir is emptying.

CASE STUDY

This section presents a case study of an SMP conducted for the selection of Stage 2 DBPR monitoring sites to guide utilities in using the information provided in the above sections. Utility D's information was used as an example to develop this case study. In addition, this section outlines the information utilities need to provide in their IDSE report, as described above.

Standard Monitoring Plan Preliminary Steps

This section of the IDSE report needs to present the following information.

System Description

Utility D is located in the western portion of the U.S. This utility distributes treated surface water from a wholesaler, supplemented with their own well water. In the section of the distribution system studied, Utility D purchases 100 percent of its distributed water. For the purpose of this exercise, the studied section of Utility D's distribution system was considered as if it were the entire distribution system.

In 2001, the average day system delivery was approximately 5.9 mgd (22,330 m^3/d), the maximum daily flow was 7.1 mgd (26,880 m^3/d), and the minimum daily flow was 4.5 mgd (17,030 m^3/d).

Source Water Information. In this distribution system, the water may come from one of three possible water treatment plants from the same wholesaler; the water can also be a blend of the three plant effluents on occasion. These plants are fed by either one of two surface water sources, or a blend of both sources.

POEs and Service Areas Under the Influence of Each POE. In this section of the IDSE report, producing systems need to discuss the POEs, and service areas under the influence of each POE. This section does not pertain to Utility D since this utility is a 100-percent purchasing system.

Treatment Description. All water treatment plants supplying Utility D's distribution system use conventional treatment consisting of coagulation with alum in a rapid mixer, flocculation, sedimentation, free chlorination, filtration on anthracite and sand, followed by free chlorination and addition of ammonia to form chloramines. The target chlorine-to-ammonia-N weight ratio is 5:1. These plants have a combined capacity of 1,670 mgd (6,321,600 m^3/d). They do not routinely switch to free chlorine. There were no significant changes at the water treatment plants during the study period.

Description of the Distribution System. Approximately 53 percent of Utility D's pipes are unlined cast iron and steel pipes installed during the 1930's, 40's and 50's. The remainder of the pipes are ductile iron, PVC and transite. Utility D's distribution system is a residential area fed by a 14-inch (35-cm) main. There are two possible POEs to this area, although they both distributed water of similar quality. The distribution system contains no storage reservoir or booster chlorination station. Flushing is performed regularly.

Schematic of the Distribution System. In this section, utilities need to show a schematic of their distribution system. For confidentiality purposes, the schematic of Utility D's distribution system is not shown in this section.

Standard Monitoring Program Monitoring Requirements

Utility D's characteristics can be summarized as follows:

Population served:	6,109 customers
Source water:	100 percent surface water
Buying and selling relationships:	100 percent purchasing system.

Based on these characteristics, Table 5.1 can be used to determine the required final SMP sites and monitoring frequency:

Near POE sites:	0
Average residence time sites:	1
Sites representative of high THM4:	2
Sites representative of high HAA5:	1
Total SMP sites:	4
Monitoring frequency:	Every 90 days for one year.

SMP Site Selection Using Data Sources and Tools. For the purpose of this exercise, the following assumptions were made about Utility D's data collected during this study:

- The monitoring frequency for the data collected was assumed to be every 90 days.
- The disinfectant residual data was assumed to be collected as part of the Total Coliform Rule (TCR) monitoring program.
- Only sampling sites and dates with complete data were used to determine the SMP sites (sites with missing data points were disregarded).

Using the flow chart in Figure 5.3, it was determined that Utility D has a hydraulic model and sufficient disinfectant residual data that can be used in SMP site selection, as shown below.

Hydraulic Model. Utility D employs H2ONET as a hydraulic model, and all pipes were included in the model. The model was last updated and calibrated in 1998, although tracer studies were not conducted. Water ages were obtained by running the model for 72 hours at average flow and demand conditions, and the evaluation included minimum, average and maximum hourly flow conditions. In addition to water ages, flow distributions and directions were calculated throughout the system.

A number of sampling locations were selected to represent a cross-section of water ages, pipe diameter, pipe materials (cast iron, PVC or asbestos cement), as well as diverse geographic areas. Because this distribution system can be fed by two POEs and either one or both POEs were used throughout the sampling period, the model was run for the particular hydraulic scenario encountered during each sampling. The water ages during each sampling are shown in Table 5.4, and are discussed below to determine the required SMP sites. The high water ages are shown in bold characters in this table to represent possible SMP sites with high THM4 concentrations.

Disinfectant Residual. Total chlorine residual data collected during each sampling are shown in Table 5.5, along with the average total chlorine residual at each location. The average residence time sites are estimated by locating the sites that have disinfectant residual concentrations closer or equal to the calculated average disinfectant residual concentration of the distribution system, which is 2.2 mg/L. The two bold rows indicate the potential average residence time sites that could be used for the SMP. In addition, the sites with the lowest disinfectant residual (in this case, Sites No.10-B/11-A) can qualify as high THM4 sites, while other sites with low but measurable disinfectant residuals (at least 0.5 mg/L for chloramines) can be selected as high HAA5 sites.

Table 5.4
Utility D water ages (hours) at each sampling location and sampling date. Bold sites have high water ages

Sampling locations	Sampling Date 1	Sampling Date 2	Sampling Date 3	Sampling Date 4	Sampling Date 5	Sampling Date 6
No.1	0.0	0.0	0.0	0.0	0.0	0.0
No.2	1.6	1.6	2.6	5.2	2.0	2.0
No.3	2.4	2.4	2.7	6.3	2.5	2.5
No.4	3.3	3.3	3.8	7.5	3.3	3.3
No.5	5.7	5.7	3.7	8.3	4.0	4.0
No.6	**10.5**	**10.5**	**6.3**	**12.6**	**7.0**	**7.0**
No.7	**18.4**	**18.4**	**13.1**	**15.2**	**10.9**	**10.9**
No.10-B/11-A	**37.2**	**37.2**	**2.2**	**3.8**	**3.4**	**3.4**

Table 5.5
Disinfectant residual data at each location for each sampling date and average disinfectant residual data. Bold sites have disinfectant residuals that are close to the average disinfectant residual

Sampling location	Sampling Date 1	Sampling Date 2	Sampling Date 3	Sampling Date 4	Sampling Date 5	Sampling Date 6	Average total chlorine residual
No.1	2.20	3.00	3.4	2.80	2.50	2.4	2.72
No.2	**1.24**	**2.10**	**3.4**	**1.10**	**2.50**	**3.5**	**2.31**
No.3	**1.83**	**1.60**	**3.5**	**1.10**	**2.60**	**2.3**	**2.16**
No.4	1.72	2.13	3.5	1.00	2.90	3.4	2.44
No.5	1.89	2.19	3.7	1.10	2.40	2.8	2.35
No.6	1.36	2.04	2.4	1.50	1.30	3.3	1.98
No.7	1.58	2.05	2.8	1.90	2.05	2.3	2.11
No.10-B/11-A	0.10	0.11	3.2	2.60	1.20	3.2	1.74
Distribution system average residual disinfectant concentration							2.23

Summary of Final SMP Sites. Sampling sites were selected based on the distribution system water age, disinfectant residual data and geographic coverage. The guidelines presented in Figure 5.4 were used to select the following final SMP sites:

- Site representative of high THM4 concentration: This site corresponds to sampling location No.10-B/11-A and represents one of the high water age locations in the distribution system. This site also has the lowest total chlorine residual. As such, it can represent one of the high THM4 sites for the SMP.
- Site representative of high THM4 concentration: This site corresponds to sampling location No.6 and represents one of the high water age locations in the distribution

system. The site has a relatively low total chlorine residual as well. As such, it can represent the second high THM4 site for the SMP.

- Site representative of high HAA5 concentration: This site corresponds to sampling location No.7 and represents one of the high water age locations in the distribution system. This site also has a total chlorine residual greater than 0.5 mg/L to ensure that HAAs are not biodegraded.
- Average residence time site: This site corresponds to sampling location No.3. This site has an average total chlorine residual of 2.2 mg/L, which is the same as the overall average distribution system disinfectant residual value.

SMP Sampling Schedule. According to the SMP requirements presented in Table 5.1, the monitoring frequency for Utility D is every 90 days. The sampling dates need to include the controlling month of the year (i.e., month with the highest THM4 or HAA5 concentration, whichever of the two is the highest, or the month with the maximum temperature), and the remaining sampling dates are determined thereafter.

Distribution System Map. In this section, utilities need to present a distribution system map that shows the major transmission mains, all of the Stage 1 DBPR compliance monitoring sites and the final SMP sampling sites. Such a map is not shown here for Utility D for confidentiality purposes.

Standard Monitoring Program Results

This section of the IDSE report has two subsections:

- Introduction: In this subsection, any discrepancies between the proposed SMP sampling dates and the actual sampling dates need to be presented and explained.
- Summary of Stage 1 DBPR compliance data and SMP data: This subsection needs to present DBP results from the Stage 1 DBPR compliance monitoring and the SMP.

Proposed Stage 2 DBPR Compliance Monitoring Sites

This section of the IDSE report needs to present the following items.

Site Summary

Based on the requirements presented in Table 5.1, two Stage 2 DBPR compliance monitoring sites need to be selected from the final SMP sites and the Stage 1 DBPR compliance sites. This section needs to present the proposed sites.

Justification of Site Selection

According to Table 5.1, one high THM4 and one high HAA5 site are required for Utility D's Stage 2 DBPR compliance monitoring. Using the flow chart presented in Figure 5.5, the site with the highest THM4 LRAA value needs to be selected first. Next, a site with high HAA5 LRAAs needs to be selected.

Proposed Stage 2 DBPR Compliance Monitoring Schedule

It is suggested that utilities collect their Stage 2 DBPR compliance samples on the same dates as the Stage 1 DBPR compliance monitoring dates for consistency. Also, the sampling dates need to include the warmest month as a control month.

Map of Proposed Stage 2 DBPR Compliance Monitoring Sites

In this section, utilities need to show the proposed Stage 2 DBPR monitoring sites on their distribution system map.

SUMMARY AND CONCLUSIONS

Under the Stage 2 DBPR, systems will be required to select new compliance monitoring sites that capture peak DBP concentrations in the distribution system. The first and most important step in the Stage 2 DBPR is the IDSE during which utilities will select their new compliance monitoring sites. The following summarizes the key elements of the IDSE:

- Systems need to first determine if they are required to conduct an IDSE or not. Systems may qualify for an IDSE waiver if they serve a population of less than 500 persons, or if they have demonstrated that they have historically low distribution system DBP levels (<40 μg/L for THM4 and <30 μg/L for HAA5).
- Utilities need to determine their systems' classification to identify their IDSE and Stage 2 DBPR requirements and schedule, which includes number of monitoring sites, types of monitoring sites, and monitoring frequency. System classification is determined based on system characteristics such as source water, population served, buying or selling relationship with other systems, purchasing or treating water, or number of treatment plants.
- Water systems need to perform an SMP or an SSS to comply with the IDSE requirements. The choice between SMP and SSS is system-specific, and each utility needs to consult with its State/primacy agency regarding the selected IDSE option (SMP or SSS).
- The SMP entails a one-year distribution system monitoring effort to determine areas/locations which routinely show higher DBP concentrations. The general approach for selecting SMP monitoring sites is to use available data sources and analytical tools to select a large number of potential sites (referred to as preliminary sites). From this group of sites, systems need to narrow down preliminary sites to final SMP sites. Systems need to monitor at the final SMP sites for one year

according to their respective sampling schedule. Stage 2 DBPR sites are then selected among the final SMP sites and existing Stage 1 DBPR data.

- To use the SSS approach to select Stage 2 DBPR monitoring sites, systems must have one of the following:
  - Sufficient historical data of the regulated THM4 and HAA5 that are equivalent or superior to data that would be obtained using the SMP. When using this approach, systems must select the Stage 2 DBPR compliance monitoring sites from the Stage 1 DBPR compliance monitoring sites and historical DBP data sites.
  - A well-calibrated water distribution system model and conduct at least one round of new sampling during the month of peak historical THM4 levels (or the month of peak distribution system water temperature if peak THM4 data are not available). When using this approach, systems must confirm that the model-predicted Stage 2 DBPR compliance monitoring sites are consistent with the THM4 and HAA5 concentrations from the one round of new sampling, Stage 1 DBPR compliance monitoring results, and any other historical THM4 and HAA5 data.
  - Systems can use alternative options to conduct an SSS (e.g., historical data combined with new data), but they need to evaluate their methodology very carefully.
- Before deciding to conduct an SSS, systems should carefully check the adequacy of their data and information available, and discuss their plan with their State/primacy agency.
- After conducting their IDSE, systems need to submit their IDSE report according to their respective schedule. At this point, the early schedule requires systems to submit their report within 2 years of Rule promulgation, while the late schedule gives systems 4 years after Rule promulgation to submit their report. This schedule is subject to change as the final rule is promulgated.

The USEPA recognizes that DBP formation and decay is system specific and the guidance provided will not apply to every system. The operational experience and knowledge of system personnel and all available information should be considered in selecting sampling sites. Best professional judgment should be exercised in the application of the manual guidelines to select sampling sites. Several findings were drawn from this study that could be used as guidelines when selecting distribution system monitoring locations to comply with the Stage 2 DBPR. These findings were presented above and are summarized here:

- The time of sampling should be carefully considered when planning monitoring programs or when using historical data to account for seasonal variations in DBP concentrations. On a larger scale, climatic changes such as high precipitation and runoff events are likely to distort DBP data as well. Thus data obtained in a non-typical year should be carefully assessed before selecting DBP monitoring sites.
- Care should be taken when relying solely on water age or chlorine residual to determine potential locations with high THM4 concentrations in a free chlorinated system. Correlation analyses with the data obtained during this study indicated that in general, correlations were relatively weak between THM4 concentrations and water

age or free chlorine residual. In chloraminated systems, high THM4 concentrations may not necessarily be found at locations with high water age when nitrification occurs.

- When determining potential locations with high HAA5 concentrations, a thorough examination of the data is warranted since the low disinfectant residual limit presented in the USEPA guidelines (at least 0.2 mg/L free chlorine or 0.5 mg/L as chlorine for chloraminated systems) may allow biodegradation to occur and may not represent the location with the highest HAA concentration. In fact, when correlation analyses were performed with the data obtained from the free chlorinated systems, HAA concentrations were poorly correlated with water age and free chlorine residuals.
- Chloraminated systems that periodically switch to free chlorine can expect peak DBP concentrations during the free chlorination period, which may result in non-compliance. In addition, the location of peak THM and HAA concentrations may change when switching disinfectant.

Important observations were also made during this study regarding the use of distribution system models. The following issues may lead to modeling errors or inaccuracies, or affect calculations of water age:

- Variations in treatment plant finished water flows, hydraulic patterns in the distribution system, distribution system features (e.g., location and status of PRVs, reservoir elevation, long transmission mains), storage within the system, and demand values and patterns
- Skeletonization of the distribution system
- Presence of specific devices such as autoflushers, or distribution system operations such as flushing
- Status of the reservoirs (i.e., filling or emptying).

CHAPTER 6
SUMMARY, CONCLUSIONS AND RECOMMENDATIONS

The goal of this study was to evaluate the formation and decay of THM4, HAA9, and NDMA in five full-scale drinking water distribution systems. Utilities were selected based on finished water characteristics, seasonal temperature changes, geographic location, type of final disinfectant utilized, minimum DBP concentrations (THM4 greater than 40 μg/L and HAA5 greater than 20 μg/L), and several characteristics specific to each distribution system. Each utility had a hydraulic model of their distribution system that was used to determine water age throughout the system. Two utilities utilized free chlorine as their secondary disinfectant, two utilities used chloramines as their secondary disinfectant, and one utility used chloramines with a one-month switch to free chlorine every year to limit nitrification occurrence. Within each distribution system, 12 to 16 sampling locations were selected on the basis of water age, pipe material and diameter, pressure zones, presence of booster chlorination stations, presence of storage reservoirs, and presence of potential interferences such as autoflushers.

Samples were collected 6 to 8 times over a one-year period, except in the system that periodically switches to free chlorine, where two samplings were conducted just before the switch to free chlorine, two samplings were conducted during the free chlorination period, and two sampling were conducted following the return to chloramines. Samples were analyzed for THM4, HAA9, NDMA, as well as several water quality parameters such as temperature, pH, free and total chlorine residuals, TOC, UV absorbance at 254 nm, AOC, HPCs, bromide, and ammonia, nitrite, and nitrate at the chloraminating utilities.

Based on the data collected during this project, the following conclusions can be drawn:

- As expected in the free chlorinated systems, THM4 concentrations increased as water age increased and free chlorine residual decreased in distribution systems. In the chloraminated systems, THM4 concentrations remained relatively constant with increasing water age, with occasional decreases in concentrations observed at MRTLs. These decreases were surprising and occurred when signs of nitrification were present, as evidence by increases in nitrite, nitrate and HPC concentrations, and decreases in ammonia and chlorine levels. Results suggested that although water age may help find locations with maximum THM4 concentration in the distribution system, it may not be a good indicator in the chloraminated systems.
- In the chloraminated system switching periodically to free chlorine, increases in THM4 concentrations from the POE to the location with the maximum THM4 concentration were approximately 32 percent during the chloramination periods. During the free chlorination period, increases in THM4 concentrations were much greater: 135 and 161 percent, with peak THM4 concentration as high as 175 μg/L. Although such trend was expected (i.e., greater increases in THM4 concentrations with water age during the free chlorinated period than during the chloraminated periods), chloraminated systems that periodically switch disinfectant can expect important peaks in DBP concentrations during the free chlorination period that may result in non-compliance.

- In the free chlorinated systems, HAA9 concentrations generally increased or remained constant as water age increased and free chlorine residual decreased, except at sampling locations with the greatest water age where a decrease in HAA concentrations was measured. In the chloraminated systems, HAA9 concentrations were relatively constant, with the exception of the MRTLs where decreases in concentrations were observed. Lower HAA concentrations were generally observed at MRTLs and at locations with lower chlorine residuals, which suggest that biodegradation was responsible for the decreases in HAA9 concentration observed. Results suggest that it may be challenging to determine the location of peak HAA5 concentrations in distribution systems considering that biodegradation may occur in some systems and during certain periods of the year only.
- In the chloraminated system periodically switching to free chlorine, HAA9 concentrations remained relatively constant with increasing water age during the chloramination periods; however, during the free chlorination period, HAA9 concentrations increased by 44 to 53 percent from the POE to the location with the maximum HAA9 concentration, with peak HAA9 concentrations of 104 to 145 μg/L. These trends were similar to those observed for THM4 in this system. HAA biodegradation was not observed in this system.
- Surprisingly high NDMA concentrations were measured in the systems studied; although, there were no obvious differences between the concentrations measured in the free chlorinated and the chloraminated systems. No particular trend in NDMA concentrations could be identified with increasing water age in the distribution systems.
- The HAA5-to-HAA9 ratios were similar in both free chlorinated systems as well as in one of the chloraminated system studied (72 to 79 percent), but this was much lower in the other chloraminated system (Utility D, 39 percent) and greater in the chloraminated system periodically switching to free chlorine (Utility E, >84 percent). These results suggest that the non-regulated HAAs comprise a significant proportion of the HAA9 in most of the systems studied. Such results are concerning considering that the brominated HAAs, which are not well represented by HAA5, are those that may present the greatest health risks.
- Seasonal variations in DBP concentrations can be significant. In this study, THM4 and HAA9 concentrations were lowest in colder months and highest in warmer months, as expected. Similarly, slightly greater NDMA concentrations were observed in warm water than in colder water in the systems studied; although, this trend was not evident at each sampling location. As such, the time of sampling should be carefully considered when planning monitoring programs or when using historical data. On a larger scale, climatic changes including months with higher precipitation and runoff are likely to distort DBP data as well.
- Pipe material, pipe diameter, and a storage reservoir did not have an appreciable effect on THM4, HAA9, nor NDMA concentrations in the distribution systems studied (Appendix D). The trends observed around the storage reservoir, however, need to be considered cautiously as they are dependant on the reservoir hydraulic scenario (i.e., filling or emptying) at the time of sampling. As such, it is difficult to examine the effect of storage reservoirs at the full scale.

- The booster chlorination station increased THM4 concentrations by approximately 16 to 101 percent (Appendix D). In the case of HAAs, concentrations either decreased when the chlorine residual decreased to low levels, increased when the chlorine residual remained high, or remained constant when the chlorine residual decreased but remained detectable. Changes in NDMA concentrations were observed during booster chlorination, but no particular trend could be distinguished. Similar to the above comment in relation to storage reservoirs, the trends observed around a booster chlorination station are dependant on the hydraulic scenario at the time of sampling. During this study, chlorine residual data collected around the booster chlorination station suggest that the hydraulic scenario might have been unexpected at times, leading to distorted DBP data.

In addition to the evaluation of DBP concentrations and speciation in full-scale distribution systems, three bacterial species were examined in bench-scale experiments with regards to their ability to degrade DBPs: a *Burkholderia* species, a *Sphingomonas* species, and *Xantherobacter autotrophicus* GJ-10. The following conclusions were drawn from this part of the study, which is presented in Appendix A:

- All three bacterial isolates were capable of degrading ClAA, Cl_2AA, BrAA, Br_2AA and BrClAA. The *Sphingomonas* species was more effective than the *Burkholderia* species in degrading HAAs, but GJ-10 appeared to be more effective than the other two species. None of the bacterial isolates were capable of effectively degrading Cl_3AA or Br_3AA.
- There appeared to be a selectivity for the X_2AAs over the XAAs by these bacteria. Similarly, the *Sphingomonas* species and GJ-10 exhibited a greater affinity for the brominated substrates over the chlorinated substrates.
- All three bacterial isolates appeared capable of degrading MCAN and DCAN. The order of reactivity was the following: MCAN > DCAN > TCAN. The *Sphingomonas* species was more effective than the *Burkholderia* species with regard to HAN degradation; although, GJ-10 appeared to be more effective than either of the other two isolates.

Bench-scale experiments were also conducted to examine the degradation mechanism of HAAs (Appendix A). In experiments with GJ-10 enzyme extracts or whole cells and Cl_2AA, two main degradation products were observed: glyoxylic and oxalic acids. Glycolic and ClAA also seemed to be formed during degradation. The proposed mechanism of degradation by GJ-10 is some form of reductive dehalogenation as well as hydrolysis. The *Burkholderia* isolate only produced detectable levels of degradation products at very high Cl_2AA concentrations (1,000 mg/L). The degradation products observed were hydroxychloroacetic acid intermediate, glyoxylic acid, and oxalic acid. No degradation products were observed in experiments using the *Sphingomonas* isolate.

RECOMMENDATIONS

The following recommendations for future studies can be drawn from this project:

- This study addressed temporal variations in DBP concentrations and speciation in distribution systems by examining the effect of seasons. Although some information is available regarding short-term (e.g., diurnal or hourly) variations in DBP concentrations in full-scale systems, studies are required to further address this issue. Temporal studies to evaluate the effects of diurnal and hourly variations in distribution systems on DBP concentrations would be useful to confirm the conclusions drawn during this study, better understand the formation and decay of DBPs in distribution systems, determine the impact of changes in flow and demand on water quality, and assess the sensitivity of THM and HAA concentrations to variations in water age at a sampling location.
- While many utilities use chloramines to control the formation of DBPs downstream of the treatment plant, chloramines are limited in their ability to truly limit THM and HAA formation. As observed during this study, high THM and HAA concentrations were measured in all of the chloraminated systems. As such, tools other than conversion to chloramines are needed to help utilities limit DBP concentrations in their distribution system and meet the Stage 2 DBPR requirements.
- This study also showed that the use of chloramines is not without problems. Two of the utilities that utilized chloramines experienced nitrification (historical data indicate that the third utility also experience nitrification; although, nitrification was not observed during this study because samplings were conducted in the winter and spring periods only). Investigations into the causes of nitrification, as well as nitrification prevention and control practices, would be highly useful for those utilities that have nitrification problems in their distribution system.
- Interestingly, this and other studies suggest that nitrification may be "beneficial" as some nitrifying bacteria can degrade THM4 via co-metabolism. Further studies are required to better address this phenomenon and to determine whether other microorganisms have the same capability.
- This study supports the proposed mechanism for HAA degradation via a hydrolytic process, with the formation of the glycolic, glyxoylic, and oxalic acids (Appendix A); however, additional studies are warranted to confirm this mechanism using analytical methods that would possess sufficient sensitivity to detect the degradation products.
- The degradation of HANs by species of *Sphingomonas*, *Burkholderia,* as well as *Xanthobacter autotrophicus* GJ-10 is a relatively new and novel observation (Appendix A) and warrants further study in view of the potential release of cyanide ion as a possible degradation by-product.
- To accommodate for the high numbers of THM and HAA samples that had to be analyzed during this study, two laboratories were used. Inter-laboratory comparison tests were conducted to ensure comparability of results. Results showed some differences, mainly when the DBP concentrations were low. The reasons for these differences are unknown, and the study was not designed to

address them. These data, however, emphasize the need to further examine the analytical methods that are suggested to comply with the Stage 2 DBPR. As observed during this study, even reputable laboratories that use strict QC procedures may show differences in THM or HAA concentrations, particularly when the concentrations are low. The source(s) of these differences need to be identified and examined.

- A statistical analysis of the parameters analyzed during this study using the ANOVA procedure is recommended. This would allow identification of the statistically significant variables that impact DBP concentrations in the distribution systems. Although basic linear correlations were calculated with the data obtained (Appendix C), this study was not designed to allow for more elaborated analyses such as ANOVA. To achieve accurate results from an ANOVA analysis, the sample size would need to be increased, and additional samplings would be required to broaden the range of conditions under which the distribution systems were analyzed.
- When examining water quality around a storage reservoir or a booster chlorination station, it is important to consider the hydraulic scenario present at the moment of sampling. In the case of a reservoir, depending on its piping configuration, aged water from the reservoir may be found at both upstream and downstream sampling locations when the reservoir is emptying. It this study, it appeared that the status of the reservoirs did not affect the change in concentration of neither of the DBPs studied; although, further investigation are warranted to confirm this observation. The hydraulic scenario also deserves attention around booster chlorination stations. In this study, greater free chlorine residuals were observed at times upstream of the booster chlorination station, which suggest unexpected hydraulic conditions at this location and the possibility of data distortion. The zone of influence of a reservoir or a booster chlorination station also needs to be addressed.

APPENDIX A
MICROBIAL DEGRADATION OF HAAs AND OTHER DBPs

The aim of this task was to identify the bacteria responsible for the degradation of HAAs and other DBPs, and to determine the degree and mechanisms of degradation. Activities conducted during this task also identified the compounds that result from the breakdown of degradable DBPs. Based on information obtained from the literature review and previous studies, this part of the study was divided into two specific tasks, which were pursued concurrently:

1. The major objective was centered on the identification of the two HAA degrading bacteria previously identified (Williams et al. 1994, 1998). Bench-scale studies were conducted to further delineate the scope of their biodegradation activity, and to isolate the enzymes that are responsible for this activity (referred to as the "bacterial study").
2. The second part focused on the mechanism of action of the specific dehalogenase enzymes with various DBPs. This portion of the research depended on the development of an analytical method that allowed one to follow the HAA degradation mechanism in order to identify the likely degradation products, and thus, establish a mechanism of action.

LITERATURE REVIEW

Microbial degradation of certain halogenated compounds has led to the discovery of several rather unique and somewhat specific dehalogenase enzymes from a variety of common bacteria, as summarized in Table A.1. The study of these microbial dehalogenase systems may be important in their possible application to bioremediation of halogen containing compounds in the environment and in drinking water. The scope and nature of the substrates for the bacterial metabolism or catabolism of these halogenated compounds ranges from a variety of haloaromatics, haloalkanes and haloalkanoic acids. In many cases, the halogenated organic compounds appear to serve as the sole energy source for growth of the bacteria.

This review summarizes the following:

- Early studies that identified dehalogenating bacteria
- The GJ-10 enzymatic system (used as a positive control in the HAA degradation experiments)
- Enzyme mechanisms for GJ-10 and other dehalogenating enzyme systems;
- Methods for the determination of dehalogenating activity, and
- Dehalogenation of HAAs in potable water systems.

Table A.1
Bacteria with dehalogenase activity

Bacteria	Substrate	Enzyme	Reference
Corynebacterium species		3B haloalkane dehalogenase	Slater et al. (1997)
Mycobacterium, GP1	1,2 dibromoethane	-	Poelarends et al. (1999)
Rhodococcus rhodochrous	-	3B haloalkane dehalogenase	Newman et al. (1999)
Ancyclobacter aquaticus	1,2 dichloroethane	3R haloalkane dehalogenase	Van den Wijngaard et al. (1992)
Xanthobacter autotrophicus, GJ-10	1,2 dichloroethane	3R haloalkane dehalogenase 1L-2-haloacid dehalogenase	Janssen et al. (1984), Van der Ploeg et al. (1991)
Bacillus species and *Corynebacterium* species	Trichloromethane, chloroacetic acid trichloroacetic acid	3B haloalkane dehalogenase - -	Olaniran et al. (2001)
Pseudomonas cepacia, MBA4	Bromoacetic acid	1L-2-haloacid dehalogenase	Tsang et al. (1988)
Flavobacterium species	Pentachlorophenol	Halo-alcohol dehalogenase	Saber and Crawford (1985)
Moraxella species B	-	1L-2-haloacid dehalogenase	Kawasaki et al. (1992)
Pseudomonas putida, AJ1/23	DL-2-chloropropionate	1L-2-haloacid dehalogenase 1D-2-haloacid dehalogenase	Smith et al. (1990)
Alcaligenes xylosooxidans	2-2-chloropropionate	2R-2-haloacid dehalogenase	Schwarze et al. (1997)
Pseudomonas putida, PP3	-	2R-2-haloacid dehalogenase 2I-2-haloacid dehalogenase	Slater et al. (1997)
Pseudomonas sp. 113	D- and L-2-chloropropionates	2I-2-haloacid dehalogenase	Slater et al. (1997)
Burkholderia sp. FA1	Fluoroacetate	Fluoroacetate dehalogenase	Kurihara et al. (2003)

Slater et al. (1979) were some of the first investigators to carry out the partial purification of two dehalogenase enzymes that were isolated from *Pseudomonas putida* which was grown specifically on 2,2'-dichloropropanoic acid (2-DCPA). One of the two dehalogenases appears to have preferred chloropropanoic acid while the second enzyme was more specific for ClAA. Later studies by Slater and Weightman (1980) suggest that a selectivity could be induced in the *P. putida* strains leading to specificity of an altered strain, which had a more rapid transport and uptake of 2-DCPA. During subsequent studies, these same authors (Weightman et al. 1982) showed that the two dehalogenase fractions from *P. putida* were capable of utilizing both the D and L forms of 2-DCPA. The dechlorination of the L-2-DCPA was found to be 80 percent of the rate of the L isomer, suggesting some degree of stereospecificity in the degradation mechanism.

Smith et al. (1990) were able to purify and characterize a specific D-2-haloacid dehalogenase from the *P. putida* strain AJ1/23, which was capable of the stereospecific dehalogenation of D-2-chloropropanic acid. These authors developed an ion-chromatographic assay which allowed them to examine a variety of 2-halocarboxylic acids. Their data suggest that this enzyme possesses a definite specificity for the brominated forms of the carboxylic acids such as bromoacetic acid and DL-2-bromopropionic acid. Maximum enzyme activity was established at a pH of 9.5 and at a temperature of 50°C. Using sodium dodecyl sulphate polyacrylamide gel electrophoresis (SDS-PAGE) gradient gels, the molecular weight for this enzyme was found to be approximately 31,800 +/- 500.

GJ-10 Enzymatic System

Earlier studies by Uchiyama et al. (1992) have shown that certain heterotrophic bacteria can degrade various chlorinated organics. Three different bacterial species were identified in this study, one of which was in the genus *Pseudomonas* and another was described as an oxidase positive, Gram-negative species which could effectively utilize Cl_2AA as its sole source. This bacterium was later identified as *Xanthobacter autotrophicus*.

Meusel and Rehm (1993) described the degradation of certain chlorinated organic acids such as Cl_2AA by immobilized *X. autotrophicus* GJ-10 via enzymatic degradation. This particular bacterium was capable of degrading Cl_2AA at levels as high as 20 mM (2.58 g/L). It was later shown that the specific enzyme that was responsible for this bacterial degradation was a haloacid dehalogenase better known as GJ-10. Apparently, there was considerable enzyme induction within the bacteria resulting from the exposure to Cl_2AA as the sole carbon source. GJ-10 is a unique enzyme system, which has been shown to degrade a wide variety of halogenated organic compounds.

The same GJ-10 enzyme was described by van der Ploeg et al. (1991) in their study of the bacterial degradation of various HAAs. This study showed substrate selectivity by the enzyme in that ClAA, Cl_2AA, BrAA and Br_2AA were effectively degraded, but Cl_3AA was not. These authors offered several mechanisms to account for the enzymatic degradation of the HAAs. The overall mechanism proposed involves direct hydrolysis of the acyl function by water to give the corresponding hydroxylchloroacetate. This proposed mechanism leads to an increase in chloride concentration during HAA degradation together with a decrease in pH. Similar responses were observed by Williams et al. (1995 and 1998) with several bacteria isolated from MRTLs.

Rehm and Heinze (1993) studied the bacterial degradation of Cl_2AA by GJ-10 that was entrapped on calcium-alginate-immobilized cells. Typically, the Cl_2AA degradation by GJ-10 led to a release of hydrochloric acid (HCl) and a drop in pH. As the HCl accumulates and the pH proceeds to drop below the optimum pH of 8.0, there was a resistance to further dehalogenation.

Williams et al. (1996, 1997, 1998) isolated a crude enzyme fraction, the haloacid dehalogenase, from commercially available *X. autotrophicus* GJ-10. This enzyme was shown to effectively degrade HAAs in bench-scale studies. Using this same experimental procedure, a crude enzyme fraction was isolated from bacteria collected at MRTLs in a drinking water distribution system, Newport News Waterworks, VA, and this enzyme was found to also degrade HAAs. This enzyme fraction was the subject of more extensive purification, which led to a purified enzyme fraction that was quite effective in the degradation of a variety of DBPs (Williams et al. 1996, 1997, 1998).

According to Keuning et al. (1985), this GJ-10 enzyme was capable of catalyzing the hydrolytic dehalogenation of C1 to C4 haloalkanes including brominated and iodinated species. Maximum activity was observed at pH of 8.2 and at 37°C. Interestingly enough, this bacterium apparently exhibits two dehalogenase enzymes. One enzyme is apparently specific for halogenated alkanes while the second enzyme is specific for halogenated carboxylic acids like ClAA and Cl_2AA. In the case of the degradation of 1,2-dichloroethane, the enzyme first coverts this substrate to 2-chloroethanol, which is then oxidized to chloroacetaldehyde and then to ClAA, which is then degraded by the second dehalogenase to glycolate and finally to oxalic acid.

Damborsky et al. (2000, 2001) have continued to investigate the substrate specificity and structure-activity relationships of GJ-10 as well as the comparative binding energy analysis of substrate specificity associated with this particular enzyme system. Uchiyama et al. (1992) have described a heterotrophic bacterium (strain DA4) from *Methylocystis sp* strain M that appears to be identical to GJ-10 and was capable of complete mineralization of trichloroethylene. The DA4 enzyme isolated in this study was also capable of degrading Cl_2AA to HCl and dihydroxyacetic acid, which is then metabolized to carbon dioxide.

Welander et al. (1995) have described the growth of an aerobic bacterium using Cl_3AA as the sole energy source. Although the specific bacterium was not positively identified, it was described as being Gram-negative, catalase-positive and cytochrome C-positive and may have been related to an *Acinetobacter* strain. More recently, Soda et al. (2000) have studied the mechanisms associated with bacterial dehalogenation of 2-haloalkanoic acids. The literature describes two distinct mechanisms operating with the various haloacid enzymes. The L-2-haloacid dehalogenase mechanism involves the participation of a carboxyl group of the enzyme serving as a nucleophile to displace the halogen from the substrate to produce an intermediate ester. This ester then undergoes direct hydrolysis to give the hydroxy acid product. The DL–2-haloacid dehalogenase enzyme from *Pseudomonas* apparently utilizes a single step mechanism that involves the attack of an enzyme directed water molecule in the halogen displacement of the haloacid.

Enzyme Mechanisms

Dehalogenating enzymes such as hydrolytic dehalogenases, glutathione transferases, monooxygenases and hydratases are classified according to their reaction type, which is derived from product analysis, cofactor requirement, DNA sequence homologies, or X-ray crystallographic data and site-directed mutagenesis studies (Janssen et al. 1994). The haloalkane dehalogenase and the haloacid dehalogenase are both considered to be hydrolytic in nature. They catalyze a nucleophilic displacement reaction with water as the sole co-substrate. The haloalkane dehalogenases are classified into two groups on the basis of substrate specificity. The first is the 3R haloalkane hydrolytic dehalogenase (dhlA), which is found in gram negative bacteria and shows a limited range of substrate specificity. GJ-10 contains the dhlA gene, and is discussed further below. The second class is the 3B haloalkane hydrolytic dehalogenase and is found in gram positive bacteria (Slater et al. 1997).

The haloacid dehalogenases consist of four specific classes, which are also based on substrate specificity:

- Class 1L 2-halogenated alkanoic acid hydrolytic dehalogenase
- Class 1D 2-halogenated alkanoic acid hydrolytic dehalogenase
- Class 2I 2-halogenated alkanoic acid hydrolytic dehalogenase
- Class 2R 2-halogenated alkanoic acid hydrolytic dehalogenase.

The 1L dehalogenases catalyzes the dehalogenation of L-2-haloalkanoic acids, while the 1D dehalogenase specifically acts on D-2-haloalkanoic acids. The 2I and 2R dehalogenases apparently acts on both enantiomers of the substrates, as illustrated in Figure A.1 (Slater et al. 1997; Kurihara et al. 2000). The 2R dehalogenase is strongly inhibited by sulfhydryl-blocking reagents, while the 2I dehalogenase is not. GJ-10 contains the 1L-2-haloacid dehalogenase (van der Ploeg et al. 1991).

(A)

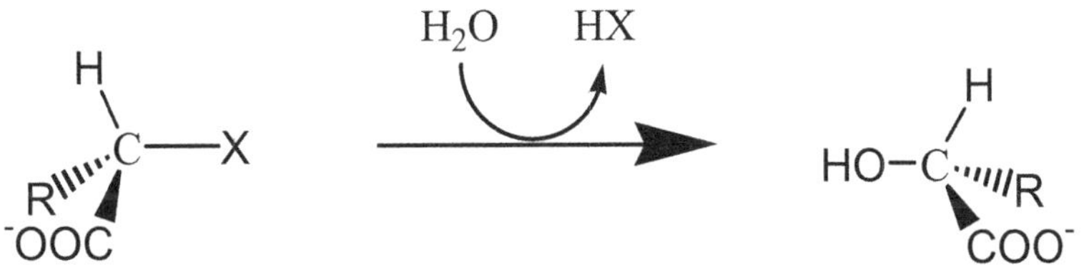

(B)

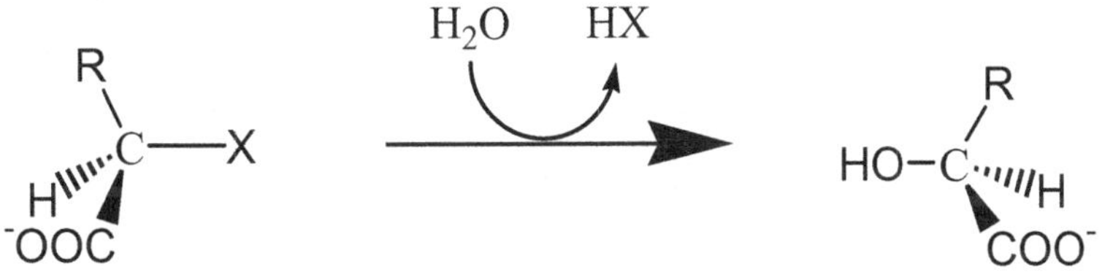

Source: Reprinted from *Journal of Molecular Catalysis B:Enzymatic,* 10(1-3):57-65, Kurihara, T., E. Nobuyoshi, and K. Soda, Bacterial 2-haloacid Dehalogenases: Structures and Reaction Mechanisms, 2000, with permission from Elsevier.

Figure A.1 Reactions catalyzed by the four classes of haloacid dehalogenases. The 1L dehalogenases catalyze the reaction shown in (A). The 1D dehalogenases catalyze the reaction shown in (B). The 2I and 2R dehalogenases catalyze both (A) and (B)

The enzymes associated with the 1L class have been described as having molecular weights ranging from 25,000 to 28,000 Daltons in various bacteria. These enzymes show a maximum reactivity in the pH range of 9 to 11, and have very similar amino acid sequences to each other within the class (Kurihara et al. 2000).

Two different reaction mechanisms have been proposed for hydrolytic dehalogenation. One mechanism involves the attack of the water molecule on the substrate (HAA) to displace the halide ion through direct nucleophilic attack (Figure A.2a). In the second mechanism (Figure A.2b) a carboxylate group, from an aspartate or a glutamate unit on the enzyme, acts as the nucleophile and attacks the alpha carbon of the substrate to release the halide ion. This results in the formation of an ester intermediate, which is then subsequently hydrolyzed by an attack of water on the carbonyl carbon as shown in Figure A.2 (van der Ploeg et al. 1991; Kurihara et al. 2000).

On the other hand, work by Soda et al. (2000) offered a very definitive discussion of the proposed reaction mechanism associated with the activity of various haloacid dehalogenases. According to this author, there appears to be three specific classifications of the 2-haloacid dehalogenases based on the stereochemistry of the substrates:

- L-2-haloacid dehalogenase that has a specificity for the L-2-haloalkanoic acids
- D-2-haloacid dehalogenase which prefers the D-haloalkanoic acids
- D,L-haloacid dehalogenases that are capable of utilizing either the D or L forms of the substrate.

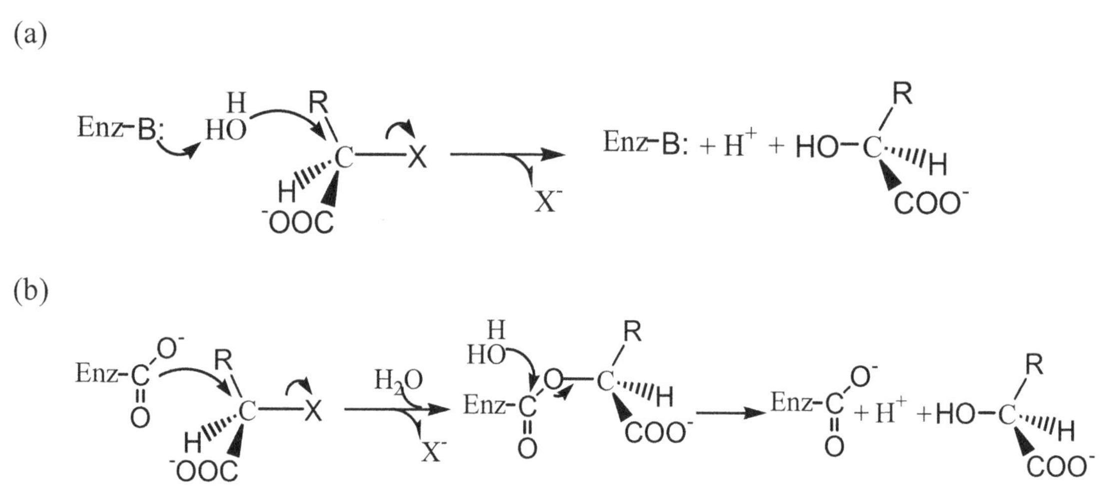

Source: Reprinted from *Journal of Molecular Catalysis B:Enzymatic,* 10(1-3):57-65, Kurihara, T., E. Nobuyoshi, and K. Soda, Bacterial 2-haloacid Dehalogenases: Structures and Reaction Mechanisms, 2000, with permission from Elsevier.

Figure A.2 Proposed mechanisms for enzymatic dehalogenation of HAAs

In the original mechanism proposed by Soda et al. (2000), both the D and L-2 haloacid dehalogenases initiate the first step of the HAA degradation by way of an ester intermediate that results from the nucleophilic attack of a specific carboxylate group that is associated with the active site in the dehalogenase enzyme (Figure A.2). It was proposed that this ester intermediate would then undergo a rapid hydrolysis step initiated or assisted by an Asp 180 unit, which was closely associated with the active site of the enzyme. Soda et al. (2000) also proposed that the D,L-2 haloacid dehalogenase enzymes carry out their degradation by a much simpler single step mechanism. This classic Sn2 mechanism involves the enzyme mediated direct attack of a molecule of water at the halogenated 2-carbon with subsequent displacement of the halide ion.

Methods for the Determination of Dehalogenase Activity

One of the earliest methods used to detect dehalogenase activity was based on the procedure described by Bergman and Sanik (1957). This method used colorimetry to measure the concentration of free chloride ions that are released by the activity of the dehalogenases. The chloride ion released in this hydrolysis would then displace a thiocyanate ion from mercuric thiocyanate, which produced a highly colored ferric thiocyanate complex in the presence of a ferric ion (Equation A.1).

$$2Cl + Hg(SCN)_2 + 2Fe^3 \rightarrow HgCl_2 + Fe(SCN)^{2+} \quad \text{(A.1)}$$

The colored solution could then be assayed in a spectrophotometer at 460 nm and intensity of the color produced was found to be proportional to the concentration of chloride ion released from the substrate. Chloride ions produced by dehalogenase enzymes may also be measured using a chloride ion-specific electrode (ISE) method. The chloride ISE is used in conjunction with a reference electrode to form a complete chemical cell. An electrical potential develops across the electrode surface, which can be compared to the fixed potential of the reference electrode. The voltage that develops between the sensing and the reference electrode represents the chloride ion concentration in the solution (Harris 2003).

In previous studies conducted with the two HAA degrading bacteria from the Newport News MRTL sites, the method of choice to follow chloride ion release was an ISE (Williams et al. 1998). In subsequent years the reproduction of these results, using the ISE, has proven to be unsuccessful. A considerable amount of drift in the ISE was experienced in the chloride measurement and protein film formation appears to occur on the electrode surface. This method has been found to equilibrate very slowly and is also subject to interference by other ions (Holloway et al. 1998).

Consequently, a colorimetric procedure based on the use of pH indicators has been implemented to follow the bacterial activity of the two bacteria isolates. Loos (1975) used bromocresol purple in bacterial growth media for the identification and isolation of microorganisms capable of degrading the chlorinated herbicide, 2,4-dicholorophenoxyacetic acid. Bromocresol indicator changes from purple to yellow as the pH drops from 7.0 to 5.4. Apparently, if the production of chloride in the pesticide degrading culture exceeds the production of cations, other than the hydronium ion, then hydrochloric acid is formed. If the acid produced was sufficient enough to overcome the buffering capacity in the media and thus lower the pH, the indicator incorporated in the media would change color from purple to yellow. The colorimetric responses usually occurred within 3 to 11 days (Loos 1975). Saber and

Crawford (1985) incorporated this same technique to isolate and characterize a purified culture of bacteria able to degrade pentachlorophenol using bromothymol blue in agar.

Strotmann and Roschenthaler (1987) also used bromothymol blue as a screening mechanism with *Pseudomonas putida* and could demonstrate that the liberation of protons is directly proportional to the liberation of chloride and reciprocally proportional to the concentration of 2-chloroethanol in the medium. In these experiments it was important to use the appropriate controls that did not contain the substrate in question in order to ensure the color change was not occurring as a result of release of other metabolic acids. One major advantage of this colorimetric method is that it reduces the time and effort associated with the steps of cell culture, harvesting, and crude enzyme extraction from bacteria that do not contain the enzyme of interest.

In addition to the use of pH indicators as a qualitative tool to detect dehalogenase activity, Holloway et al. (1998) has quantitatively used phenol red to measure activity from isolated crude GJ-10 enzyme in a 96-well plate. The dehalogenases catalyze the hydrolytic displacement of a chloride ion from the substrate and produces a subsequent drop in pH. This activity is calculated by constructing a standard curve using varying amounts of HCl in a reaction mixture containing the indicator and measuring the resulting optical densities at 540 nm. The decrease in optical density that is observed can be compared to a standard curve to obtain the activity of the enzyme (Holloway et al. 1998).

Haloacetic Acid Degradation in Potable Water Systems

Zhou and Xie (2002) conducted experiments to study the biodegradation of HAAs. Three dechlorinated samples and one sample straight from a local water distribution system were spiked with ClAA, Cl_2AA, and Cl_3AA and mixed with 10 mL of bacterial seed prepared with pieces of biofilm washed from biologically activated carbon (BAC) collected from filters at the local water treatment plant. Results showed that ClAA, Cl_2AA, and Cl_3AA in dechlorinated water were completely degraded. Conversely, there was little change in the Cl_2AA concentration in the water sample that maintained a total chlorine residual greater than 1 mg/L for eight days. Because residual chlorine restrains bacterial growth in water, the complete degradation of HAAs in the dechlorinated water samples may result from biological activity. The results also imply that biofilm on the surface of BAC contained the bacteria that degraded the HAAs. It is possible, therefore, to use BAC to biodegrade HAAs in drinking water treatment processes. Cl_3AA degraded much lower than ClAA and Cl_2AA. The majority of the Cl_3AA degraded in the last 48 hours after the ClAA and Cl_2AA had degraded completely (Figure A.3). This observation is in agreement with Williams et al.'s (1995) report, which used bacterial seed isolated from water distribution systems at the MRTLs.

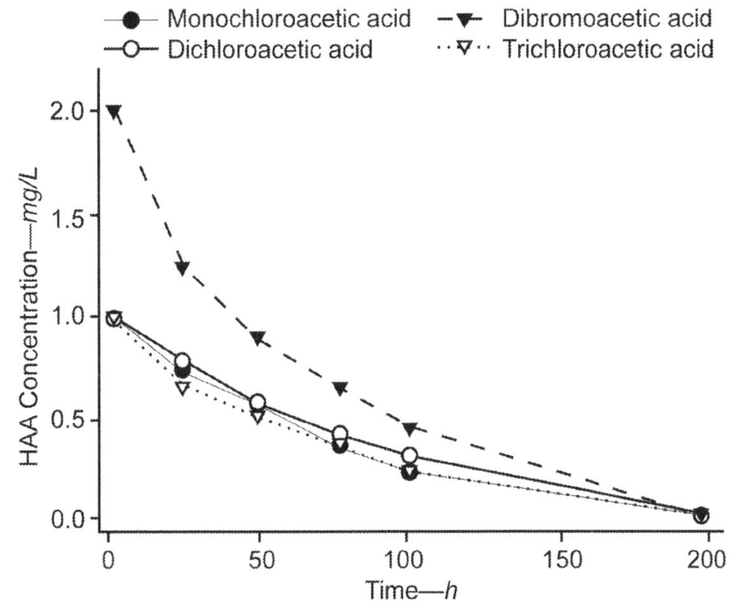

Source: Zhou and Xie. 2002. Use of BAC for HAA Removal - Part 1, Column Study. *Journal AWWA*, 94(4):194 – 200. Published with permission from AWWA.

Figure A.3 Removal of HAAs in the presence of BAC

More specific information is available regarding HAA degradation. Xie and Zhou (2002) confirmed that HAAs are biologically degraded in his investigation of HAA removal by BAC. In the first part of the study, two BAC columns packed with acclimated granular activated carbon (GAC) and two columns packed with autoclaved GAC were used. The GAC in both kinds of filters had been online for 4 and 26 months respectively. The acclimated GAC and the autoclaved GAC columns were fed synthetic water samples (originally contained 4 μg/L ClAA, 11 μg/L Cl_2AA, and 8 μg/L Cl_3AA) that were spiked with 50 μg/L of each of the five HAAs. The water also contained 1-2 mg/L of total chlorine residual which was not quenched in these column studies. In both BAC columns, ClAA, Cl_2AA, BrAA, and Br_2AA were completely removed during a 7-hour run. Cl_3AA, however, was detected in effluents at 10 μg/L. The removal efficiency for HAAs was significantly reduced in the columns with autoclaved GAC. Both Br_2AA and Cl_3AA were removed to a moderate degree in the column with autoclaved 26-month GAC. These results indicate that all HAA5 are biologically degradable, although carbon adsorption was also involved in removal of Br_2AA and Cl_3AA.

In a 76-day experiment conducted by Xie and Zhou (2002), results indicated that both carbon adsorption and biological degradation are involved in HAA removal. The HAA removal at the beginning of the run was from carbon adsorption. After GAC adsorption capacity for HAAs was exhausted and before the bioactivity for the HAAs was developed, no HAA removal was observed. Once the bioactivity was fully developed, HAAs were completely biodegraded. This mechanism was illustrated by the removal of ClAA, BrAA, and Cl_2AA. For Br_2AA and Cl_3AA, a combination of adsorption and biodegradation explained their removal as the bioactivity was developed before the GAC adsorption capacity was completely exhausted. The time required for the development of the bioactivity to degrade each HAA also varied, with the bioactivity of degradation of X_2AAs lagging behind that of XAAs.

MATERIALS AND METHODS

To conduct the tasks listed at the beginning of this chapter, the following approach was used.

Isolation and Dehalogenase Screening

Bacterial samples presumed to be capable of HAA degradation were isolated from water samples collected from Utility C's MRTLs. The bacteria were evaluated for dehalogenase activity using the colorimetric indicator plate method described earlier. The bacterial isolates were streaked on Difco™ R2A agar (BD; Franklin Lake, NJ) plates containing 10 mM Cl_2AA (Fluka-Sigma-Aldrich; St Louis, MO) with 3 percent bromocresol purple (Sigma-Aldrich) or 20 μg/mL phenol red (The Coleman & Bell Co.; Norwood, Ohio)(Loos 1975; Holloway et al. 1998). These plates were incubated at 33°C for 2 to 7 days and carefully monitored for a color change from purple or red to yellow, indicating the release of HCl and dehalogenase activity with this substrate. The bacteria that presented dehalogenase activity were then tested further for substrate specificity by using indicator plates containing HAAs such as ClAA, Cl_3AA, BrAA, Br_2AA, Br_3AA or BrClAA.

Growth of Bacteria

Once the dehalogenase activity was confirmed, these same bacteria were then grown on R2A agar plates. Many bacterial strains found in drinking water may have been stressed by recent passage through the disinfection process and may be starved because of the minimal concentrations of available nutrients in water (Geldreich 1996). R2A agar was chosen because it is a low nutrient medium that stimulates the growth of stressed and chlorine-tolerant bacteria. A medium rich in nutrients supports the growth of faster growing bacteria and may suppress the growth of the slower growing or stressed bacteria found in drinking water.

One-inch (2.5-cm) portions of the R2A agar were then removed from the culture plate and placed into 25 mL of sterilized R2A broth (BBL) in sterile 50-mL conical tubes and allowed to incubate at 33°C under rotary shaking in a Lab Line Orbit Environ-Shaker (Barnstead International; Dubuque, IA) at 1000 rpm. The growth rate was estimated by measuring the optical density (OD) at 600 nm (OD_{600}) using a Milton Roy Spectronic 20D spectrophotometer (Spectronic Instruments; Rochester, NY). OD_{600} is commonly accepted as the wavelength to estimate the number of cells within a population (Trun and Trempy 2003). The end of the log phase, which occurs at the end of an exponential growth period, was determined by plotting OD_{600} versus time.

When the culture had reached the end of its log phase, the contents of the conical tube were added to 200 mL of fresh R2A broth and placed into a shaker bath (Precision Scientific Co.; Winchester, VA) at 33°C and approximately 72 rpm. Once the end of the log phase was reached, 50 mL of the culture was added to 500 mL of fresh R2A broth containing 1 mM Cl_2AA as the substrate. *Xanthobacter autotrophicus*, GJ-10, which had been obtained from the Metropolitan Water District of Southern California was chosen as internal control. The same growth procedure was followed for *X. autotrophicus*, except tryptose broth was used in place of R2A broth.

Identification of HAA Degrading Bacteria

Samples of the two bacteria of interest were sent to the Bacterial Identification and Fatty Acid Analysis Laboratory at the University of Florida where the MIDI Sherlock Microbial Identification System and Biolog analyses were performed. The Sherlock Microbial Identification System identifies bacteria based on a gas chromatographic analysis of the whole-cell fatty acid content followed by comparison of these data to an extensive databank. The Biolog system measures cellular metabolism by introducing the bacteria to a variety of carbon sources in a 96-well microplate containing tetrazolium violet. The test yields a characteristic pattern of purple wells which constitutes a "metabolic fingerprint" unique to each bacteria. These patterns can then be compared to a databank for identification. The bacteria were also sent to the Microbial ID laboratory in Newark, Delaware, for confirmation of identification using the MIDI system and polymerase chain reaction (PCR). PCR testing uses primer sets to amplify genomic DNA obtained from the bacteria, resulting in a product of approximately 500 base pairs (bp) in size. The PCR product is then sequenced and analyzed by comparison to a database for identification. The region of DNA amplified is the 16S rRNA gene, which is used to successfully differentiate microbial genus and/or species by bp variances.

Whole-Cell Dehalogenase Colorimetric Assays

An initial dehalogenase activity assay was developed in whole cells based on a method introduced by Chang et al. (1999). Cells were suspended in a buffer consisting of 1 mM HEPES (Fisher Biotech; Pittsburg, PA) at pH 8.2, 20 mM sodium sulfate (Mallinckrodt Baker, Inc; Phillipburg, NJ), and 1 mM EDTA (Fisher Scientific International, Inc.; Pittsburg, PA). The absorbance was determined for a 6-mL sample at 600 nm. This sample was divided into two 3-mL samples and 20 μg/mL of phenol red indicator was added to each aliquot. One sample served as the negative control, while the other samples were exposed to 10 mM Cl_2AA at pH 9.0. The reaction time started upon addition of the Cl_2AA. Once a color change was observed, samples were centrifuged to remove the cells. The absorbance of the supernatant was then determined at 570 nm to calculate the rate of change and the relative dehalogenase activity. Eventually, this procedure was adapted to a microplate format. Equation A.2 was used:

$$\text{One unit of dehalogenase activity} = 1000 \times (OD_{570(-)} - OD_{570(+)}) / (T \times V \times OD_{600}) \quad (A.2)$$

Where:
$OD_{570(-)}$: OD_{570} of the negative control
$OD_{570(+)}$: OD_{570} of the sample containing substrate
T: Elapsed reaction time after adding substrate (minutes)
V: Volume of suspended culture (mL)

Preparation of Crude Cell Extracts

Once the dehalogenase containing bacteria had been grown up to a proper optical density at 600 nm, the crude bacteria were harvested by centrifugation at 10,000xg for 30 minutes at 20ºC in a Beckman centrifuge Model J2-21 (Beckman; Miami, FL). The cells were washed with 10 mM Tris (Sigma), 1 mM EDTA, 1 mM mercaptoethanol (Fisher Scientific) buffer (pH 7.8), and recentrifuged at 10,000xg. The pellet was then resuspended in the same buffer (1 mL/g of wet cells), and cells were disrupted by sonication using a Branson Sonifier Cell Disruptor 185 (Branson Ultrasonics Co.; Danbury, CT) and a macroprobe for five 30-second bursts separated by 60-second cooling periods on ice. The sonicated phase was centrifuged at 10,000xg and 4ºC for 30 minutes to remove cell debris. The supernatant, referred here as the crude extract, was stored at 4ºC or used immediately in further analyses. The pellet was reconstituted and then vortexed in 1 mL of 0.01-percent Triton X-100 (Fisher Scientific), and then centrifuged at 10,000xg for 15 minutes to dislodge any possible enzymes located in the bacterial cell membrane.

Protein Estimation

The protein concentration, in the cell extracts, was determined by measuring the absorbance at 280 nm. Aromatic proteins, such as tryptophan and tyrosine together with cystine typically absorb UV light with an absorbance maximum at 280 nm. The relationship of absorbance to protein concentration is linear. Because different proteins have widely varying absorption characteristics, considerable error may occur, especially for unknowns or protein mixtures. It is commonly accepted to use Equation A.3 to roughly estimate protein concentration (Caprette 1997).

$$\text{Concentration (mg/mL)} = \frac{\text{Absorbance at 280 n}}{\text{Path length (cm)}} \qquad \text{(A.3)}$$

Equation A.3 assumes that protein has an extinction coefficient of 1 for a 1-mg/mL sample. The extinction coefficient of most proteins ranges from 0.5 to 1.5 for a 1-mg/mL sample (Kirschenbaum 1975). To avoid possible errors, a study performed by Gill and von Hippel (1989) discovered that by using the aromatic amino acids and cystine components of a sequence, Equation A.4 could be used to accurately determine (+/-5 percent) the molar extinction coefficient.

$$\varepsilon\,(\text{protein}) = \#(\text{Tyr}) * \varepsilon\,(\text{Tyr}) + \#(\text{Trp}) * \varepsilon\,(\text{Trp}) + \#(\text{Cystine}) * \varepsilon\,(\text{Cystine}) \qquad \text{(A.4)}$$

Where:
ε: extinction coefficient
#: number of given amino acid
Tyr: Tyrosine
Trp: Tryptophan

The ExPASy (Expert Protein Analysis System) proteomics server (SIB 2004) has a tool named ProtParam that allows for the computation of various physical and chemical parameters for a given protein, including the molar extinction coefficient (Appel et al. 1994). Once the extinction coefficient is determined the Beer-Lambert law may be used to determine the protein concentration (Equation A.5).

$$A = \varepsilon * c * l \quad \text{(A.5)}$$

Where:
A: Absorbance
ε: Extinction coefficient
c: Concentration (mol/L)
l: Optical path length

Colorimetric methods for protein estimation such as Bradford or other dye-binding assays (Caprette 1997) are generally less reliable unless previously calibrated for the specific protein under investigation.

Enzyme Purification

Enzyme purification was initiated first by precipitation using saturated ammonium sulfate (Mallinckrodt Baker, Inc.; Phillipburg, NJ) for a final concentration of 35 percent to the crude bacterial extract at 4°C on a stir plate. This mixture was removed from the stir plate, kept on ice for 30 minutes and then centrifuged at 10,000g at 4°C for 30 minutes. The resulting pellet was removed, and the remaining supernatant was treated with ammonium sulfate to a final concentration of 65 percent. As the salt concentration was increased, a point of minimum protein solubility was reached and the crude enzyme precipitated from solution. After centrifugation, the precipitant was resuspended in 2 to 3 mL of 50 mM Tris, 1 mM EDTA, and 1 mM mercaptoethanol (TEM) buffer (pH 7.2). This suspension was then dialyzed against a 50 mM TEM buffer for 6 hours on ice to remove any residual ammonium sulfate bound to the concentrated protein. A 4.0 cm x 10.0 cm dialysis tubing (Fisher Scientific) with a molecular weight cutoff of 6,000 to 8,000 was used for dialysis.

Enzyme Characterization

The dialyzed enzyme fraction was applied to a diethylaminoethyl-cellulose chromatography (DEAE) column (10 cm x 1 cm). It was equilibrated with 1 mM HEPES, 20 mM sodium sulfate and 1 mM EDTA buffer at pH 8.2 to allow binding of the enzyme. The enzyme in the column was eluted at 0.84 mL/minute with a 0.1 to 0.5 M gradient of ammonium sulfate (Keuning et al. 1985), and monitored at 280 nm with a LKB Bromma UV source (Stockhom, Sweden). While sodium chloride or potassium chloride is usually used for elution, these salts present a source of chloride ion contamination, and therefore, were not used in this study. Aliquots of 2 mL were collected with a Bio-Rad 2110 Fraction Collector (Bio-Rad Laboratories; Hercules, Calif.) and the individual fractions were then tested for dehalogenase activity by the colorimetric procedure (Holloway et al. 1998).

Gel filtration chromatography, also called size exclusion chromatography, is effective in the separation of proteins of different sizes, and was used to determine the molecular weight of the dehalogenase enzyme. The Sephacryl S-200 gel-filtration column (42.5 cm x 3 cm) has a stationary phase consisting of a crossed-linked polyacrylamide, and was equilibrated with 1 mM HEPES, 20 mM sodium sulfate and 1 mM EDTA buffer at pH 8.2. The active DEAE fractions were combined and placed on the column. Fractions were eluted by step gradient with this same buffer at a flow rate of 1.3 mL/minute and monitored at 280 nm with a LKB Bromma UV source. Molecules are eluted in order of decreasing size and the column was calibrated using Dextran Blue 2000 (MW 2,000,000) (Pharmacia Fine Chemicals; Upsala, Sweden), myoglobin (MW 17,000), and Vitamin B_{12} (MW 1350). Aliquots of 2 mL were collected with a BioRad 2110 Fraction Collector. Fractions containing the dehalogenase were pooled and dialyzed against a new salt buffer. The void volume (V_o) of the column was taken as the elution volume of Blue Dextran 2000. The relative retention of each protein was measured and the elution volume/void volume (V_e/V_o) was calculated and used to prepare a calibration curve.

The molecular weight and purity of the enzyme was determined by running the enzyme sample on a 10-percent discontinuous sodium dodecyl sulfate (SDS) polyacrylamide slab gel by the method of Laemmli (1970). The separating gel was prepared by mixing 1.66 mL of 30-percent acrylamide-bisacrylamide (Promega Co.; Madison, WI), 50 µL 10-percent SDS (Mallinckrodt) and 1.25 mL of 3.0 M Tris-HCl (pH 8.8), and brought to a final volume of 11 mL with deionized water. 50 µl of 1.5-percent ammonium persulfate (Sigma) and 5 µL N,N,N',N'-tetramethylethylenediamine (TEMED) (BioRad) were added to the final volume to initiate polymerization of the acrylamide. The stacking gel was prepared using 0.4 mL of a solution of 30-percent acrylamide and 0.8-percent bisacrylamide, 20 µL of 10-percent SDS and 0.6 mL of 0.5 M Tris-HCl (pH 6.8), and brought to a final volume of 2 mL with deionized water. Polymerization was initiated by the addition of 20 µL of ammonium persulfate and 15 µL of TEMED.

The separating gel was added between the two glass plates and covered with water saturated butanol (EM Science). After polymerization, the butanol was rinsed off with deionized water and the stacking gel was added on top of the separating gel to the top of the glass plate. A 10-well comb was placed between the plates. After, about 30 minutes, the polymerization was complete and the plates containing the gel were placed in a Bio-Rad Mini-Protein II Cell gel apparatus (Bio-Rad Laboratories). A buffer, made up of a 1:10 dilution of 30.3 g of 0.25 M Tris at pH 8.3, 144 g of 1.92 M glycine (Fisher Scientific; Far Lawn, NJ), 10 g of SDS, and deionized water to make 100 mL, was added to the reservoir tank and the comb was removed from the plates. Tracking dye, bromophenol blue in 50 percent glycerol, 50 mM Tris, and SDS buffer at pH 7.8, at a volume of 20 µL was added to 20 µL of enzyme extracts and were set in an 80°C water bath for 3 minutes to allow denaturation. Next, 20 µL of enzyme sample and 10 µL of a molecular weight ladder (Bio-Rad Laboratories) were each added to separate wells in the gel. The gel was run for approximately 2 hours at 100 V on an E-C 500 power source. Once the tracking dye reached 0.5 to 1 cm from the bottom of the gel, the gel was stained with a methanolic Coomassie brilliant blue R250 solution, made of 0.25 g Coomassie brilliant blue R250 (Sigma), 300 mL methanol (EDM Chemical; Gibbstown, NJ), 100 mL glacial acetic acid (Fisher Scientific) and 600 mL deionized water. Next, the gel was destained in the same solution without the Coomassie brilliant blue R250. If the bands were difficult to visualize at this point, the gel was then stained with a silver nitrate solution. The silver nitrite staining technique is generally 50 times more sensitive than the typical Coomassie staining and results in much clearer bands.

The gel was soaked in fixative, (same solution as the destaining solution mentioned above), for at least one hour. The gel was rinsed with 20 percent ethanol for 20 minutes and in Milli-Q water for 10 minutes. The gel was then soaked in a sensitizer solution made up of 0.02 percent sodium thiosulfate for one minute, followed by three, 20-second rinses in Milli-Q water. Next, the gel was soaked in 0.2 percent silver nitrate for 45 minutes and then rinsed with Milli-Q water for 5 to 10 seconds. Finally, the gel was soaked in a developer solution made up of 0.3 percent sodium carbonate, 0.025 percent formaldehyde, and 10 mg/L sodium thiosulfate. When the bands were adequately visualized, the gel was soaked in a stop solution consisting of a 2.5 percent Tris and 0.4-M acetic acid solution for 5 minutes (Blum et al. 1987).

Dehalogenase Studies

After the enzyme had been partially purified, it was tested for substrate specificity. Cl_2AA was substituted in the colorimetric buffer used in the method of Holloway et al. (1998) with ClAA, Cl_3AA, BrAA, Br_2AA, Br_3AA, and BrClAA. The relative activities at varying substrate concentrations in the range of 1 μM to 50 mM were also calculated in order to determine the Michaelis-Menten constant (K_M), which was done by plotting the velocity of the reaction (μmol/minute) versus the substrate concentration. The ensuing plot should plateau off due to the enzyme becoming saturated with substrate; this resulting maximum velocity (V_{max}) can then be halved to determine the K_M. By comparing the V_{max}/K_M ratio, the activity of different enzymes could be compared. The larger the V_{max}/K_M ratio, the higher the enzyme efficiency (Campbell 1999).

Study of the Haloacid Degradation Mechanism

One of the objectives of the research described in this report was the study of this dehalogenase mechanism and the delineation of the degradation products. To follow the degradation pathway associated with the dehalogenase reaction, it was necessary to develop a suitable analytical method that would allow the detection of the possible metabolites as the reaction proceeded.

Earlier work by Narayanan et al. (1999) described a rather interesting method using high-performance liquid chromatography (HPLC) for the detection of low levels of Cl_2AA and its metabolites in blood and urine as related to the consumption of Cl_2AA in drinking water by mice and rats. Three major metabolites were detected using this HPLC method and were identified as glycolic acid, glyoxylic acid and oxalic acid. These are the same three degradation products that have been proposed in earlier research studies. Narayanan et al. (1999) utilized a Model DX-300, Dionex liquid chromatographic system and a Dionex AG11 analytical anion-exchange column and a conductivity detector. The mobile phase consisted of a 0.01 mM sodium hydroxide solution in 40 percent methanol, which was then followed by a linear gradient from 0.01 mM to 60 mM sodium hydroxide in 40 percent methanol for 30 minutes. Excellent separations of the various metabolites and the parent Cl_2AA were accomplished using this method.

The analytical method used in this study was more direct as it utilized a Bio-Rad FAST ACID (Aminex HPX-87H, 300 mm x 7.8 mm) column, which could be used to separate and identify simple organic acids including acetic, formic, succinic, malic, citric and oxalic acids. The typical mobile phase for the separation of these acids as well as the HAAs was 4 mM sulfuric acid with a flow rate of 0.60 mL/minute. The acids were detected using a set wavelength of 210 nm. The various HAAs of interest to this study could be easily separated over a 20-minute period and gave the following retention times:

Br_3AA:	6.5 minutes
Cl_3AA:	8.3 minutes
Cl_2AA:	8.3 minutes
BrClAA:	8.9 minutes
Br_2AA:	10.9 minutes
ClAA:	15.8 minutes
BrAA:	18.4 minutes.

In general, this method provided a detection limit for the various HAAs in the range of 1 to 5 mg/L. ClAA was especially difficult to detect at low concentration, which presented a problem since ClAA was to be the first target HAA in the mechanistic studies.

Using this FAST ACID column, it was anticipated that the enzymatic degradation of ClAA could be followed and that the general mechanism associated with this substrate would be basically the same for the other HAAs. The likely degradation products of ClAA have already been described and include, in a stepwise fashion, glycolic acid, glyoxylic acid and finally oxalic acid. Using this FAST ACID column, the elution order for these three degradation products was established as follows: 6.68 minutes for oxalic acic, 9.32 minutes for glyoxylic acid, and 12.4 minutes for glycolic acid. There were no major problems distinguishing the positions of the various HAAs and the three possible metabolites of ClAA. However, Cl_3AA and oxalic acid did have very similar retention times (6.4 minutes and 6.57-6.63 minutes, respectively) and BrClAA and gyloxylic acid had very similar retention times (9.11 and 9.29 minutes, respectively). Unfortunately, the proposed degradation or metabolic products of the more highly substituted HAAs, such as Cl_2AA and Br_2AA, are not commercially available and their HPLC profile could not be established.

In view of the sensitivity of the crude enzyme fractions to the conditions used in these studies (pH and temperature), the HPLC method was later adapted to the study of the bacterial dehalogenation of HAAs using whole bacterial cells from GJ-10 and a *Burkholderia* species isolated from one of the distribution systems. These whole-cell studies, using Cl_2AA as the substrate, gave very similar results as those with the crude enzymes, but the operation was much easier and is perhaps more relevant or more easily related to the actual situation in the distribution system.

Initial experiments investigated the degradation products of ClAA, as the degradation products are known. However, no degradation products were observed, due to minimal degradation, as observed in the previous experiments. Further experiments were conducted on Cl_2AA using GJ-10, as well as the two bacterial isolates. As GJ-10 exhibited greater degradation of Cl_2AA than the two bacterial isolates, the majority of the experiments used GJ-10.

The degradation pathway experiments were conducted as follows. Samples of the bacteria were suspended in dilution water and centrifuged. The supernatant was removed and the cells were resuspended in dilution water and centrifuged a second time to remove any elements of the culture media that might interfere with the HPLC analysis. The cells were then resuspended in dilution water and the pH of the reaction media was adjusted to 9.04. A sample of Cl_2AA (concentration ranging from 10 to 100 mg/L) was then added to the cell suspension and the reaction was stirred at room temperature for a period of approximately 90 minutes. At the end of the reaction period, the reaction media was centrifuged and aliquots of the supernatant were analyzed by HPLC on the FAST-ACID column.

RESULTS AND DISCUSSION

This part of the study was divided into two tasks: (1) identification of selected HAA degrading bacteria, and (2) determination of the mechanism of action of dehalogenase enzymes with various DBPs.

Part 1 – Bacterial Study

The specific objectives of the bacterial study were as follows:

1. Isolate HAA degrading bacteria from Utility C's MRTL sites.
2. Culture and identify the bacteria isolated from these samples.
3. Isolate and purify the dehalogenase enzymes from these bacteria.
4. Determine the physical character (e.g., molecular weight) of the enzymes.
5. Determine the scope of activity and optimum conditions for the bacterial degradation of certain HAAs and other DBPs.

Throughout the bacterial study, the bacterium *Xanthobacter autotrophicus* GJ-10 was used as a positive control, based on its known ability to dehalogenate HAAs.

Bacterial Characteristics

Initially, two bacteria were isolated from Utility C's distribution samples that were apparently capable of degrading HAAs. One was a white bacterium, while the other was a yellow-pigmented bacterium. A sample of the yellow bacteria was isolated from a plate inoculated with a water sample collected at a location with a water age of 8 hours on June 11, 2003. The sample (1-mL aliquot) was spread plated onto a bromocresol-purple indicator plate containing 10 mM Cl_2AA and then incubated at 33°C, as described in the Materials and Methods section. Dehalogenase activity, as shown by a distinct change in color from purple to yellow, was not observed until 17 days later. This same bacterium sample was then purified by subsequent streaking on four sets of plates, as follows: fresh indicator media, R2A media, R2A media, and indicator media to isolate Cl_2AA degrading bacteria and confirm yellow pigmentation. On the final plate, dehalogenating activity was confirmed by a change in color within 8 days. A sample of this bacterium was sent out for identification and a bulk growth culture was initiated for enzyme isolation.

From a sample collected on June 11, 2003 in Utility C's distribution system at a location with a water age of 21 hours, a white bacterium was isolated on a nutrient agar plate. Various different bacterial colonies were observed on the plate and they were streaked onto an indicator plate. One of the active colonies was then plated onto a plain R2A and an indicator plate. Significant growth was observed after approximately 4 days on the R2A plate, with dehalogenating activity observed by a color change at 10 days on the indicator plate. Both bacteria were streaked onto R2A plates containing indicator without Cl_2AA, to ensure the bacteria were not causing a false positive color change from natural acids released during metabolic processes. At this point, when both bacteria continued to test positive for dehalogenase activity and apparently were free from other bacterial colonies, these purified samples were then sent out for identification.

Bacteria Identification

Bacterial identification was conducted at the University of Florida using both MIDI and Biolog analyses, as well as at the Microbial ID Laboratory in Newark, Delaware, using MIDI and PCR analyses, as described in the Materials and Methods section. MIDI results are expressed as a similarity index (SI), representing how similar the fatty acid composition of the unknown sample is to that of reference strains. An SI of 0.500 or higher is considered a good match; and exact match would have an SI of 1.000.

The yellow isolate was identified as a *Sphingomonas* sp. The isolate was initially identified as *Flavobacterium johnseniae* by MIDI at the University of Florida (SI of 0.605); Biolog results were inconclusive. The same isolate was also identified as *Mycobacterium mucogenicum*, by both MIDI and PCR analyses at the Microbial ID Laboratory. However, when the culture was plated to check for contamination, it was concluded that the *M. mucogenicum* was a contaminant and not responsible for dehalogenating activity. A second sample was identified by MIDI at the Microbial ID Laboratory as *Novosphingobium aromaticivorans*, also known as a *Sphingomonas* species. PCR analysis confirmed that the bacterium was indeed a *Sphingomonas*, not *Flacobacterium*, species.

In the case of the white isolate, the results of MIDI analyses by the University of Florida were inconclusive. MIDI analysis at the Microbial ID Laboratory identified the isolate as *Burkholderia cepacia* (SI of 0.742). PCR analyses at the same laboratory were inconclusive. The identification was supported by earlier work by the authors of this report, in which this bacterium was identified as *B. cepacia* using the simpler API system (this bacterium previously was referred to as *Pseumonas cepacia*). *B. cepacia* has also been documented in various dehalogenase studies (Tsang 2000, 2002).

The GJ-10 material that was used throughout these studies as an internal positive control was also submitted to the Microbial ID laboratory and was identified as *Xanthobacter flavus*.

Substrate Specificity Testing with Indicator Plates

The two bacteria together with GJ-10 were tested for substrate specificity by incorporating various HAAs into the purple indicator plates in place of the Cl_2AA. The following substrates were tested: ClAA, Cl_2AA, Cl_3AA, BrAA, Br_2AA, Br_3AA and BrClAA. The production of a yellow color in the indicator method was considered as an indication of HAA degradation activity. Results obtained are summarized in Table A.2, and show that GJ-10 was the most reactive of the three bacteria as it produced positive results within 1 to 2 days, whereas the *Sphingomonas* sp. required 2 to 3 days to give positive results, and the *Burkholderia sp.* required 6 to 7 days.

Table A.2
Substrate/isolate combinations for which dehalogenating activity was observed

Bacteria	Substrates giving a positive response
GJ-10	Cl_2AA, Br_2AA, BrClAA
Sphingomonas species	Cl_2AA, Br_2AA
Burkholderia species	Cl_2AA

GJ-10 was observed to degrade Cl_2AA, Br_2AA and BrClAA, but the *Sphingomonas* isolate only degraded the Cl_2AA and the Br_2AA, and the *Burkholderia* isolate reacted with only Cl_2AA. The bromocresol purple used in this study is purple at a pH of 6.8 and above, and yellow at a pH of 5.2. Typically, the reactions were begun at a pH of 8.2. To potentially accelerate the screening process for dehalogenase activity, phenol red was substituted for the bromocresol purple in subsequent studies. Phenol red is red at a pH of 8.2 and yellow at a pH of 6.8.

Haloacid dehalogenase activity has been shown to occur between pH 9 and 11. As such, it is likely that the activity of the enzyme may be minimal and perhaps even nonexistent at a pH of 5.2. The negative results for the BrClAA for the *Sphingomonas* isolate and the Br_2AA and BrClAA for the *Burkholderia* isolate may represent false negatives, based on the results of GJ-10 and the low pH levels.

Bacterial Growth Studies

Growth curves were developed for the *Burkholderia* isolate and GJ-10 (Figures A.4 and A.5). The *Burkholderia* isolate was cultured in R2A broth as it did not grow well in in tryptose or Luria broth. GJ-10 was cultured in nutrient-rich tryptose broth. The growth was monitored in an autoclaved cuvette containing 3 mL of growth medium based on absorbance at 600 nm, as presented in the Materials and Methods section. The cuvettes were inoculated with bacteria from agar plates in triplicate and monitored for absorbance readings.

The *Burkholderia* isolate grew much more slowly than either GJ-10 or the yellow *Sphingomonas* isolate. The growth of GJ-10 experienced a 4-hour lag phase and reached the end of its exponential growth phase, at an optical density of about 0.700, in approximately 24 hours. The *Burkholderia* species exhibited a much longer lag phase of about 48 hours. It reached the end of its exponential growth phase at the same optical density as GJ-10 (0.700), however, this point was not reached for approximately one week. The time required for bacteria to complete a growth cycle is highly variable and is dependent on both nutritional and genetic factors (Madigan et al. 2000). Upon harvesting, approximately, 2.5 and 1.3 g (wet weight) of GJ-10 and the *Burkholderia* isolate, respectively, were isolated per 500 mL of broth.

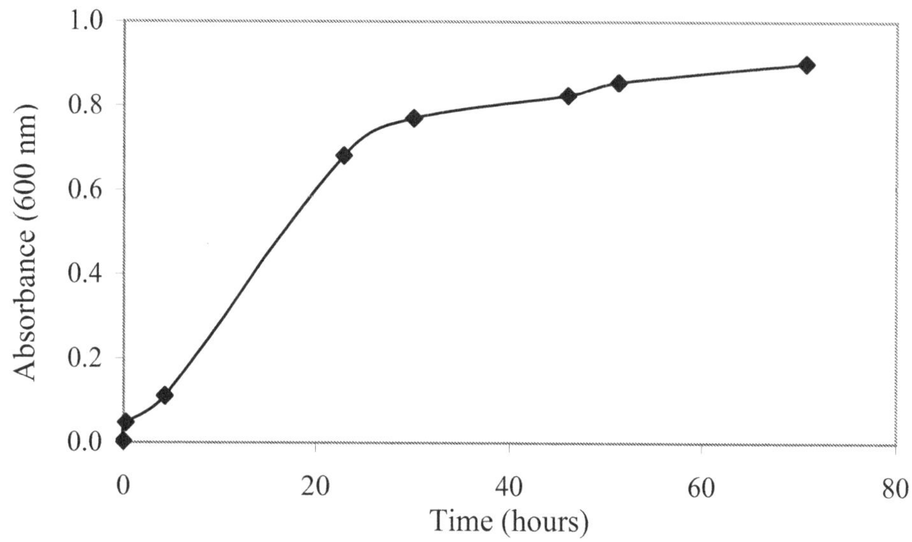

Figure A.4 The growth curve for *Xanthobacter autotrophicus*, GJ-10

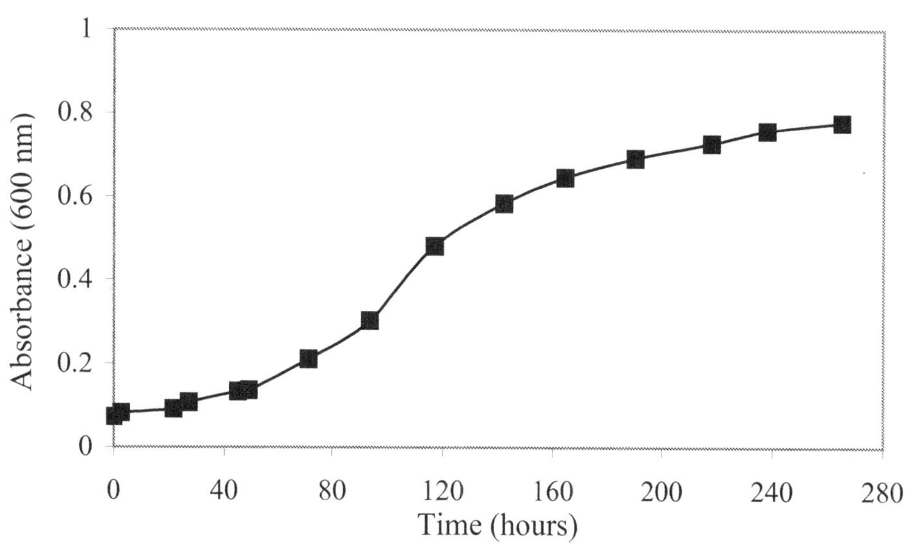

Figure A.5 The growth curve for the *Burkolderia* isolate

Protein Purification

The purification of the GJ-10 haloacid enzyme was attempted first since it was to serve as the control in subsequent studies. During the DEAE column step, the GJ-10 enzyme eluted with 0.1 M ammonium sulfate, as per Keuning et al. (1985). The active fractions were combined and tested on the S-200 size exclusion column. The eluants were placed in a 96-well plate with reaction buffer containing 10 mM Cl_2AA to determine which samples contained activity, as described in the Materials and Methods section. The anticipated color change of the indicator from red to yellow was observed in the fraction that eluted at 76.4 mL, which could be associated with the GJ-10 dehalogenase enzyme. Figure A.6 depicts a size exclusion calibration curve, using blue dextran, myglobin and Vitamin B_{12} as standards.

From this calibration curve, a rough estimate of the native molecular weight of this enzyme was determined to be approximately 30,000 Daltons. van der Ploeg has characterized the actual size of the native GJ-10 enzyme as being 36,000 Daltons (Slater 1997). It is suspected that one source of error might be due to the size of the column used in this experiment. For purposes of calculating total protein for GJ-10, the subunit molecular weight of 27,469 Daltons was used since it is the current value listed with the National Center for Biotechnology Information (NCBI) database (2004).

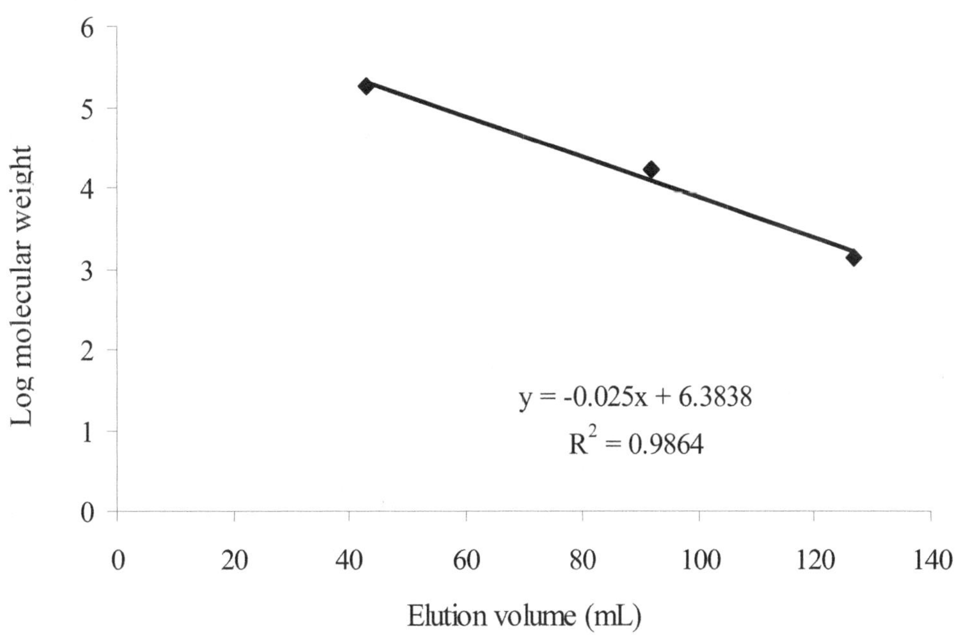

Figure A.6 Size exclusion for GJ-10 enzyme

The purification scheme and concentration at each step for haloacid dehalogenase from GJ-10 are shown in Table A.3, and in Table A.4 for *Burkholderia* isolate. The initial eluant fraction is defined as the crude extract. The parameters described in Tables A.3 and A.4 were defined as follows:

- The total protein at each step was calculated as the molar concentration of the enzyme (absorbance divided by the extinction coefficient) multiplied by the volume of the eluant and the molecular weight of the enzyme.
- The total activity is expressed in milliunits (mU), where a unit is defined as the amount of enzyme required to catalyze the formation of 1 μmol of halide per minute.
- The specific activity is defined as the amount of activity (mU) per mass of enzyme (mg).
- The yield is defined as the ratio of the total activity of the purified and crude extracts.
- The purification factor is defined as the ratio of the specific activities of the purified and crude extracts.

Table A.3 suggests that a significant amount of enzyme was lost in the first purification step (using 65-percent ammonia sulfate) and the specific activity (i.e., effective concentration) actually decreased, as shown by the fairly low yield (39 percent) and the purification factor less than one (0.83). However, in the next two purification steps, the decreases in yield were moderate and the specific activity greatly increased. In particular, a great increase in the specific activity resulted from the DEAE-cellulose step. The final purification step resulted in a complete loss of the enzyme, as shown by the lack of detectable activity. The purification procedure was repeated for the proteins isolated from the *Burkholderia* isolate. To calculate the total protein concentrations, the molecular weight of 29,939 Daltons was obtained from the NCBI database for *Burkholderia cepia*. The extinction coefficients for both GJ-10 and the *Burkholderia* species could be estimated using their published NCBI protein sequences for the haloacid enzyme and the ExPASy proteomics server using Equation A.5, which resulted in values of 29,870 and 22,310 $cm^{-1}M^{-1}$, respectively.

Table A.3
Purification scheme of 2-haloacid dehalogenase from GJ-10

Step	Total protein (mg)	Total activity (mU)	Specific activity (mU/mg)	Yield (percent)	Purification factor
Crude extract	54.8	5.94	0.11	100	
Ammonium sulfate 65 percent	25.9	2.33	0.09	39	0.83
Ammonium sulfate 35 percent	6.79	2.04	0.30	34	2.8
DEAE-cellulose	0.72	1.37	1.9	23	17.7
Sephacryl S-200	0.25	ND*	-	-	-

*ND indicates no detectable enzyme activity

Table A.4
Purification scheme of 2-haloacid dehalogenase from the *Burkholderia* isolate

Step	Total protein (mg)
Crude extract	24.9
Ammonium sulfate 65 percent	9.47
Ammonium sulfate 35 percent	8.72
Dialysis	3.16
DEAE-cellulose (0.3 M)	0.24
DEAE-cellulose (0.4 M)	0.12
DEAE-cellulose (0.5 M)	0.19

Standard Curve for Indicator Activity Analysis

A standard curve was produced following the method of Holloway et al. (1998) by adding increasing amounts of HCl, as presented in the Materials and Methods section, using bromocresol purple as the indicator. This method was then tested in triplicate with the partially purified GJ-10 enzyme and 4 to 30 mM Cl_2AA in order to establish an ideal substrate concentration for activity. For the GJ-10 enzyme, the highest activity that was observed was 0.0074 Unit which occurred using 10 mM Cl_2AA. To limit pH variations in this method, the indicator was changed to incorporate phenol red and a new standard curve was constructed. The absorbance was plotted versus the amount of added HCl. The resulting equation was the following:

$$y = -6.3(+/-0.4)\ x + 0.95(+/-0.04) \qquad (R^2 = 0.98) \qquad (A.6)$$

Control experiments were conducted, with only the reaction buffer, indicator, enzyme, but without Cl_2AA. No color change was observed in any of these tests. This same procedure was then used for the analysis of the haloacid enzyme from the *Burkholderia* isolate.

Protein Purification of Crude Dehalogenase Enzymes from GJ-10 and the Burkholderia Species

In the DEAE purification steps, the *Burkholdera* haloacid dehalogenase enzyme eluted at 0.3 to 0.5 M ammonium sulfate, which was slightly higher than the 0.1-M elution for the GJ-10 enzyme. This slight difference would suggest that the *Burkholderia* haloacid enzyme has a different genetic make-up than the GJ-10 enzyme, even though they are both in the 1L 2-halogenated alkanoic acid hydrolytic dehalogenase class (Slater et al. 1997).

Aliquots of 4 μL of the eluted fractions were placed into a 96-well plate containing reaction buffer with phenol red as the indicator and 10 mM Cl_2AA, as described in the Materials and Methods section. Three distinct regions from the eluted fractions gave positive color changes from red to yellow, corresponding to the elution from 0.3, 0.4, and 0.5 M ammonium sulfate. The occurrence of three specific regions would suggest the presence of three distinctly different dehalogenase enzymes. These three regions corresponding to potentially active fractions were combined for each of the individual areas and treated as separate samples. The following enzyme fractions were further characterized using SDS-PAGE:

- Samples of the crude GJ-10 enzyme
- Samples of the GJ-10 enzyme from the DEAE elution
- Samples of the GJ-10 from the size exclusion study
- Samples of the crude *Burkholderia* species enzyme
- Samples of the three *Burkholderia* fractions collected from the DEAE purification

The lanes for the DEAE eluant that were collected from the *Burkholderia* species were very difficult to visualize when stained with Coomassie due to low concentrations. The gel was then restained with silver nitrate (Figure A.7). In the silver nitrate gel, Lanes 1 and 2 (GJ-10) show a band at a level consistent with the 28,200 Dalton standard soybean trypsin inhibitor. van der Ploeg found the molecular mass of the purified enzyme from GJ-10 to be approximately 28,000 Daltons (Slater et al. 1997).

Interestingly, there were no bands present for any of the samples from the *Burkholderia* isolate (Lanes 5 to 8) in the range of 25,000 to 28,000 Daltons, which is the mass range reported for the 1L 2-halogenated alkanoic acid hydrolytic dehalogenase class (Kuihara et. 2000). From the multiplicity of other bands in the gel, it was rather obvious that the enzymes from the *Burkholdera* isolate were still only partially purified.

Whole-Cell Assay for Haloacid Speciation Using the Colorimetric Method

A whole-cell assay for GJ-10, the *Sphingomonas* isolate and the *Burkholderia* isolate gave positive results for specific activity with ClAA, Cl_2AA, BrAA, Br_2AA, BrClAA, MCAN, and DCAN. GJ-10 also gave positive results for TCAN as the substrate but Cl_3AA and Br_3AA were not suitable substrates for any of the three bacterial species. GJ-10 exhibited significantly higher activities for Cl_2AA, Br_2AA and BrClAA than the *Sphingomonas* and *Burkholderia* isolates at a 10 mM concentration (Table A.5). The colorimetric test using GJ-10 with various substrates was rather rapid and easily observed visually (Figure A.8), whereas the *Sphingomonas* and *Burkholderia* isolate responses were more difficult to visualize apparently due to a lower rate of reaction with the substrates. GJ-10's activity for Cl_2AA, Br_2AA and BrClAA was 57.3, 95.3 and 162 Units, respectively, compared to the 3 Units or less of activity observed for the *Sphingomonas* and *Burkholderia* isolates. One unit of dehalogenase activity was defined as the change of absorbance at OD_{570} per minute per mL, per absorbance at OD_{600}. This trend in activity for the various haloacids suggests that these bacterial enzymes prefer the X_2AAs to the XAAs or X_3AAs.

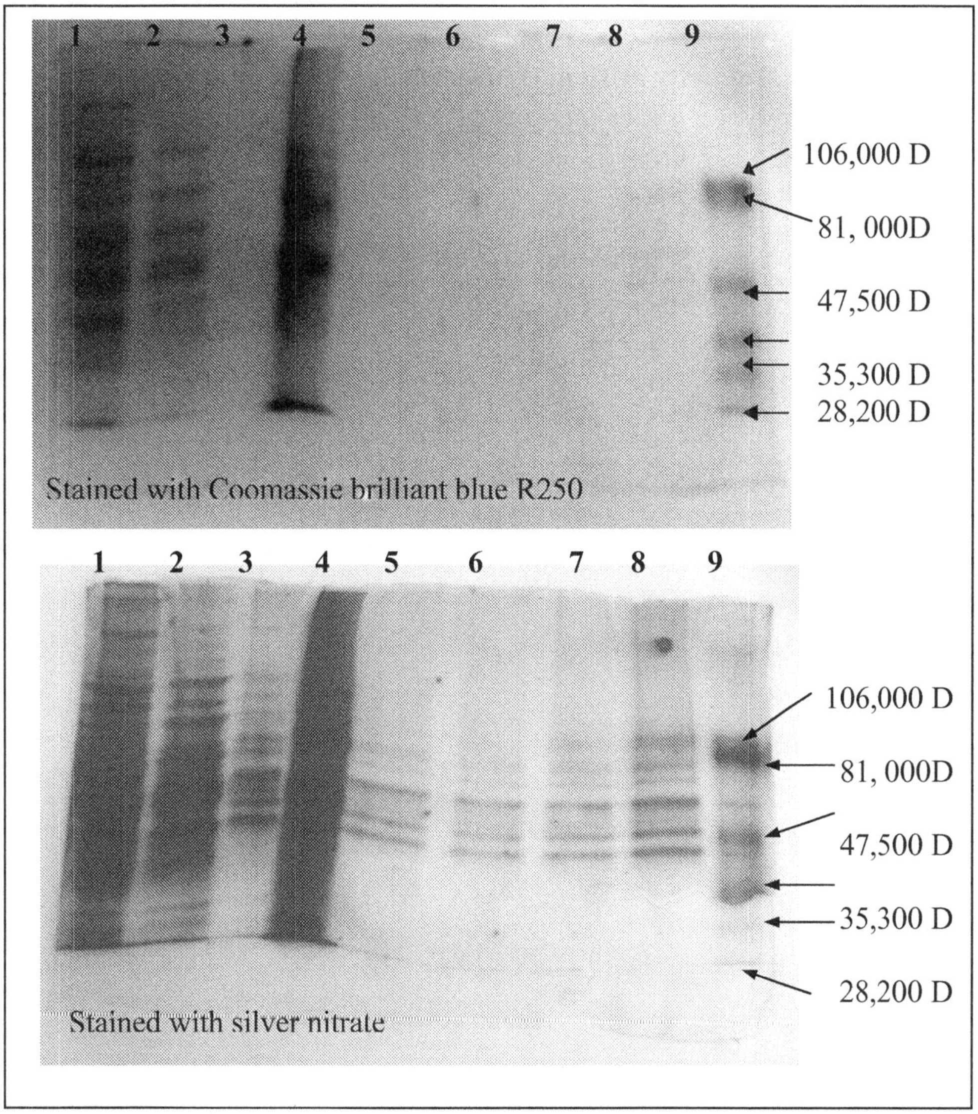

Figure A.7 SDS-PAGE gel. The wells contain the following:

1. Crude GJ-10 enzyme
2. GJ-10 eluent from DEAE
3. GJ-10 eluent from size exclusion chromatography
4. Crude enzyme from *Burkholderia* species
5. 0.3 M ammonium sulfate DEAE eluent for *Burkholderia* species
6. 0.4 M ammonium sulfate DEAE eluent for *Burkholderia* species

7 & 8. 0.5 M ammonium sulfate DEAE eluent for *Burkholderia* species

9. Molecular weight ladder, Phosphorylase B 106,000 Daltons; Bovine serum albumin 81,000 Daltons; Ovalbumin 47,500 Daltons; Carbonic anhydrase 35,300 Daltons; Soybean trypsin inhibitor 28,200 Daltons; Lysozyme 20,800 Daltons.

Table A.5
Comparison of whole-cell activity at 10 mM of substrate (Units of dehalogenase activity)

	GJ-10 activity	*Burkholderia* isolate activity	*Sphingomonas* isolate activity
ClAA	5.5	0.1	1.6
Cl_2AA	57	1.5	2.0
BrAA	2.5	1.1	3.6
Br_2AA	95	1.2	2.0
BrClAA	162	0	0.5
MCAN	3.1	0.9	1.1
DCAN	1.8	1.0	3.0
TCAN	1.4	-	-

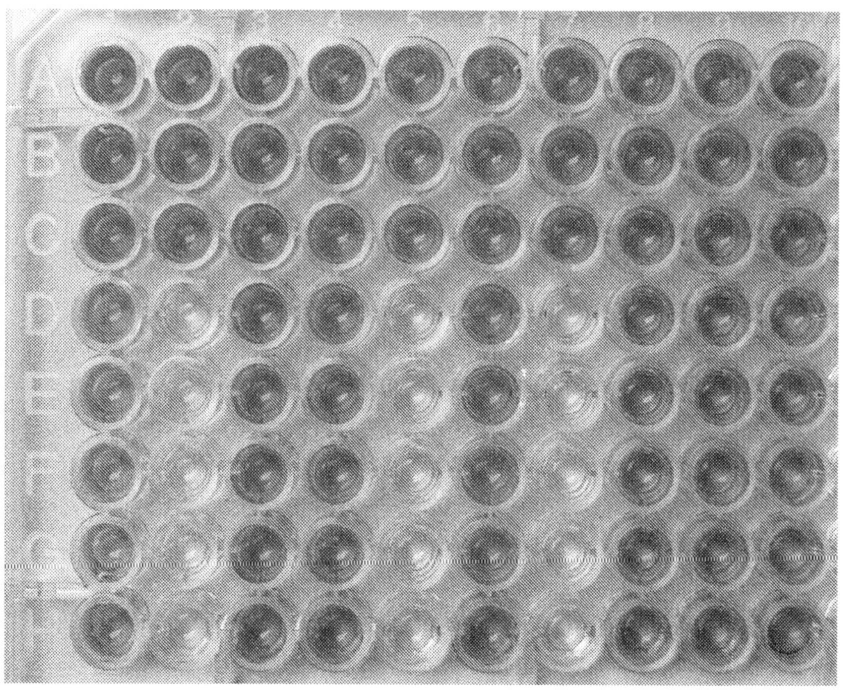

Figure A.8 Typical substrate specificity test for GJ-10. Rows A, B, and C contain bacteria in buffer with phenol red. Rows D thru H contain test substrate reactions. The substrates, at 10 mM concentrations, are in the columns as follows: 1 ClAA, 2 Cl_2AA, 3 Cl_3AA, 4 BrAA, 5 Br_2AA, 6 Br_3AA, 7 BrClAA, 8 MCAN, 9 DCAN, 10 TCAN

Stereochemically speaking, the X_3AAs such as Cl_3AA and Br_3AA may present a more difficult site for enzymatic hydrolysis due to the sheer bulk of the three halogens attached to the alpha carbon. Assuming that a typical Sn2 type of reaction is involved in the hydrolysis of the HAA substrates, this might be considered to be a stereochemically controlled reaction which may constitute considerable stearic hindrance to the approaching enzyme.

The XAAs, such as ClAA, appear to be less suitable to this phenomenon compared to the X_2AAs such as Cl_2AA and Br_2AA. This may be explained by the added inductive effect of the two halogens, which would make the alpha carbon of the HAA more electron deficient, and thus, more attractive to the enzyme's nucleophilic attack.

GJ-10 was shown to exhibit a higher degree of activity in Br_2AA compared to Cl_2AA, which is in agreement with the study performed by van der Ploeg et al. (1991) with GJ-10. The stability of the carbon-halogen bond increases with increasing electronegativity of the halogen and therefore the ease of dehalogenation proceeds in the following order: I > Br > Cl > F (Hardman 1991). In addition, bond dissociation energy for the chlorine bonded to the carbon of the ClAA is 311 kJ/mol and the bromine bonded to the chlorine of BrAA is 257 kJ/mol (Lagoa et al. 2001). The larger the bond dissociation energy, the stronger the bond.

However, BrClAA had a higher activity than either Br_2AA or Cl_2AA in the GJ-10 study. This observation may result from a combination of the inductive effect of the chlorine atom and the greater leaving group potential of the bromine in the case of BrClAA.

GJ-10 was more effective than either of the two isolates (*Sphingomonas* and *Burkholderia*) for dehalogenation of HAAs; and the *Sphingomonas* isolate was more effective with all HAAs than the *Burkholderia* isolate. Trends were noted that may indicate subtle differences in mechanisms between the GJ-10 haloacid dehalogenase and those of the two other species. The trend for GJ-10 suggests a combination of inductive and stearic effects. With GJ-10, the order of reactivity appears to be: BrClAA > Br_2AA > Cl_2AA > ClAA > BrAA. The order of reactivity of the *Sphingomonas* isolate was quite different: BrAA > Br_2AA = Cl_2AA > ClAA > BrClAA, whereas the *Burkholderia* isolate exhibited an even more diverse order of substrate specificity: Cl_2AA > Br_2AA = BrAA > ClAA > BrClAA.

The potential for bacterial degradation of certain HANs was also examined. Interestingly, it appeared that the HAAs were better substrates than the HANs for all three bacterial species. The hydrolytic mechanism that one would suppose for the HANs should be quite similar to that proposed for the haloacid dehalogenases. However the final metabolic product would not be oxalic acid, as it is the case in the HAA degradation, but rather one might see an increase in available cyanide. Considering DCAN as a possible substrate and assuming that the degradation mechanism is similar to that for the HAAs, then a likely reaction scenario might be represented in the reaction scheme presented in Figure A.9.

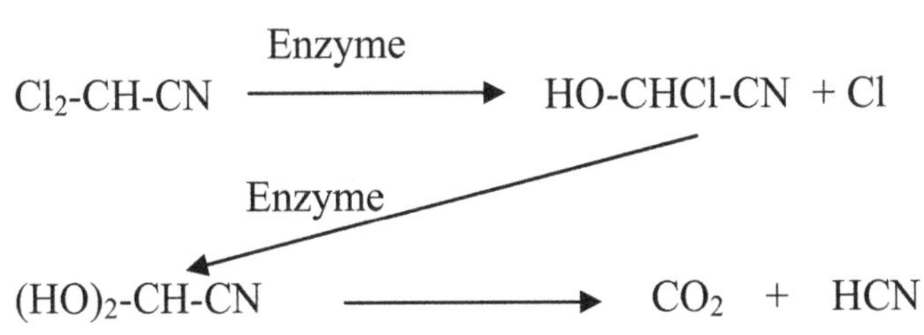

Figure A.9 Proposed reaction scheme for DCAN degradation

From this whole-cell activity data, it would appear that GJ-10 is more effective at degrading the HANs than either of the two isolates, with the *Sphingomonas* isolate exhibiting greater efficacy than the *Burkholderia* isolate. GJ-10's order of reactivity was found to be MCAN > DCAN > TCAN, whereas the *Burkholderia* and *Sphingomonas* isolates exhibited the same relative order of reactivity, which was DCAN > MCAN. GJ-10 was apparently capable of using TCAN as a suitable substrate, whereas the other two bacterial species were ineffective in the degradation of this substrate. Once again the stearic hindrance associated with the TCAN may explain the ineffectiveness of the other two bacterial species, but this same stearic assertion is apparently not completely effective against GJ-10.

Rate Studies

Kinetic studies were conducted on the GJ-10 control and the *Burkholderia* isolate to determine the Michaelis constant using Michaelis-Menton and Lineweaver-Burk plots, as described in the Materials and Methods section. These studies were not conducted on the *Sphingomonas* isolate; sufficient dehalogenase enzyme could not be extracted from the isolate, likely due to the smaller quantities of this bacteria that could be cultured.

Kinetic experiments were first done with GJ-10 in order to optimize the reaction conditions. The Michaelis-Menton kinetics suggest that GJ-10 exhibited the same order of activity trend that was observed during the specificity testing: BrClAA > Br_2AA > Cl_2AA. The decreased activity with Cl_2AA may result from the greater leaving group potential associated with the bromine component in each substrate, due to decreased electronegativity and lower bond energy for the brominated compounds, as discussed above. Increased rates were observed as the substrate concentration increased and then a constant was observed as the maximum velocity, V_m, was approached (ranging from 49 to 73 Units). The estimated V_m occurred at substrate concentrations of 30, 1 and 10 mM for BrClAA, Br_2AA and Cl_2AA, respectively. The Lineweaver-Burk plot for the X_2AAs for GJ-10 is presented in Figure A.10, and provided for more reliable V_m values.

The maximum velocity appeared to decrease significantly in the GJ-10 studies with the XAAs (ClAA and BrAA). This may be due to the lesser inductive effect afforded by the single halogen attached to the alpha carbon. This study shows the V_m to occur at about 50 mM for both ClAA and BrAA with activities of 22 and 11 Units, respectively, as compared to V_m values of 49 Units or greater for the X_2AAs. Values of K_m based on these data were 10 and 20 mM for ClAA and BrAA, respectively (Figure A.11). The Lineweaver-Burk plot however gives a K_m of 3.4 and 11.6 mM for ClAA and BrAA, respectively (Figure A.12). The trend appears to be consistent, but these K_m values were higher than those of the X_2AAs suggesting that perhaps the substrate binding may be better for these compounds, but the reaction occurs at a much slower rate.

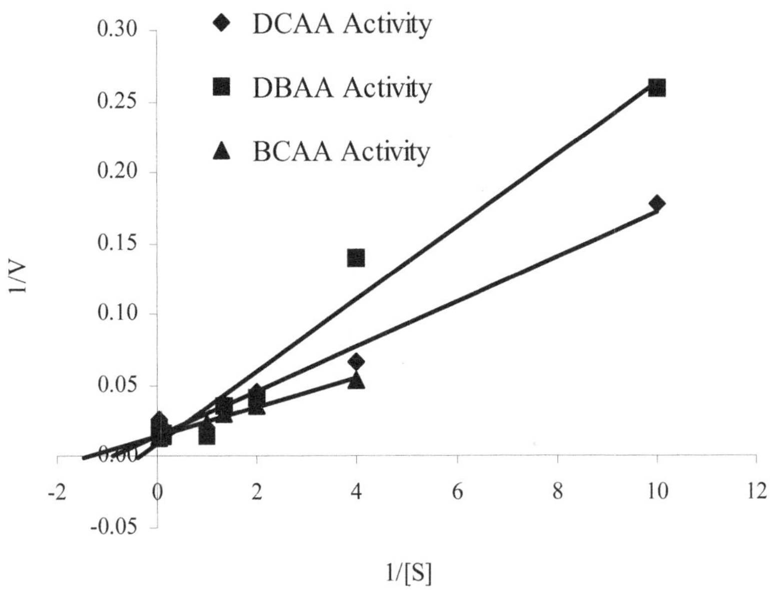

GJ-10	Data range (mM)	Slope	R^2 value	V_m (Unit)	K_m (mM)
Cl_2AA	0.1 to 30	y=0.0158x + 0.0148	0.9812	67.57	1.07
Br_2AA	0.1 to 50	y=0.0254x + 0.0093	0.9657	103.1	2.61
BrClAA	0.25 to 50	y=0.0103x + 0.0143	0.9937	69.90	0.72

Figure A.10 Lineweaver-Burk plot of GJ-10 activity with X_2AAs

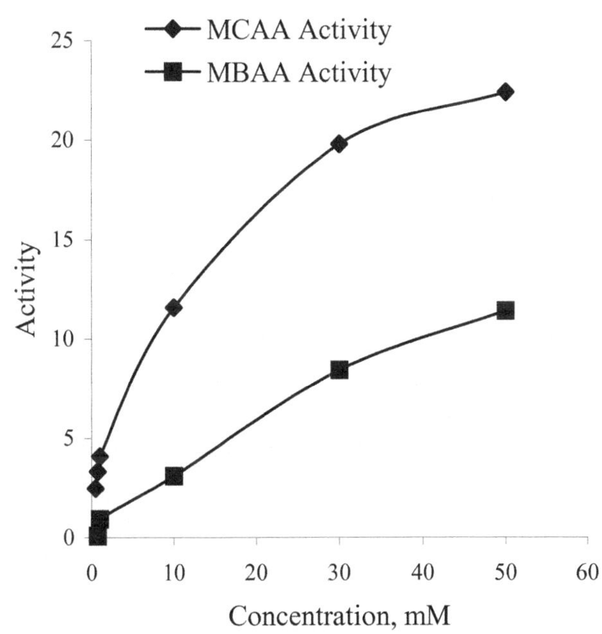

GJ-10	V_m (Unit)	$1/2V_m$ (Unit)	K_m (mM)
ClAA	22	11	10
BrAA	11	6	20

Figure A.11 Michaelis-Menton plot of GJ-10 activity with XAAs

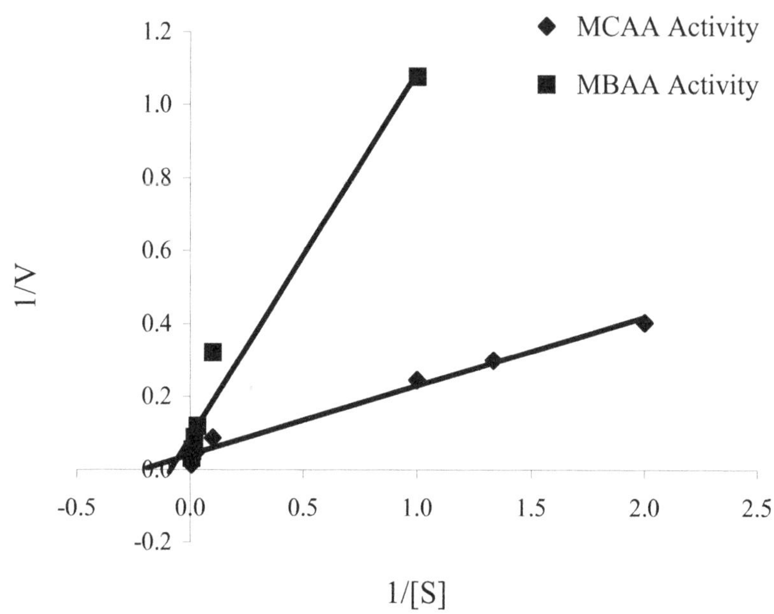

GJ-10	Data range (mM)	Slope	R^2 value	V_m (Unit)	K_m (mM)
ClAA	0.5 to 50	y=0.1799x + 0.0537	0.9935	18.62	3.400
BrAA	1 to 50	y=1.0035x + 0.0861	0.9704	11.61	11.66

Figure A.12 Lineweaver-Burk plot of GJ-10 activity with XAAs

The activity of the *Burkholderia* species was then examined through the use of Michaelis-Menton and Lineweaver-Burk plots. The Michaelis-Menton plot exhibited maximum velocity at 10, 30 and 10 mM for BrClAA, Br_2AA and Cl_2AA, respectively (Figure A.13). These data correspond to activities that are significantly lower than GJ-10's at 2.9, 5.3 and 5.2 Units for BrClAA, Br_2AA and Cl_2AA, respectively. The resulting K_m values for BrClAA, Br_2AA and Cl_2AA obtained by the Michaelis-Menton plot were 0.90, 0.85 and 1.25 mM, respectively. The Lineweaver-Burk plot for these X_2AAs gave K_m values of 1.11 mM for BrClAA, 0.56 mM for Br_2AA and 5.37 mM for Cl_2AA (Figure A.14). Although the magnitudes of the K_m values differed, the trends were the same (i.e., activity for Cl_2AA > BrClAA > Br_2AA).

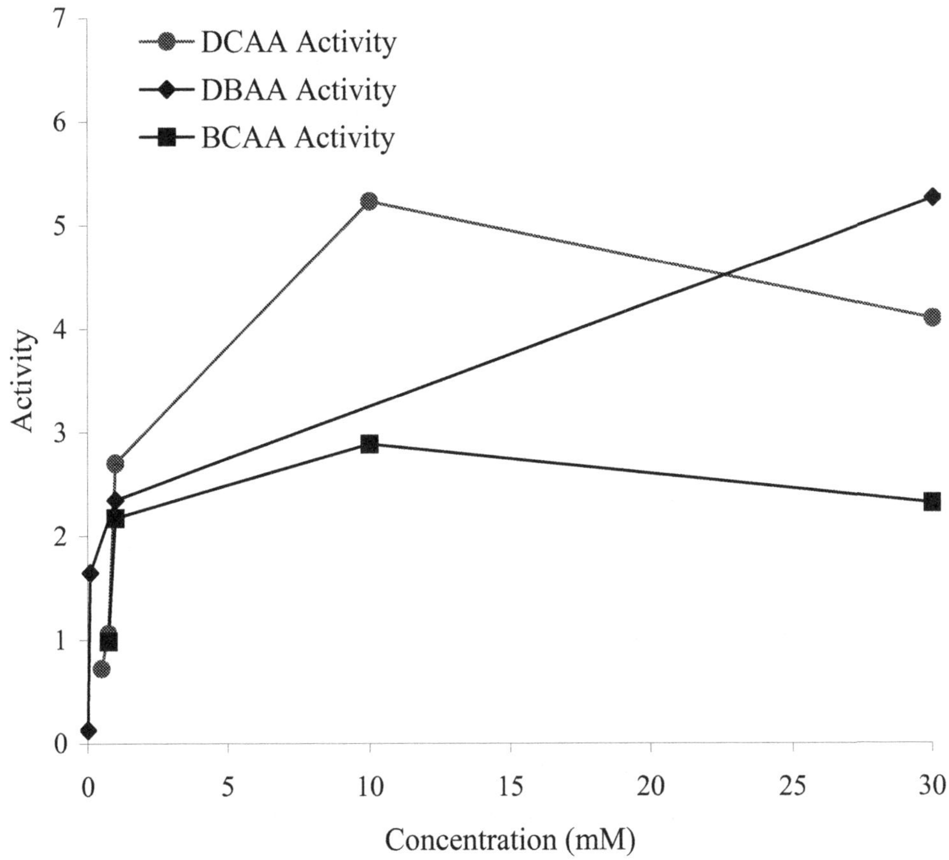

Burkholderia species	V_m (Unit)	$1/2V_m$ (Unit)	K_m (mM)
Cl_2AA first phase	5.23	2.75	1.25
Br_2AA first phase	5.26	2.63	0.85
BrClAA first phase	2.89	1.45	0.90

Figure A.13 Michaelis-Menton plot of *Burkholderia* isolate activity with X_2AAs

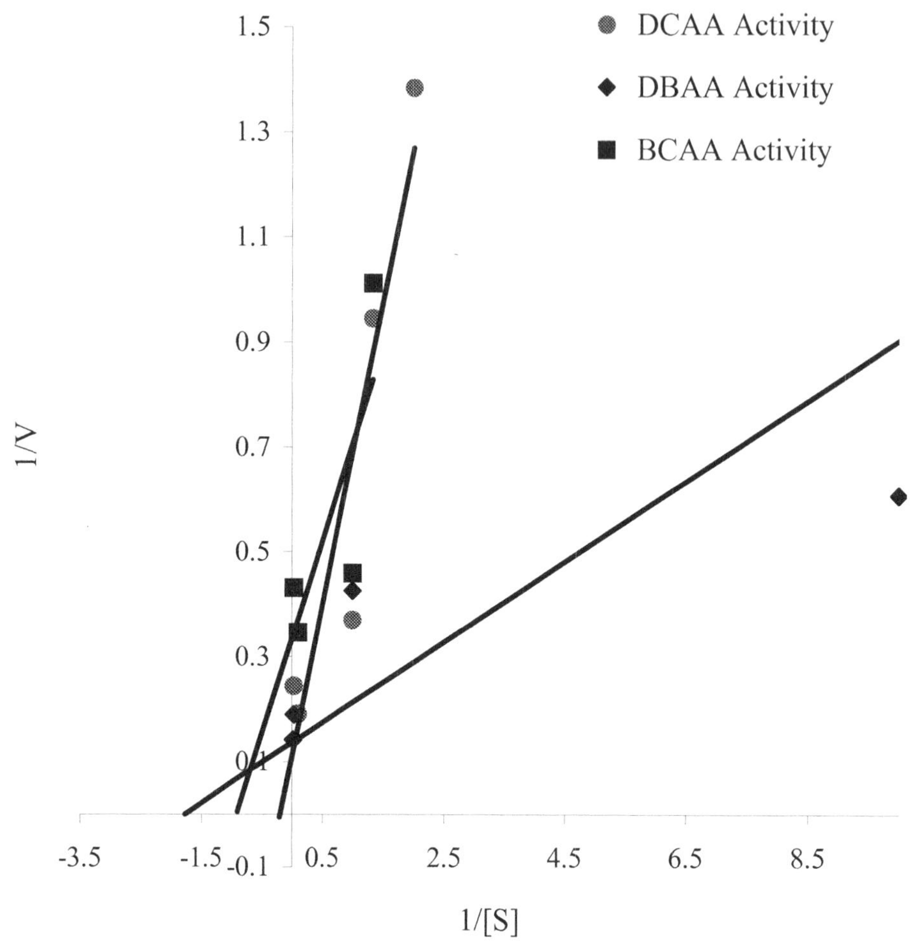

Burkholderia species	Data range (mM)	Slope	R^2 value	V_m (Unit)	K_m (mM)
Cl_2AA	0.5 to 30	y=0.5805x + 0.1082	0.8766	9.24	5.37
Br_2AA	0.01 to 50	y=0.0766x + 0.1363	0.9970	7.34	0.56
BrClAA	0.75 to 30	Y=0.3708x + 0.3335	0.6316	3.00	1.11

Figure A.14 Lineweaver-Burk plot of *Burkholderia* species activity with X_2AAs

The activity with ClAA and BrAA was lower than for the X_2AAs with the *Burkholderia* isolate, which may once again be due to the lesser inductive effect in these XAAs. The Michaelis-Menton shows V_m to occur at 1 mM for ClAA and 10 mM for BrAA, which translated to activities of 1.7 and 7.4 Units, respectively. This gives K_m values of 0.9 mM for ClAA and 5.2 mM for BrAA (Figure A.15). However, the Lineweaver-Burk plot gives K_m values of 0.41 mM for ClAA and 0.28 mM for BrAA (Figure A.16). This exceptionally low level of activity and variance between the K_m values suggest that this method of detection for such minimal activity may be beyond an accurate limit of detection. Tsang et al. (1988) reported K_m values of 120 μM for ClAA and 92 μM for BrAA for a dehalogenase extracted from *Burkholderia cepacia* MBA4.

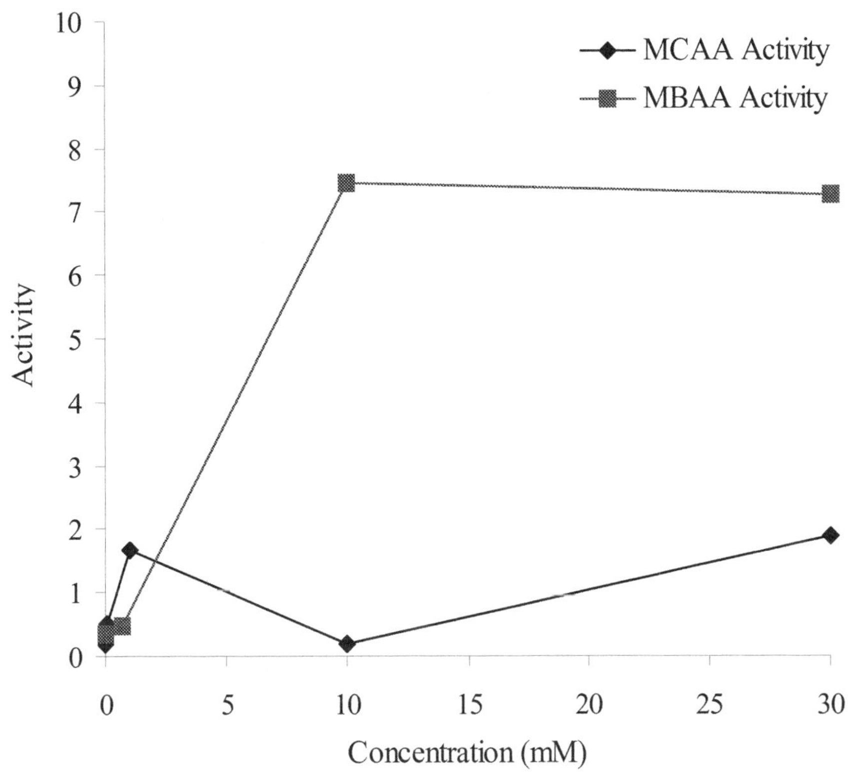

Burkholderia species	V_m (Unit)	$1/2V_m$ (Unit)	K_m (mM)
ClAA	1.67	0.84	0.9
BrAA	7.4	3.7	5.20

Figure A.15 Michaelis-Menton plot of *Burkholderia* species activity with XAAs

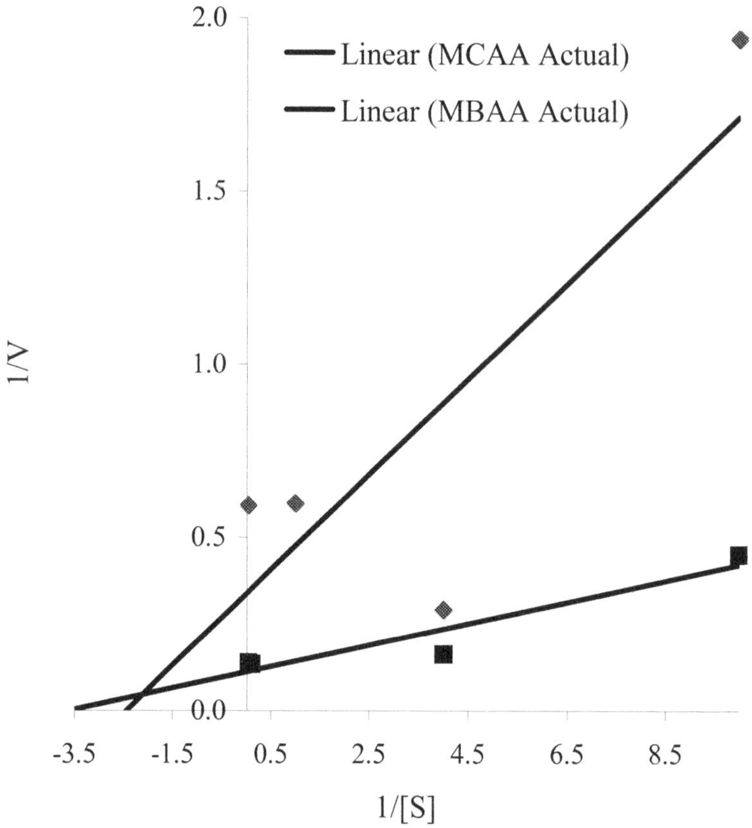

Burkholderia species	Data range (mM)	Slope	R^2 value	V_m (Unit)	K_m (mM)
ClAA	0.1 to 30	y=0.1375x + 0.3381	0.7022	2.96	0.41
BrAA	0.1 to 30	y=0.0309x + 0.1120	0.8986	8.93	0.28

Figure A.16 Lineweaver-Burk plot of *Burkholderia* species activity with XAAs

MCAN, DCAN and TCAN were examined to determine the level of activity with the dehalogenase enzymes of both GJ-10 and the *Burkholderia* isolate. The Michaelis-Menton plot exhibited maximum velocity at 10 mM for MCAN and DCAN with both bacteria (Figure A.17). GJ-10 did not appear to have activity with DCAN at relatively low concentrations, but produced an activity of 8.11 Units. No K_m values could be calculated for GJ-10 using DCAN as the substrate. GJ-10 exhibited an activity of 4.44 Unit with MCAN, and the *Burkholderia* isolate produced activities of 1.58 Unit with MCAN and 2.40 Unit with DCAN. Surprisingly, the K_m values calculated from the Michaelis-Menton plot were all 0.85 mM. The values from the Lineweaver-Burk plot gave K_m values of 6.73 mM for GJ-10 with MCAN, and 4.30 mM with MCAN and 0.54 mM with DCAN for the *Burkholderia* species (Figure A.18).

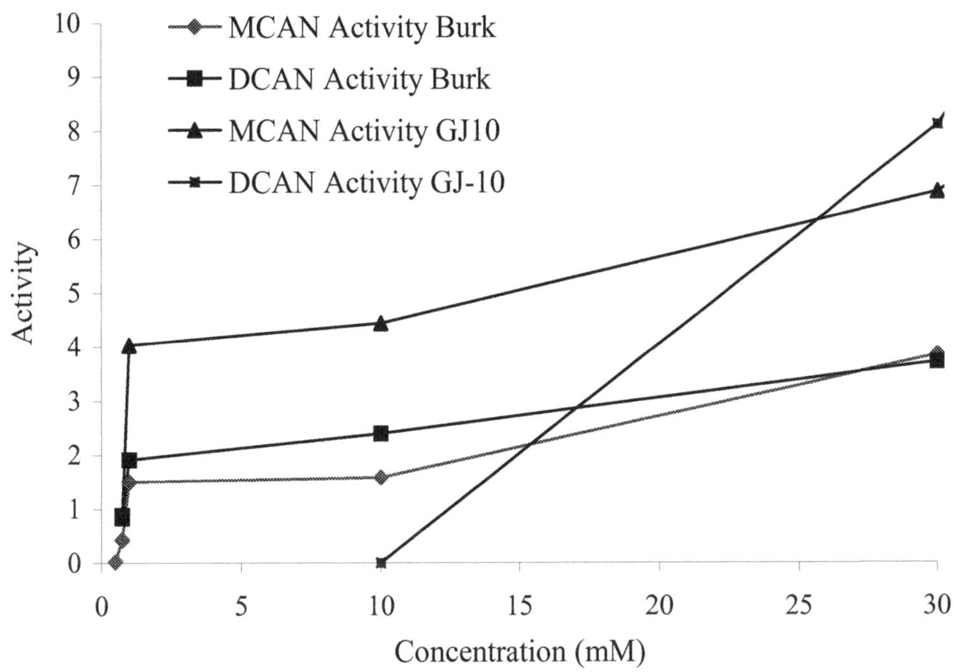

	V_m (Unit)	$1/2V_m$ (Unit)	K_m (mM)
ClAN *Burkholderia*	1.58	0.79	0.85
Cl_2AN *Burkholderia*	2.40	1.20	0.85
MCAN GJ-10	4.44	2.22	0.85
DCAN GJ-10	-	-	-

Figure A.17 Michaelis-Menton plot of *Burkholderia* isolate and GJ-10 activity with HANs

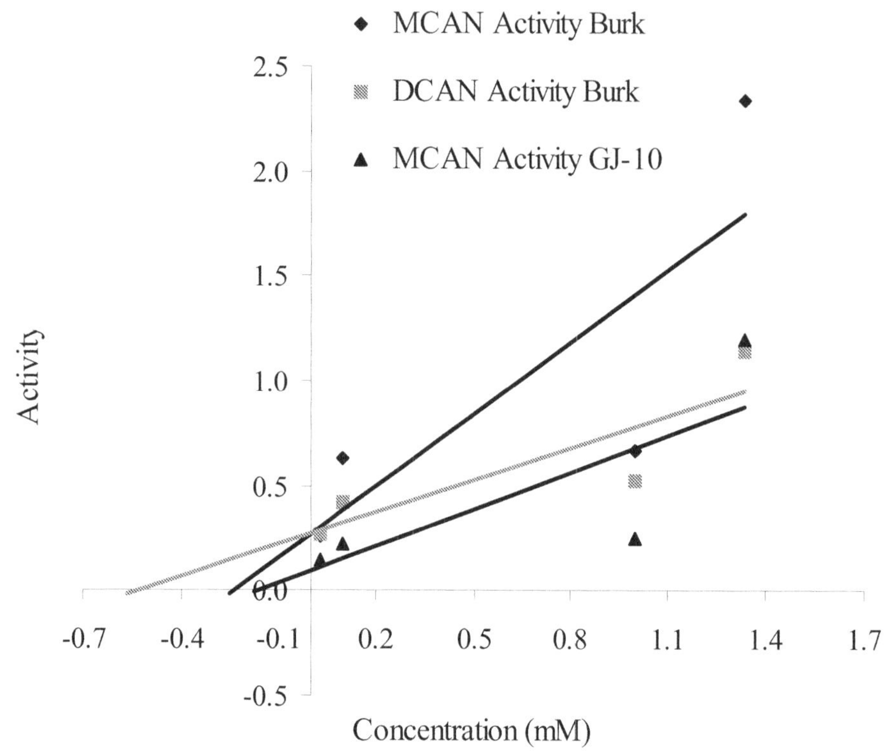

	Data range (mM)	Slope	R^2 value	V_m (Unit)	K_m (mM)
MCAN *Burkholderia*	0.75 to 30	y=1.1484x + 0.2671	0.6445	3.74	4.30
DCAN *Burkholderia*	0.75 to 30	y=0.5085x + 0.2735	0.7472	3.66	0.54
MCAN GJ-10	0.75 to 30	y=0.5937x + 0.0882	0.6005	11.34	6.73
DCAN GJ-10	-	-	-	-	-

Figure A.18 Lineweaver-Burk plot of *Burkholderia* isolate and GJ-10 activity with HANs

Differences in K_m values between the Michaelis-Menton plot and the Lineweaver-Burk plots may be a result of the plots themselves. The K_m values for the Michaelis-Menton plots were visually estimated, which undoubtedly added a source of error. Since these studies were generally conducted with relatively small concentrations of substrate, small errors in measured activity may have lead to large errors when the inverse was plotted, which may be reflected in errors in K_{max} and V_{max} values in the Lineweaver-Burk plots.

Part 2 – Study of the Haloacid Degradation Mechanism

One of the objectives of the research described in this report was to study the dehalogenase mechanism and the delineation of the degradation products. The mechanisms were studied using an HPLC method adapted from Narayanan et al. (1999), but using a Bio-Rad FAST ACID column, as described in the Materials and Methods section. The method was able to detect seven HAAs (HAA7: XAAs, X_2AAs, Br_3AA, and Cl_3AA), as well as the three main degradation products of ClAA (i.e., glycolic acid, glyoxylic acid and oxalic acid). Unfortunately, the proposed degradation or metabolic products of the more highly substituted HAAs, such as Cl_2AA and Br_2AA, are not commercially available and their HPLC profiles could not be established. In general, this method provided a detection limit for the various HAAs in the range of 1 to 5 mg/L.

This analytical method was first used to assess the hydrolytic potential of the various HAAs, which could be significant to this study. Samples of each of the HAA7 were examined for any extent of hydrolysis over a 72-hour period at pH 8.5 in dilution water. The only compound that exhibited extensive hydrolysis was Br_3AA, which degraded quite rapidly to give $CHBr_3$. All of the other HAAs were relatively inert to hydrolysis under these conditions for up to 72 hours.

In previous studies conducted by the authors of this report, samples of the crude enzyme fractions were used to follow the degradation pathway with various HAAs. In the current study, the sensitivity of the crude enzyme to the reaction conditions, especially the change in pH that accompanies the dehalogenation, created considerable problems. The use of the crude enzymes was eventually abandoned to focus on a whole-cell assay for these degradation mechanism studies. The use of whole cells may create some problems inherent in the absorption of the substrate and the release of any degradation metabolites back into the reaction media for analysis. The use of whole bacterial cells however seemed appropriate if not very relevant in view of their presence, as such, in the distribution system.

Experiments were conducted to validate the ability of the HPLC method to work on whole-cell extracts. Initially, the bacterial isolates were reacted with 20 mg/L of Cl_2AA in dilution water and the level of Cl_2AA in the reaction was monitored using HPLC until a significant decrease in the concentration was observed. The time required for significant decrease in Cl_2AA was 11 and 15 days, for the *Sphingomonas* and *Burkholderia* isolates, respectively. In addition, whole cells of the *Sphingomonas* isolate were stirred with a composite of 5 mg/L of each of the HAA7. There was a significant decrease in the levels of Cl_2AA, BrClAA and Br_2AA, as detected by HPLC.

Since the possible metabolites for the X_2AAs and X_3AAs were not commercially available, ClAA was first examined, which should give rise to glyoxylic, glycolic and oxalic acids as the enzymatic dehalogenation reaction proceeded. However, no degradation products were detected after reaction with whole cells of GJ-10 or the two bacterial isolates. This result was not surprising in view of the relatively low levels of activity (Table A.5) that were observed for ClAA and these three different bacteria.

In experiments using GJ-10, considerable amounts of unreacted Cl_2AA were present following a 90-minute reaction time. The relevant peaks detected included glyoxylic acid (9.46 minutes) and oxalic acid (6.28 minutes). There was also a peak at approximately 12.68 minutes that may be attributed to the hydroxychloroacetic acid, which would be the first step in the degradation of Cl_2AA. The low levels of these metabolites observed in the HPLC may reflect the problems inherent in the release of these metabolites from the bacterial cell.

There was also a very interesting peak at 16.02 minutes, which is consistent with the position of ClAA in the HPLC profile. This suggests that GJ-10 may also exhibit some level of reductive dehalogenation in the conversion of Cl_2AA to ClAA in the bacterial cells. ClAA was also observed as a metabolite in the HPLC studies of Narayanan et al. (1999) in the metabolism of Cl_2AA in rats and mice.

In the degradation studies with the *Burkholderia* isolate, the concentration of the HAA substrate needed to be considerably increased to see any levels of metabolites, possibly due to the apparent lower level of activity of *Burkholderia* species compared to GJ-10. Even at 1,000 mg/L of Cl_2AA, peaks of hydroxychloroacetic acid intermediate (12.59 minutes), glyoxylic acid (9.35 minutes) and oxalic acid (6.39 minutes) were barely detected.

Although it was anticipated that the *Sphingomonas* isolate would show a similar level of activity in the whole-cell assay, Cl_2AA degradation did not occur even at very high substrate concentrations. Many of the problems encountered during the studies with the *Sphingomonas* isolate may be traced to the purity of the culture.

SUMMARY AND CONCLUSIONS

Bacterial Studies

The major emphasis of this research has centered on the isolation, identification and properties of the bacteria that were found to be associated with distribution system locations that exhibited unexpected low levels of HAAs. These locations had low disinfectant residuals and high HPCs. From the many different bacterial isolates that were cultured from these sites, two specific bacteria were shown to be quite effective in the degradation of various HAAs. These two bacteria were cultured repeatedly in laboratory and have now been identified as a *Burkholderia* species and a *Sphingomonas* species. Another bacterial species, *Xantherobacter autotrophicus*, known as GJ-10, was used a control in this study and was shown to effectively degrade various HAAs and other halogenated compounds.

Research from this study showed limited success in isolating the dehalogenase enzymes from either the *Burkholderia* or the *Sphingomonas* isolates, primarily due to problems encountered in the scale up of the culture of both species. Because of the lack of significant amounts of these enzymes and certain problems that arose with the use of the ISE method for the detection of dehalogenase activity, the focus was re-centered on whole-cell studies to establish the speciation and scope of the activity of these two bacteria to degrade various HAAs.

A relatively simply and effective colorimetric screening method was developed to detect bacteria that were capable of degrading HAAs. This method was proven useful in determining the whole-cell activity and substrate specificity of all three bacterial species with various HAAs. The following conclusions regarding the relative effectiveness of the different bacteria to degrade certain halogenated DBPs were drawn using the colorimetric method:

- All three bacterial isolates were capable of degrading ClAA, Cl_2AA, BrAA, Br_2AA and BrClAA.
- None of the bacterial isolates were capable of effectively degrading Cl_3AA or Br_3AA, which may demonstrate a distinct stearic effect that inhibits the dehalogenase enzyme activity.
- There appeared to be a selectivity for the X_2AAs over the XAAs by these bacteria. This may be due to the basic nature of the enzymes affinity for the active site. The nucleophilic substitution center would be inductively activated by two halogens.
- The *Sphingomonas* species was more effective than the *Burkholderia* species in degrading HAAs. However GJ-10 appeared to be much more effective than either of the other two bacterial species in the degradation of these same HAAs.
- It appeared that the *Sphingomonas* species preferred the brominated substrates over the chlorinated substrates. The substrate specificity for the *Burkholderia* species would appear to be more diverse and less selective.
- GJ-10 exhibited a greater affinity for Br_2AA over Cl_2AA, which may be explained by the greater leaving-group potential of the bromide over the chloride in the rate determining step of the dehalogenation mechanism.
- All three bacterial isolates appeared capable of degrading MCAN and DCAN. The order of reactivity was the following: MCAN > DCAN > TCAN. The *Sphingomonas* species was more effective than the *Burkholderia* species with regard to HAN degradation, however GJ-10 appeared to be more effective than either of the other two isolates.
- The degradation of HANs by these bacteria is a relatively new and novel observation and warrants further study in view of the potential release of cyanide ion as a possible degradation by-product.
- The simple colorimetric method used in this study could be adaptable to future laboratory procedures involving the detection of bacterial species capable of HAA degradation in the distribution system. This would be useful to utilities for selecting monitoring locations for the IDSE requirements of the Stage 2 DBPR.

Haloacetic Acid Degradation Mechanism Studies

One of the objectives of this research was to examine the mechanism of action associated with HAA degrading bacteria, which was accomplished using an HPLC method. Hydrolysis was not taken into account in the degradation study, as a separate set of experiments demonstrated that hydrolysis was a concern only for Br_3AA, which was not included in the mechanistic study.

The majority of the experiments used Cl_2AA as the substrate, as no degradation products were observed during initial tests with ClAA. In experiments with GJ-10 enzyme extracts and Cl_2AA, two expected degradation products (glyoxylic and oxalic acids) were observed. There were two additional peaks that that may have been glycolic and ClAA. These very preliminary data suggest that the GJ-10 bacteria may be capable of some form of reductive dehalogenation as well as the expected hydrolytic mechanism. The glycolic acid peak could be arising from the hydrolysis of the ClAA produced in the reductive dehalogenation step.

In view of the sensitivity of the crude enzyme fractions to the conditions used in these studies (pH and temperature), the HPLC method was later adapted to the study of the bacterial dehalogenation of HAAs using whole bacterial cells from GJ-10 and the bacterial isolates. The results with GJ-10 were very similar to those for the crude enzyme extracts. The *Burkholderia* isolate only producted detectable levels of degradation products at very high Cl_2AA concentrations (1,000 mg/L). The degradation products observed were hydroxychloroacetic acid intermediate, glyoxylic acid and oxalic acid. No degradation products were observed in experiments using the *Sphingomonas* isolate.

In retrospect, it appeared the HPLC method was not sufficiently sensitive for the detection of degradation products. However, these studies did support the proposed mechanism for HAA degradation via a hydrolytic process, as the formation of the glycolic, glyxoylic and oxalic acids was observed.

APPENDIX B
BATCH INCUBATIONS IN BOTTLES

This appendix presents detailed results of the laboratory batch incubations in bottles that were conducted during several samplings to distinguish the effect of pipes (in the distribution system only) from the effect of water age (in the distribution system and bottle incubation) on disinfectant decay and DBP formation. These batch incubations also allow to distinguish between the effect of the suspended and fixed biomass (in the distribution system) from the effect of the suspended biomass alone (in the bottle incubation).

MATERIALS AND METHODS

For the batch incubations, water was collected in 1-L amber glass bottles at the POE to the distribution system and kept in darkness, head-space free. The bottles were kept at the same temperature as the water temperature measured at the POE by placing them in the laboratory in a sink, and letting tap water flow around them during the incubation period. The batches were incubated at NNWW and UNC for the samples collected in the distribution systems of Utilities A, C and E. For Utility B, the batches were incubated at the participating utility, whereas Utility D's batches were incubated at Carollo Engineers. After a pre-defined incubation period, one of the bottles was sacrificed and aliquoted in duplicate THM4 and HAA9 vials (containing the appropriate dechlor(am)inating and preservative agents), placed on ice packs, and sent to the respective laboratories for analysis. Temperature, pH, and chlorine residuals were also measured at the time of sampling. The same procedure was repeated after longer incubation periods to span the range of water ages observed in the distribution system.

RESULTS AND DISCUSSION

Free Chlorinated Systems

Two free chlorinated distribution systems were examined, Utilities A and B. Results are presented separately for THM4 and HAA9.

Free Chlorine Residual and Trihalomethanes

In Utility A, batch incubations were analyzed after 24, 72 and 168 hours (1, 3 and 7 days). These batch incubation results could not be directly compared with results observed from the distribution system as the incubation periods did not exactly match. Nonetheless, the free chlorine residuals decreased during batch incubation, generally at the same rate as in the distribution system, or at a slightly greater rate during some samplings. These results are surprising as the opposite would be expected, i.e., greater stability of the chlorine residual in the incubation bottles. Increases in THM4 concentrations during batch incubations are summarized in Table B.1, and results indicate that concentrations increased significantly during batch incubation of Utility A. THM4 concentrations measured in the batch incubations were consistent with the distribution system results, suggesting that the distribution system pipes did not affect the THMs in Utility A. One exception was noticed: during the February 2003 sampling, the batch incubation THM results were much higher than in the distribution system for no apparent reasons. The chlorine residuals measured in this batch incubation sample were also much lower than in the distribution system during this sampling, which may make this batch incubation questionable.

Batch incubations were analyzed in Utility B after 24 and 48 hours (1 and 2 days) of incubation. Free chlorine residuals decreased in the bottles with increasing incubation time, generally at the same rate as in the distribution system, or, surprisingly, at a slightly greater rate during some samplings. Increases in THM4 concentrations during batch incubations of Utility B are summarized in Table B.1. Results show that these increases were much smaller in Utility B than in Utility A, consistent with the smaller increases in THM4 concentrations observed in Utility B's distribution system. Results were at times similar to the concentrations measured in the distribution system, or different at other times. No clear trends could be distinguished.

Haloacetic Acids

Changes in HAA9 concentrations during batch incubation are summarized in Table B.2. Results indicate that the increases in HAA9 concentrations during batch incubation were much smaller than corresponding increases in THM4 (Table B.1). HAAs tend to form faster than THMs; as a result, HAAs are expected to increase to a lesser degree than THMs. In Utility A, these increases in HAA9 concentrations cannot be directly compared with those observed in the distribution system as the incubation periods did not exactly match. Nonetheless, HAA9 concentrations measured in the batch incubations seemed slightly higher than in the distribution system. This could be explained by the more limited potential for biodegradation in the batch incubations (without biofilm) compared with the distribution system.

In the batch incubations of Utility B, HAA9 concentrations sometimes increased and sometimes decreased with increasing incubation time, with no particular trend (Table B.2). Also, batch incubation HAA9 results were consistent with distribution system concentrations, which confirm that biodegradation was not as pronounced in Utility B, as mentioned in Chapter 4.

Table B.1
Increases in THM4 concentrations (percent) measured in the batch incubations of the free chlorinated systems (from Hour 0)

	Utility A			Utility B		
	Minimum	Average	Maximum	Minimum	Average	Maximum
24-hour batch	149%	195%	271%	22%	29%	41%
48-hour batch				32%	44%	56%
72-hour batch	176%	222%	312%			
168-hour batch	167%	244%	398%			

Table B.2
Changes in HAA9 concentrations (percent) measured in the batch incubations of the free chlorinated systems

	Utility A			Utility B		
	Minimum	Average	Maximum	Minimum	Average	Maximum
24-hour batch	37%	49%	63%	-24%	2.0%	22%
48-hour batch				-33%	1.1%	23%
72-hour batch	46%	57%	70%			
168-hour batch	53%	74%	92%			

Positive percentages indicate increases in HAA9 concentrations from Hour 0, whereas negative percentages indicate decreases in HAA9 concentrations from Hour 0

Chloraminated Systems

Two chloraminated distribution systems were examined, Utilities C and D.

Total Chlorine Residual and Trihalomethanes

In Utility C, batch samples were usually collected after 24 and 48 hours (1 and 2 days) of incubation. During one of the samplings (April 2003), samples were collected after 48 and 72 hours (2 and 3 days). As expected, total chlorine residuals decreased slightly during batch incubation, at a similar or slightly lower rate than in the distribution system. Changes in THM4 concentrations during batch incubations are summarized in Table B.3. Results suggested that in general, THM4 concentrations decreased slightly with increasing incubation time in the batch samples of Utility C. In the distribution systems, THM4 concentrations tended to remain constant or decreased at MRTLs. As a result, THM4 concentrations measured in the batch incubations tended to be lower than those measured in the distribution system. This observation was also made by Meyer et al. (1993), who suggested that the organic material presents on distribution system pipe surface may contain THM precursors.

In Utility D, batch samples were generally sampled after ~12 and ~36 hours of incubation, except during two samplings when batch incubations were sampled after 24 and 57 hours. In the batch incubations of Utility D, total chlorine residuals decreased slightly during batch incubation, at a similar or slightly lower rate than in the distribution system. THM4 concentrations remained relatively stable (Table B.3). Batch incubation samples showed THM4 concentrations that were at times higher and at other times lower than the distribution system samples. No particular reasons were found to explain these differences.

Table B.3
Changes in THM4 concentrations (percent) measured in the batch incubations of the chloraminated systems

	Utility C			Utility D		
	Minimum	Average	Maximum	Minimum	Average	Maximum
9 to 12-hour batch				-3.1%	-0.78%	1.8%
24-hour batch	-25%	-14%	-2.8%		-2.8%	
32 to 36-hour batch				-2.6%	2.0%	3.9%
48-hour batch	-16%	-7.7%	8.3%			
57-hour batch					-3.2%	
72-hour batch		-8.3%				

Positive percentages indicate increases in THM4 concentrations from Hour 0, whereas negative percentages indicate decreases in THM4 concentrations from Hour 0

Haloacetic Acids

Changes in HAA9 concentrations during batch incubation are summarized in Table B.4. HAA9 concentrations generally increased with increasing incubation time in the batch incubations. When comparing batch incubation with distribution system samples, HAA9 batch incubation results showed different trends, sometimes lower than distribution system samples, sometimes higher. Little conclusions could be drawn from the data obtained.

Table B.4
Changes in HAA9 concentrations (percent) measured in the batch incubations of the chloraminated systems

	Utility C			Utility D		
	Minimum	Average	Maximum	Minimum	Average	Maximum
9 to 12-hour batch				-9.0%	-0.85%	6.2%
24-hour batch	-12%	18%	44%		6.2%	
32 to 36-hour batch				2.6%	15%	35%
48-hour batch	3.6%	18%	34%			
57-hour batch					30%	
72-hour batch		1.2%				

Positive percentages indicate increases in HAA9 concentrations from Hour 0, whereas negative percentages indicate decreases in HAA9 concentrations from Hour 0

SUMMARY AND CONCLUSIONS

This appendix presents results of batch incubations in bottles that were conducted to distinguish the effect of pipes from the effect of water age on the fate of THM4 and HAA9. Overall, the batch incubations did not provide clear results. Below are the key conclusions that could be drawn from this task:

- In the free chlorinated systems, THM4 concentrations measured in the batch incubations were at times consistent with the distribution system results (such as at Utility A), suggesting that the distribution system pipes did not affect the THMs, or results were inconsistent (such as at Utility B). In one of the chloraminated systems (Utility C), THM4 concentrations measured in the batch incubations tended to be lower than those measured in the distribution system, but were inconsistent at Utility D. Overall, no clear trend could be distinguished from the batch incubations with regards to THM4.
- At Utility A, HAA9 concentrations measured in the batch incubations seemed slightly higher than in the distribution system, but were consistent with distribution system concentrations at Utility B. These trends can be explained by the greater biodegradation of HAAs observed in Utility A's distribution system compared to Utility B.
- When comparing batch incubation with distribution system samples in the chloraminated systems, HAA9 batch incubation results showed different trends, sometimes lower than distribution system samples, sometimes higher. Little conclusions could be drawn from the data obtained with regards to HAA9 in the chloraminated systems.

APPENDIX C
CORRELATIONS BETWEEN DBP CONCENTRATIONS, WATER AGE, AND WATER QUALITY PARAMETERS

Linear correlations were calculated between DBP concentrations and water age or selected water quality parameters, in attempt to identify one or more parameters that could be used as surrogate(s) of the DBPs examined. These results are presented in this appendix.

TRIHALOMETHANES AND HALOACETIC ACIDS

Correlations were calculated in the free chlorinated systems only (Utilities A and B). Correlation calculations were generally not done in the chloraminated systems because of the stability of the THM and HAA concentrations under these conditions, although some calculations were done in the chloraminated system periodically switching to free chlorine (Utility E).

Free Chlorinated Systems

Results of linear correlations (R^2) calculated for THM4 and HAA9 are summarized in Tables C.1 and C.2, respectively. Minimum, average and maximum R^2 calculated at each utility shown at the bottom of the tables, and correlations with an R^2 greater than 0.50 are underlined. At Utility A, the MRTL (water age of 323 hours) was removed from the correlation calculations as data obtained at this location were too different from the other distribution system locations, impacting the calculations too drastically.

Results obtained from some of the samplings showed correlations between THM4 concentrations and water ages or free chlorine residuals. As expected when correlations were observed, they were positive between THM4 concentrations and water age (increases in THM4 concentrations with increasing water age, as shown on Figure C.1), and negative between THM4 concentrations and chlorine residual (increases in THM4 concentration with decreasing chlorine residual, as shown on Figure C.2). Correlations were stronger in Utility A than in Utility B for reasons that are not clear. In general, HAA9 concentrations were not correlated with water age or chlorine residual (Table C.2). At Utility A, TOC and UV absorbance data throughout the entire distribution system were available for the first two samplings only, which limited the ability to determine possible correlations with these parameters at this utility. Nonetheless, results suggested that there were no correlations between DBP concentrations and TOC or UV absorbance, even though a conservative approach was used for the calculation (removal of MRTL in Utility A). There was generally no correlation between HAA9 concentrations and HPCs in the free chlorinated systems studied.

Table C.1
Correlations (R^2) between THM4 concentrations, water age and selected water quality parameters calculated in the free chlorinated systems (Utilities A and B)

Sampling dates	Water temperature at POE (°C, °F)	Water age	Free chlorine residual	TOC concentration	UV absorbance
October 2002:					
Utility A	13.3°C (56°F)	0.80	0.84	0.85	0.12
Utility B	26°C (79°F)	0.41	0.06	0.01	0.01
November/ December 2002:					
Utility A	5.6°C (42°F)	0.58	0.73	0.00	0.00
Utility B	17°C (63°F)	0.33	0.00	0.21	0.09
January 2003:					
Utility A	6.1°C (43°F)	0.99	0.34	Not available	Not available
Utility B	11°C (52°F)	0.38	0.57	0.02	0.07
February 2003:					
Utility A	4.0°C (39°F)	0.34	0.37	Not available	Not available
Utility B	12°C (54°F)	0.68	0.17	0.37	0.00
April 2003:					
Utility B	22°C (72°F)	0.37	0.64	0.24	0.06
June 2003:					
Utility A	15.0°C (59°F)	0.80	0.42	Not available	Not available
Utility B	26°C (79°F)	0.68	0.43	0.01	0.11
August 2003:					
Utility A	20.1°C (68°F)	0.37	0.91	Not available	Not available
Utility B	27°C (81°F)	0.61	0.73	0.67	0.56
September/ October 2003:					
Utility A	16.6°C (62°F)	0.66	0.43	Not available	Not available
Utility B	22°C (72°F)	0.04	0.02	0.07	0.02
Utility A:					
Minimum		0.34	0.34	0.00	0.00
Average		0.65	0.58	0.43	0.06
Maximum		0.99	0.91	0.85	0.12
Utility B:					
Minimum		0.04	0.00	0.01	0.00
Average		0.44	0.33	0.20	0.12
Maximum		0.68	0.73	0.67	0.56

Correlations with an R^2 greater than 0.50 are underlined.

Table C.2
Correlations (R^2) between HAA9 concentrations, water age and selected water quality parameters calculated in the free chlorinated systems (Utilities A and B)

Sampling dates	Water temperature at POE (°C, °F)	Water age	Free chlorine residual	TOC concentration	UV absorbance	HPCs
October 2002:						
Utility A	13.3°C (56°F)	0.36	0.19	0.42	0.26	Not available
Utility B	26°C (79°F)	0.00	0.00	0.07	0.01	0.01
November/ December 2002:						
Utility A	5.6°C (42°F)	0.36	<u>0.79</u>	0.08	0.27	0.11
Utility B	17°C (63°F)	0.06	0.09	0.00	0.02	0.09
January 2003:						
Utility A	6.1°C (43°F)	<u>0.62</u>	0.23	Not available	Not available	<u>0.65</u>
Utility B	11°C (52°F)	0.22	0.43	0.03	0.03	0.33
February 2003:						
Utility A	4.0°C (39°F)	0.15	0.00	Not available	Not available	0.19
Utility B	12°C (54°F)	0.04	0.01	0.01	0.43	0.00
April 2003:						
Utility B	22°C (72°F)	0.47	<u>0.66</u>	0.34	0.05	<u>0.62</u>
June 2003:						
Utility A	15.0°C (59°F)	0.10	0.17	Not available	Not available	0.18
Utility B	26°C (79°F)	0.02	0.13	0.12	0.26	0.02
August 2003:						
Utility A	20.1°C (68°F)	0.20	0.26	Not available	Not available	0.24
Utility B	27°C (81°F)	0.11	0.08	0.10	0.00	0.00
September/ October 2003:						
Utility A	16.6°C (62°F)	<u>0.76</u>	<u>0.73</u>	Not available	Not available	0.19
Utility B	22°C (72°F)	0.12	0.19	0.10	0.15	0.12
Utility A:						
Minimum		0.10	0.00	0.08	0.26	0.11
Average		0.36	0.34	0.25	0.27	0.26
Maximum		<u>0.76</u>	<u>0.79</u>	0.42	0.27	<u>0.65</u>
Utility B:						
Minimum		0.00	0.00	0.00	0.00	0.00
Average		0.13	0.20	0.10	0.12	0.15
Maximum		0.47	<u>0.66</u>	0.34	0.43	<u>0.62</u>

Correlations with an R^2 greater than 0.50 are underlined.

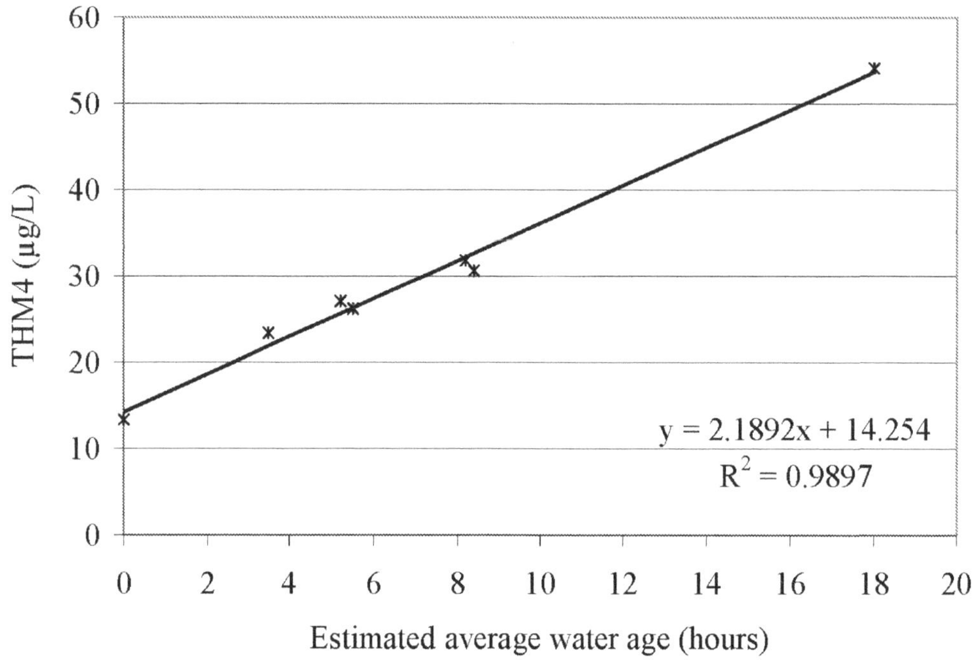

Figure C.1 Correlation between THM4 concentrations and water ages measured during the January 7, 2003 sampling in Utility A (free chlorinated system) (water temperature at the POE was 6.1°C, 43°F)

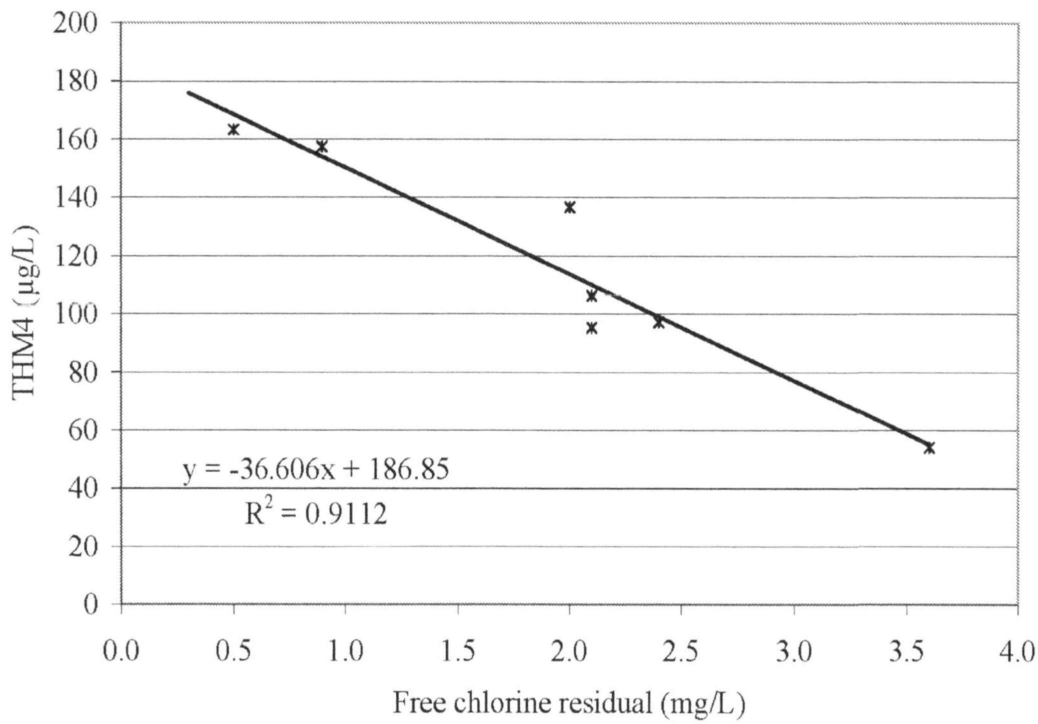

Figure C.2 Correlation between THM4 concentrations and free chlorine residual measured during the August 12, 2003 sampling in Utility A (free chlorinated system) (water temperature at the POE was 20.1°C, 68°F)

Chloraminated System Periodically Switching to Free Chlorine

Linear correlations (R^2) were calculated with the data obtained during the free chlorination period. Some correlations were also calculated with the data collected during the chloramination periods, although no specific trends were expected because of the stability of the THM and HAA concentrations under these conditions, as mentioned above. Results are summarized in Table C.3 and typical correlations are illustrated on Figures C.3 and C.4.

During the free chlorination period, positive correlations were observed between DBP concentrations and water age, and negative correlations between DBP concentrations and chlorine residual, as expected. Correlations were generally stronger between DBP concentrations and chlorine residual than between DBP concentrations and other water quality parameters. Correlations seemed stronger with THMs than with HAAs. No correlation was observed between HAA9 concentrations and HPCs. As expected, all correlations were stronger during the free chlorination period than during the chloramination periods.

N-NITROSODIMETHYLAMINE

NDMA concentrations were measured in one of the chloraminated systems (Utility C), in the chloraminated system that periodically switches to free chlorine (Utility E), and in one of the free chlorinated systems (Utility A) as a control.

Chloraminated System

Results of linear correlations (R^2) calculated with the NDMA data obtained from the chloraminated system are summarized in Table C.4, with minimum, average and maximum R^2 shown at the bottom of the table. The MRTL (water age of >120 hours) was removed from the correlation calculations as this location results that were too different from the other distribution system locations (absence of disinfectant residual, very high nitrite concentration, etc.). In general, NDMA concentrations were not correlated with the parameters examined. Some correlations were observed with total chlorine residual, but the direction of correlation varied from sampling to sampling. Free ammonia concentrations were measured during two of the samplings conducted. Correlations between NDMA and free ammonia concentrations (R^2) were 0.52 and 0.32, and the correlations were negative.

Table C.3
Correlations (R^2) between THM4 concentrations, HAA9 concentrations, total chlorine residuals and HPCs calculated in Utility E

		THM4		HAA9		
Sampling dates	Water temperature (°C, °F)	Water age	Chlorine residual	Water age	Chlorine residual	HPCs
Chloramines (prior to switch):						
February 11, 2003:	6.4°C (44°F) at POE 1 9.0°C (48°F) at POE 2	0.61	0.20	0.29	0.03	0.01
February 28, 2003:	7.0°C (45°F) at POE 1 8.0°C (46°F) at POE 2	0.55	0.04	0.42	0.17	Not available
Free chlorine:						
March 11, 2003	9.9°C (50°F) at POE 1 10.2°C (50°F) at POE 2	0.66	0.84	0.50	0.51	0.27
March 31, 2003	13.3°C (56°F) at both POEs	0.65	0.89	0.75	0.87	0.47
Chloramines (after switch):						
April 29, 2003	15.6°C (60°F) at POE 1 15.5°C (60°F) at POE 2	0.65	0.18	0.61	0.16	0.33
May 20, 2003	17.7°C (64°F) at POE 1 17.5°C (64°F) at POE 2	0.48	0.07	0.53	0.23	0.01
Average		0.60	0.37	0.52	0.33	0.22

Correlations with an R^2 greater than 0.50 are underlined.

a) THM4 – March 11, 2003 sampling (water temperature at the POEs was 9.9 and 10.2°C, 50°F)

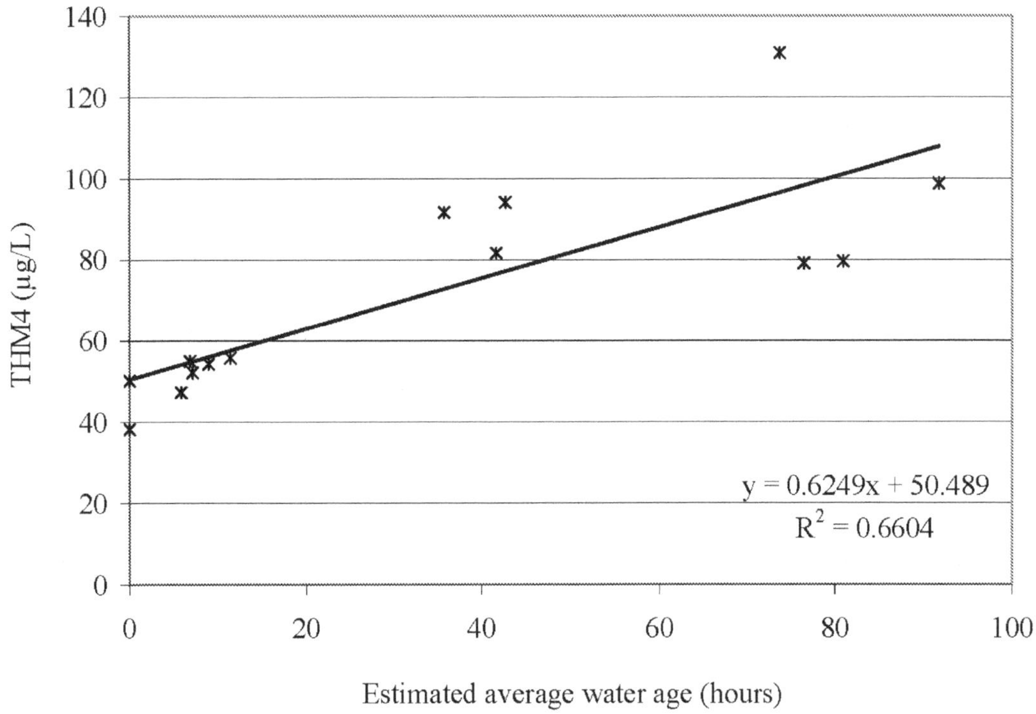

b) HAA9 – March 31, 2003 sampling (water temperature at the POEs was 13.3°C, 56°F)

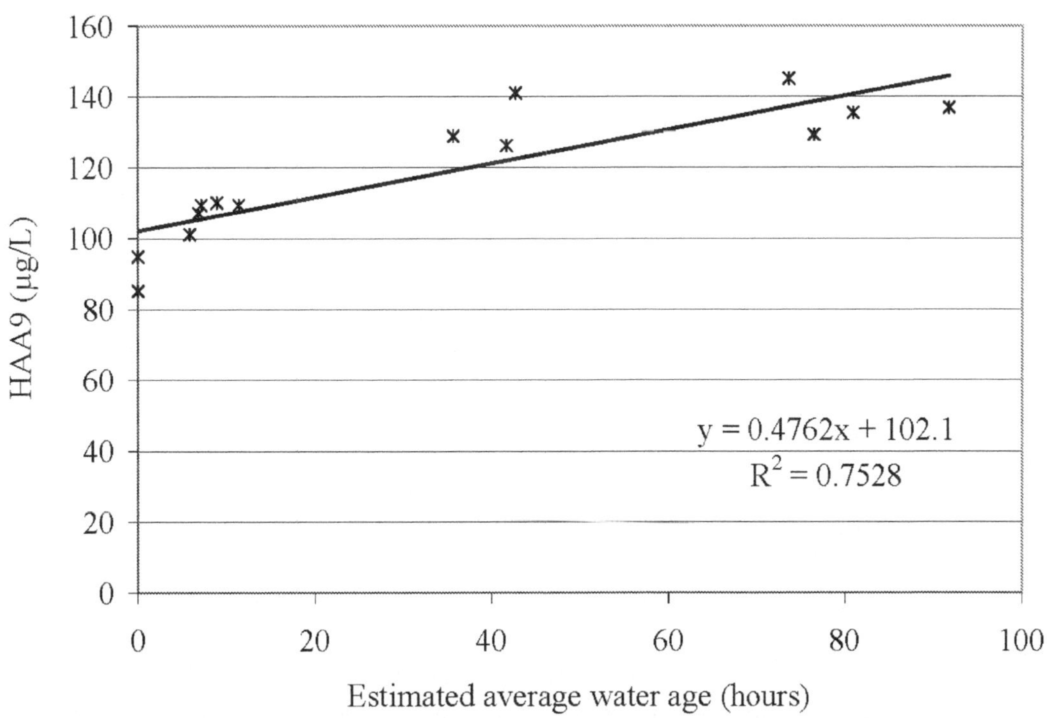

Figure C.3 Correlation between THM4 or HAA9 concentrations and water age measured during the free chlorination period in Utility E

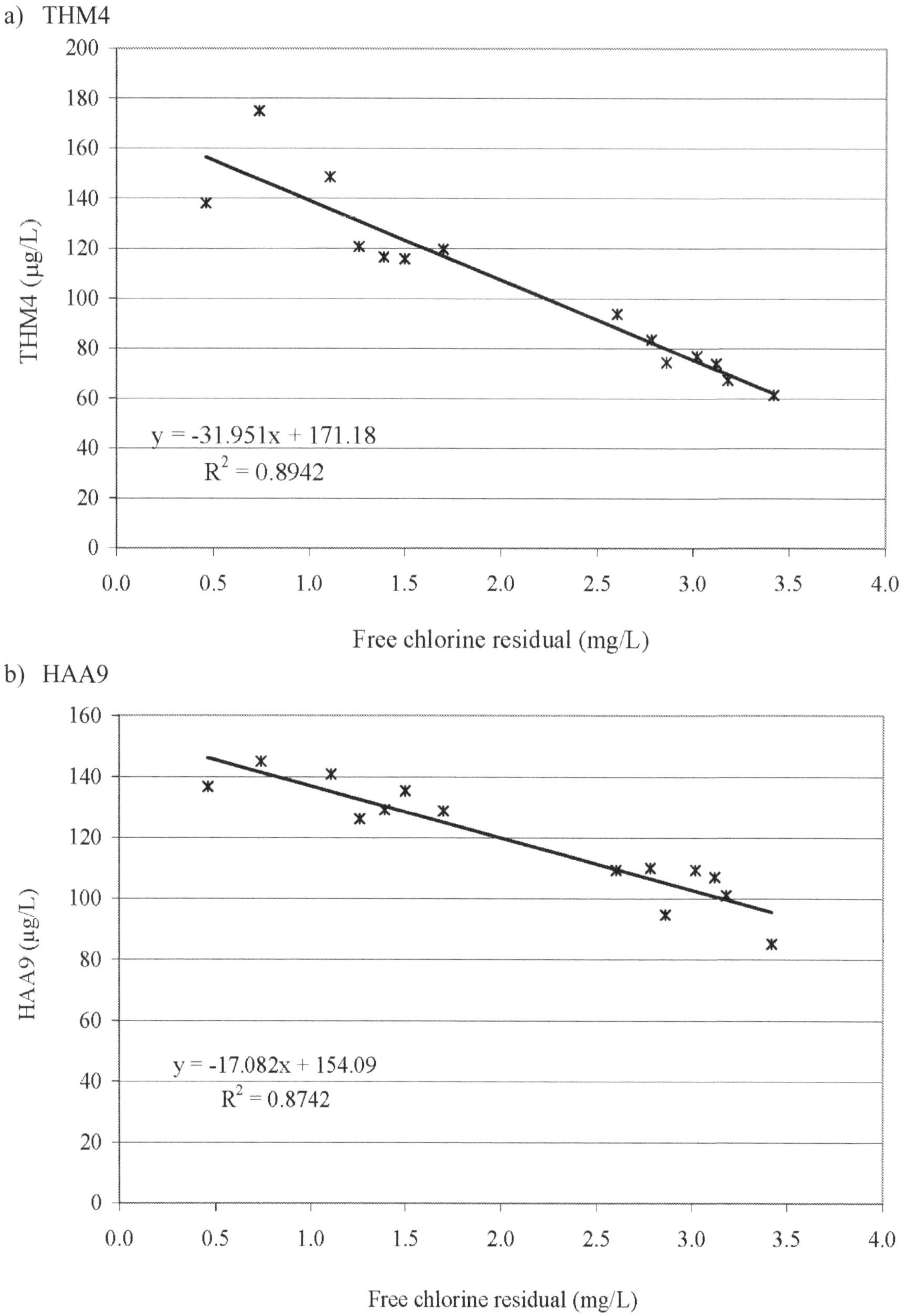

Figure C.4 Correlation between THM4 or HAA9 concentrations and chlorine residual measured during the free chlorination period in Utility E (March 31, 2003, sampling) (water temperature at the POEs was 13.3°C, 56°F)

Table C.4

Correlations (R^2) between NDMA concentrations, water age, and water quality parameters calculated in Utility C

Sampling dates	Water temperature (°C, °F)	Water age	Total chlorine	TOC	UV absorbance	Nitrite	Nitrate	Total ammonia
November 2002	16°C (61°F)	0.04	<u>0.64</u>	0.03	0.02	0.20	0.13	<u>0.53</u>
January 2003	10.5°C (51°F)	0.11	0.00	0.03	0.00	0.00	<u>0.52</u>	0.00
February 2003	8°C (46°F)	0.11	<u>0.54</u>	0.20	0.05	0.04	0.04	0.43
June 2003	25°C (77°F)	0.01	0.00	0.02	0.06	0.13	0.43	0.01
August 2003	27°C (81°F)	0.08	0.18	0.01	0.07	0.04	0.04	0.02
October 2003	19°C (66°F)	0.00	0.07	0.01	0.16	0.12	0.14	0.07
Minimum		0.00	0.00	0.01	0.00	0.00	0.04	0.00
Average		0.06	0.24	0.05	0.06	0.09	0.22	0.18
Maximum		0.11	<u>0.64</u>	0.20	0.16	0.20	<u>0.52</u>	<u>0.53</u>

Correlations with an R^2 greater than 0.50 are underlined.

Free Chlorinated System

Results of linear correlations (R^2) calculated with the NDMA data obtained from the free chlorinated system are summarized in Table C.5. The MRTL (water age of 323 hours) was removed from the correlation calculations as this location showed results that were too different from the other distribution system locations. Results showed that none of the correlations was significant (R^2<0.50). Only one sampling allowed calculation of correlations between NDMA concentrations and TOC or UV absorbance, and the R^2 was 0.03 in both cases.

Table C.5
Correlations (R^2) between NDMA concentrations, water age and free chlorine residuals calculated in Utility A

Sampling dates	Water temperature (°C, °F)	Water age	Free chlorine residual
December 2002	5.6°C (42°F)	0.25	0.49
January 2003	6.1°C (43°F)	0.04	0.01
February 2003	4.0°C (39°F)	0.03	0.13
August 2003	20.1°C (68°F)	0.29	0.01
September 2003	16.6°C (62°F)	0.07	0.19
Minimum		0.03	0.01
Average		0.14	0.17
Maximum		0.29	0.49

Chloraminated System Periodically Switching to Free Chlorine

Results of linear correlations (R^2) calculated with the NDMA data obtained from the chloraminated system that periodically switches to free chlorine are summarized in Table C.6. Overall, results showed that none of the correlations was significant (R^2<0.50). Correlations with free ammonia were not calculated during the free chlorination period as free ammonia concentrations were then below the MDL of 0.05 mg/L NH_3-N. It was not possible to correlate NDMA and nitrite as nitrite concentrations were always below the MDL of 0.1 µg/L NO_2^--N. As for nitrate, concentrations remained stable in the distribution system, with variations less than 0.1 mg/L NO_3^--N.

Table C.6
Correlations (R^2) between NDMA concentrations, water age, chlorine residuals and free ammonia concentrations calculated in Utility E

Sampling dates	Water temperature (°C)	Water age	Chlorine residual	Free ammonia
February 11, 2003	6.4°C (44°F) at POE 1 9.0°C (48°F) at POE 2	0.01	0.00	0.06
February 28, 2003	7.0°C (45°F) at POE 1 8.0°C (46°F) at POE 2	0.07	0.04	0.30
March 11, 2003	9.9°C (50°F) at POE 1 10.2°C (50°F) at POE 2	0.02	0.14	---
March 31, 2003	13.3°C (56°F) at both POEs	0.07	0.21	---
April 29, 2003	15.6°C (60°F) at POE 1 15.5°C (60°F) at POE 2	0.17	0.00	0.15
May 20, 2003	17.7°C (64°F) at POE 1 17.5°C (64°F) at POE 2	0.21	0.13	0.02
Minimum		0.01	0.00	0.02
Average		0.09	0.09	0.13
Maximum		0.21	0.21	0.30

Correlations were calculated with total chlorine residuals for the samplings of February 11, February 28, April 29 and May 20, 2003, and with free chlorine residuals for the samplings of March 11 and 31, 2003.

SUMMARY AND CONCLUSIONS

Attempts were made to find surrogate parameter(s) for THM4 and HAA9 to help predict the locations of maximum DBP concentrations in distribution systems. Results of linear correlations between DBP concentrations and several parameters however showed weak and inconsistent correlations in general.

In the free chlorinated systems, where important changes in DBP concentrations occurred, positive correlations were observed at times between THM4 concentrations and water age, and negative correlations between THM4 concentrations and free chlorine residuals. Correlations were not significant (R^2 <0.50) between HAA9 concentrations and water age, free chlorine residual or HPCs. No correlations were observed between DBP concentrations and TOC concentration or UV absorbance.

During the free chlorination period in Utility E, positive correlations were observed between DBP concentrations and water ages, and negative correlations between DBP concentrations and chlorine residuals. Correlations were strongest between DBP concentrations and chlorine residual. Correlations seemed stronger with THMs than with HAAs. No correlation was observed between HAA9 concentrations and HPCs.

Correlations were not calculated from the data obtained in the chloraminated systems because of the stability of the DBPs under these conditions.

In all three systems where NDMA concentrations were measured, NDMA concentration did not show any correlations with water age, total chlorine residual, TOC concentration, UV absorbance, nitrite, nitrate, nor total ammonia concentrations.

These results suggest that, as one could have expected, linear correlations cannot predict the fate of THM4, HAA9 and NDMA with increasing water age in full-scale distribution systems. An ANOVA procedure is recommended instead, as it would allow identification of the statistically significant variables that impact DBP concentrations in the distribution systems. However to achieve accurate results, the sample size would need to be increased, and additional samplings would be required to broaden the range of conditions under which the distribution systems are analyzed.

APPENDIX D
EFFECTS OF PIPE MATERIAL, PIPE DIAMETER, BOOSTER CHLORINATION STATION, AND STORAGE RESERVOIR

A review of the impact of pipe material and pipe diameter on DBP concentrations was performed with the data obtained from the full-scale distribution systems. When selecting the sampling locations within each system, consideration was made, as much as possible, to select locations with similar water age and pipe diameter but different pipe material, as well as locations with similar water age and pipe material but different pipe diameter. The resulting classification of each sampling location within each system is presented in the Methodology section below.

In Utility A, samples were also collected before and after a storage reservoir and a booster chlorination station to allow examination of the effect of these facilities on DBP concentrations. Again, more information regarding these sampling locations is presented in the Methodology section.

This appendix presents each topic separately: effect of pipe material, pipe diameter, booster chlorination station and storage reservoir. The DBPs examined included THM4, HAA9 and NDMA. Only the most representative graphs and tables are shown in this chapter; all other graphs and tables are presented on the CD-Rom.

METHODOLOGY

To address the effect of pipe material and diameter on the fate of DBPs, the longest possible section of the specific pipe diameter and material listed for each sampling location was used. Sampling locations were selected to ensure flow through continually decreasing pipe diameter and the same pipe material from the POE into the distribution system to the point of sample collection. For the sampling locations listed as cast iron pipe, water may have flowed through sections of lined and unlined cast iron pipe as well as lined ductile iron pipe prior to the section of cast iron pipe from which the sample was collected.

Tables D.1 through D.5 summarize the characteristics of each sampling location with regards to pipe material and diameter. The sampling locations selected at Utility B did not allow this evaluation, as can be seen from Table D.2. Because all sampling locations were in unlined cast iron pipes in Utility C, it was not possible to examine the effect of pipe material in this system.

Table D.6 presents the characteristics of the sampling locations selected in Utility A to examine the effect of a storage reservoir and a booster chlorination station on DBP concentrations. Utility A's Reservoir No.3 is an elevated reservoir with a capacity of 0.1 MG (380 m^3). This station boosts water pressure and chlorine residual. The booster chlorination station is used to maintain free chlorine residual of 1.2 mg/L in the distribution system.

Table D.1
Sampling locations sorted by pipe material and pipe diameter for Utility A

Pipe diameter	Estimated average water age (hours)			
	Low (≤9 hours)		Medium (18 hours)	High (323 hours)
	DI-TR	CI	DI - TR	DI
Small (8-10 in; 20-25 cm)	RD013 (5.2 hours)	RD007 (5.5 hours) RD002 (8.4 hours)	RD004 (18 hours)	RD025 (323 hours)
Large (16 in; 40 cm)	RD005 (8.2 hours)	RD018 (3.5 hours)		

CI:Cast Iron
DI:Ductile Iron
TR:Transite

Table D.2
Sampling locations sorted by pipe material and pipe diameter for Utility B

Pipe diameter	Estimated average water age (hours)					
	Low (≤16 hours)		Medium (19-29 hours)		High (>33 hours)	
	DI	PVC	DI	PVC	DI	PVC
Small (4 in; 10 cm)		X-3 (8.8 hours) X-11 (12 hours)		R-1 (27 hours)		D-6 (34 hours)
Large (30 in; 75 cm)	308 (4.7 hours) 883 (11 hours) 975 (15 hours)		979 (19 hours) 208 (24 hours) 232 (29 hours)		100 (34 hours) 116 (39 hours) 251 (44 hours) 227 (47 hours)	

DI:Ductile iron
PVC:Polyvinyl chloride

Table D.3

Sampling locations sorted by pipe diameter for Utility C. The pipe material was the same at all locations; unlined cast iron

Pipe diameter	Estimated average water age (hours)		
	Low (≤24 hours)	Medium (32-43 hours)	High (>120 hours)
Small (6-8 in; 15-20 cm)	Site 20 (8 hours) Site 13 (11 hours) Site 10 (21 hours) Site 2 (24 hours)	Site 9 (32 hours) Site 8 (39 hours) Site 4 (43 hours)	Site 15 (>120 hours)
Medium (10-12 in; 25-30 cm)	Site 1721 (24 hours)		
Large (12-14 in; 30-35 cm)	Site 1 (6 hours)	Site 3 (32 hours)	

Table D.4

Sampling locations sorted by pipe material and pipe diameter for Utility D

Pipe diameter	Estimated average water age (hours)			
	Low (≤10 hours)		High (>10 hours)	
	AC - PVC	CI	CI	AC
Small (6-8 in; 15-20 cm)	No.10-A (8.3 hours) No.10-B (2.2-3.8 hours)	No.3 (2.4-6.3 hours) No.4 (3.3-7.5 hours) No.5 (3.7-8.3 hours) No.6 (6.3-12.6 hours) No.8-B (7.9-15.4 hours) No.11-B (1.5 hours)	No.7 (13.1-18.4 hours) No.8-A (50 hours) No.9-A (15.3 hours) No.9-B (12.7-21.1 hours) No.11-C (10.0-19.6 hours) No.12-A (11.1 hours)	No.11A (37 hours)
Large (12 in; 30 cm)	No.2 (1.6-5.2 hours)			

AC:Asbestos cement
CI:Cast iron
PVC:Polyvinyl chloride

Table D.5
Sampling locations sorted by pipe material and pipe diameter for Utility E

Pipe diameter	Estimated average water age (hours)				
	Low (≤12 hours)		Medium (32 to 45 hours)		High (≥64 hours)
	DI	CI	DI	CI	DI
Small (2-4 in; 5-10 cm)	Site 8 (7.1 hours)	Site 7[1] (8.9 hours)	Site 5 (42.7 hours)		Site 14 (76.5 hours)
Medium (6-8 in; 15-20 cm)	Site 9 (6.8 hours)	Site 13 (11.4 hours)		Site 6 (41.7 hours)	Site 12 (91.8 hours)
Large (12-16 in; 30-40 cm)	Site 10 (5.6-6.0 hours)		Site 4* (33.3-38.2 hours)		Site 3[1] (73.7 hours) Site 11 (80.9 hours)

*These sites were fed with water from Site 2 POE; all others were fed with Site 1 POE
DI:Ductile iron
CI:Cast iron

Table D.6
Sample locations around the storage reservoir and booster chlorination station in Utility A (free chlorinated system). The pipe material was the same at all locations; ductile iron

Site identification number	Estimated water age, based on average daily flow (hours)			Pipe diameter (in, cm)
	Minimum	Average	Maximum	
RD027 - Before the reservoir	89	100	95	12 in, 30 cm
RD028 - After the reservoir, but before rechlorination station	197	201	198	12 in, 30 cm
RD020 - After the reservoir and rechlorination station	197	201	198	12 in, 30 cm
RD025 – Further downstream from the reservoir and rechlorination station	286	323	315	8 in, 20 cm

RESULTS AND DISCUSSION

The effect of pipe material, pipe diameter, booster chlorination station and storage reservoirs is discussed below in separate sections.

Effect of Pipe Material

Trihalomethanes

The effect of pipe material was studied in one of the free chlorinated systems, Utility A, and in one of the chloraminated systems, Utility D. The effect of pipe material was also examined in Utility E, the chloraminated system that periodically switches to free chlorine. Tables D.7 through D.9 present the variations in THM4 concentrations between sampling locations with similar water age and pipe diameter, but different pipe material. Figure D.1 illustrates an example of representative results obtained; all other figures are presented on the enclosed CD-Rom. These results suggest that pipe material did not have an appreciable effect on THM4 concentrations in the distribution systems studied as variations generally remained below 20 percent. THM4 concentrations seemed to be lower in the cast iron pipes of Utility A, but this trend was not observed in the other systems studied. Some important differences were observed at times, but these differences could result from variations in water age within each water age subcategory, or from differences between model-predicted and actual water ages.

Table D.7

Variations in THM4 concentrations (percent) between sampling locations with similar water age and pipe diameter, but different pipe material in Utility A (free chlorinated system)

Sampling date	Water temperature °C (°F)	Water age: 5.2 and 5.5 hours Pipe diameter: 8 in (20 cm) Pipe material: transite* and CI	Water age: 8.2 and 3.5 hours Pipe diameter: 16 in (40 cm) Pipe material: DI and CI
October 16, 2002	13.3°C (56°F)	+5.1 percent	-0.29 percent
December 3, 2002	5.6°C (42°F)	-17 percent	-66 percent
January 7, 2003	6.1°C (43°F)	-3.3 percent	-30 percent
February 25, 2003	4.0°C (39°F)	-0.32 percent	-62 percent
June 10, 2003	15.0°C (59°F)	-17 percent	-8.6 percent
August 12, 2003	20.1°C (68°F)	-2.0 percent	+18 percent
September 23, 2003	16.6°C (62°F)	+1.6 percent	-20 percent
Average		6.6 percent	29 percent

*Transite (lined asbestos cement)

Negative variations indicate that the THM4 concentration was lower in the cast iron pipe than in the transite or ductile iron pipe

CI: Cast iron

DI: Ductile iron

Table D.8

Variations in THM4 concentrations (percent) between sampling locations with similar water age and pipe diameter, but different pipe material in Utility D (chloraminated system)

Sampling date	Water temperature °C (°F)	Water age: 6.3-13 hours Diameter: 6 and 8 in (15 and 20 cm) Material: PVC and CI	Water age: 2.2-6.3 hours Diameter: 6 and 8 in (15 and 20 cm) Material: AC and CI	Water age: 50 and 37 hours Diameter: 6 and 8 in (15 and 20 cm) Material: CI and AC
November 18, 2002	21.1°C (70°F)	+76 percent		+62 percent
January 7, 2003	15.6°C (60°F)	+50 percent		+22 percent
April 8, 2003	18.3°C (64°F)		-2.6 percent	
June 10, 2003	17.8°C (64°F)		+49 percent	
July 22, 2003	21.1°C (70°F)		-2.4 percent	
	23.3°C (74°F)			
September 23, 2003	21.1°C (70°F)		+4.2 percent	
	22.2°C (72°F)			
Average		63 percent	14 percent	42 percent

Negative variations indicate that the THM4 concentration was lower in the cast iron pipe than in the PVC or asbestos-cement pipe
PVC: Polyvinyl chloride
AC: Asbestos cement
CI: Cast iron

Table D.9

Variations in THM4 concentrations (percent) between two sampling locations with similar water age and pipe diameter, but different pipe material in Utility E (chloraminated system that periodically switches to free chlorine)

Sampling date	Water temperature °C (°F)	Water age: 7.1 and 8.9 hours Pipe diameter: 2 and 4 in (5 and 10 cm) Pipe material: DI and CI	Water age: 6.8 and 11 hours Pipe diameter: 8 in (20 cm) Pipe material: DI and CI
During chloramination:			
February 11, 2003	6.4°C (44°F) 9.0°C (48°F)	+7.2 percent	-3.7 percent
February 28, 2003	7.0°C (45°F) 8.0°C (46°F)	+11 percent	+7.2 percent
During free chlorination period:			
March 11, 2003	9.9°C (50°F) 10.2°C (50°F)	+3.9 percent	+1.4 percent
March 31, 2003	13.3°C (56°F)	-12 percent	+4.0 percent
Following return to chloramines:			
April 29, 2003	15.6°C (60°F) 15.5°C (60°F)	-1.7 percent	+5.0 percent
May 20, 2003	17.7°C (64°F) 17.5°C (64°F)	-3.7 percent	-5.1 percent
Average		6.6 percent	4.4 percent

Negative variations indicate that the THM4 concentration was lower in the cast iron pipe than in the ductile iron pipe
CI: Cast iron
DI: Ductile iron

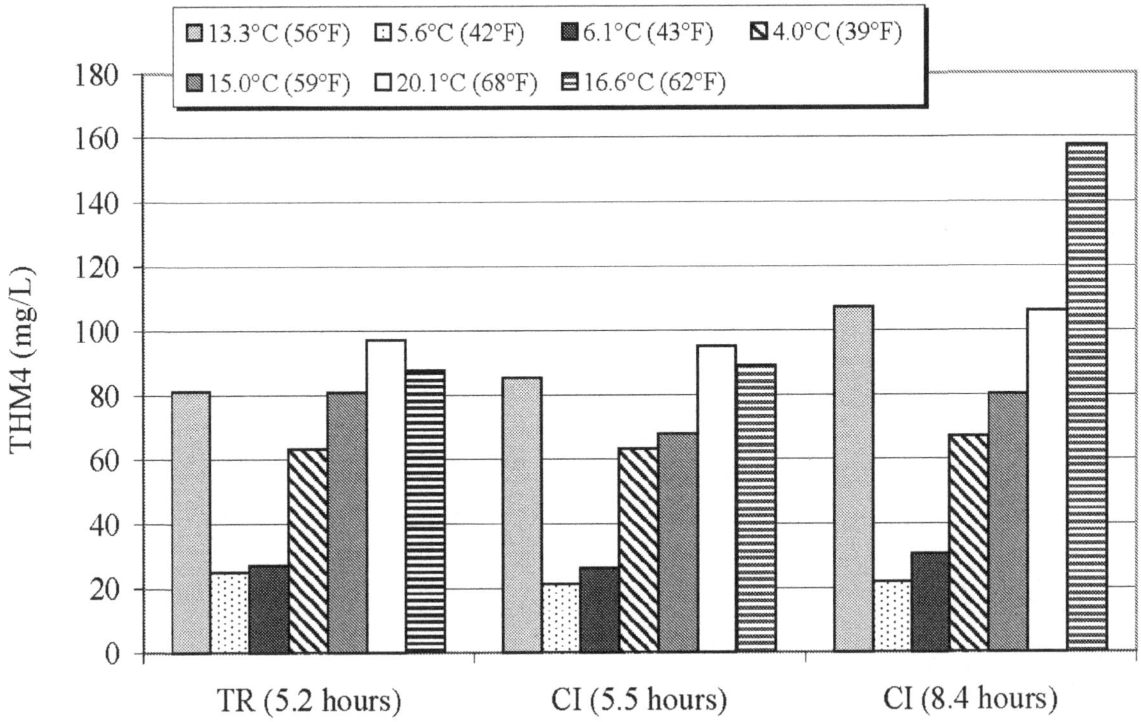

Figure D.1 Variation in THM4 concentration between sampling locations with similar water age and pipe diameter (8 or 10 inch, 20 or 25 cm), but different pipe material (transite, TR [lined asbestos cement], and cast iron, CI) in Utility A (free chlorinated system)

Haloacetic Acids

Tables D.10 through D.12 present the variations in HAA9 concentrations between sampling locations with similar water age and pipe diameter but different pipe material in Utilities A, D and E, and Figure D.2 illustrates examples of results obtained. As observed for THM4, the variations in HAA9 concentrations were generally less than 20 percent, suggesting no effect of pipe material. Greater variations were observed at times, but these differences could result from variations in water age within each water age subcategory, or from differences between model-predicted and actual water ages.

Table D.10

Variations in HAA9 concentrations (percent) between sampling locations with similar water age and pipe diameter, but different pipe material in Utility A (free chlorinated system)

Sampling date	Water temperature (°C, °F)	Water age: 5.2 and 5.5 hours Pipe diameter: 8 in (20 cm) Pipe material: transite* and CI	Water age: 8.2 and 3.5 hours Pipe diameter: 16 in (40 cm) Pipe material: DI and CI
October 16, 2002	13.3°C (56°F)	-6.7 percent	+4.7 percent
December 3, 2002	5.6°C (42°F)	-17 percent	-52 percent
January 7, 2003	6.1°C (43°F)	-5.6 percent	-30 percent
February 25, 2003	4.0°C (39°F)	-4.1 percent	+9.5 percent
June 10, 2003	15.0°C (59°F)	+24 percent	+13 percent
August 12, 2003	20.1°C (68°F)	-16 percent	-
September 23, 2003	16.6°C (62°F)	-1.9 percent	+0.45 percent
Average		11 percent	18 percent

*Transite (lined asbestos cement); CI: Cast iron; DI: Ductile iron

Negative variations indicate that the HAA9 concentration was lower in the cast iron pipe than in the transite or ductile iron pipe

Table D.11

Variations in HAA9 concentrations (percent) between sampling locations with similar water age and pipe diameter, but different pipe material in Utility D (chloraminated system)

Sampling date	Water temperature (°C, °F)	Water age: 6.3-13 hours Diameter: 6 and 8 in (15 and 20 cm) Material: PVC and CI	Water age: 2.2-6.3 hours Diameter: 6 and 8 in (15 and 20 cm) Material: AC and CI	Water age: 50 and 37 hours Diameter: 6 and 8 in (15 and 20 cm) Material: CI and AC
November 18, 2002	21.1°C (70°F)	+84 percent		+126 percent
January 7, 2003	15.6°C (60°F)	+100 percent		+0.69 percent
April 8, 2003	18.3°C (64°F)		-5.4 percent	
June 10, 2003	17.8°C (64°F)		+19 percent	
July 22, 2003	21.1°C (70°F)		-3.8 percent	
	23.3°C (74°F)			
September 23, 2003	21.1°C (70°F)		-0.4 percent	
	22.2°C (72°F)			
Average		92 percent	4.1 percent	63 percent

Negative variations indicate that the HAA9 concentration was lower in the cast iron pipe than in the PVC or asbestos-cement pipe

PVC: Polyvinyl chloride; AC: Asbestos cement; CI: Cast iron

Table D.12

Variations in HAA9 concentrations (percent) between two sampling locations with similar water age and pipe diameter, but different pipe material in Utility E (chloraminated system that periodically switches to free chlorine)

Sampling date	Water temperature (°C, °F)	Water age: 7.1 and 8.9 hours Pipe diameter: 2 and 4 in (5 and 10 cm) Pipe material: DI and CI	Water age: 6.8 and 11 hours Pipe diameter: 8 in (20 cm) Pipe material: DI and CI
During chloramination:			
February 11, 2003	6.4°C (44°F) 9.0°C (48°F)	+3.1 percent	-6.5 percent
February 28, 2003	7.0°C (45°F) 8.0°C (46°F)	+7.8 percent	+6.0 percent
During free chlorination period:			
March 11, 2003	9.9°C (50°F) 10.2°C (50°F)	+3.5 percent	+0.50 percent
March 31, 2003	13.3°C (56°F)	+0.64 percent	+2.1 percent
Following return to chloramines:			
April 29, 2003	15.6°C (60°F) 15.5°C (60°F)	-7.7 percent	+2.5 percent
May 20, 2003	17.7°C (64°F) 17.5°C (64°F)	-4.4 percent	-4.0 percent
Average		4.5 percent	3.6 percent

Negative variations indicate that the HAA9 concentration was lower in the cast iron pipe than in the ductile iron pipe
CI: Cast iron
DI: Ductile iron

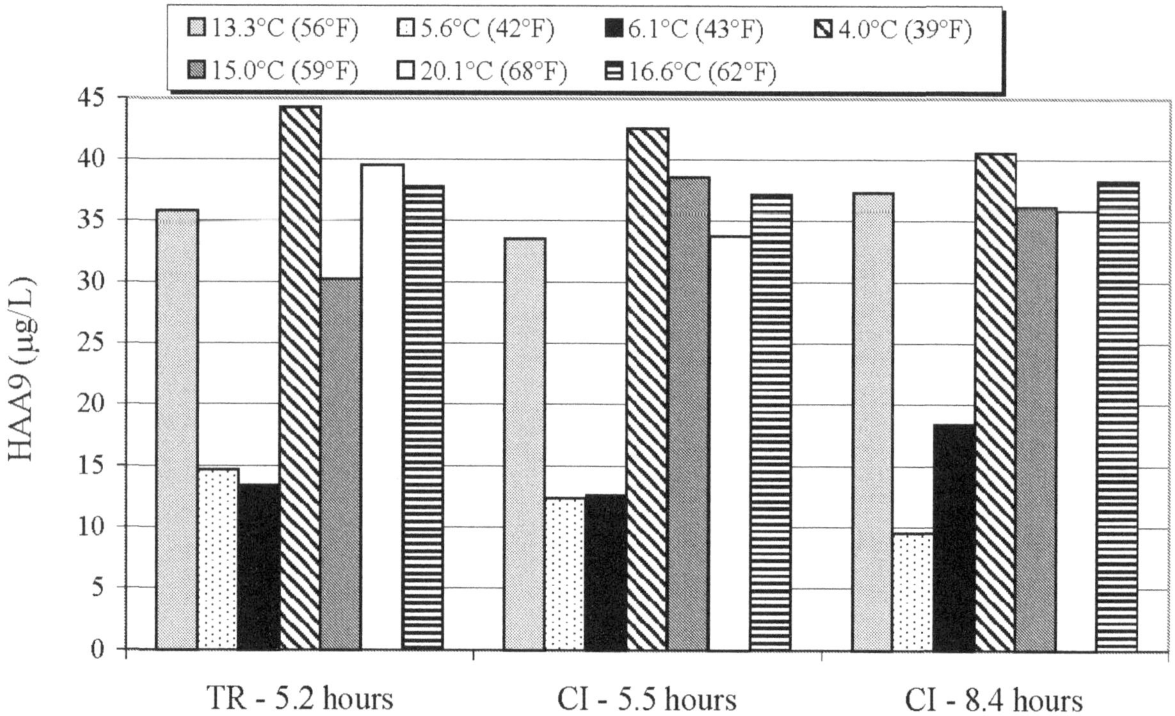

Figure D.2 Variation in HAA9 concentration between sampling locations with similar water age and pipe diameter (8 or 10 inch, 20 or 25 cm), but different pipe material (transite, TR [lined asbestos cement], and cast iron, CI) in Utility A (free chlorinated system)

N-nitrosodimethylamine

Tables D.13 and D.14 present the variations in NDMA concentrations between sampling locations with similar water age and pipe diameter but different pipe material in Utilities A and E, and Figure D.3 illustrates and example of representative results obtained. As opposed to the trends observed with THM4 and HAA9, high variations in NDMA concentrations (>20 percent) were observed between sampling locations with different pipe materials, but no specific trend could be distinguished. At times, NDMA concentrations were higher in cast iron pipes, whereas at other times, concentrations were higher in transite or ductile iron pipes. Although all samples with NDMA concentrations over 20 ng/L in Utility A occurred in the cast iron pipes, the opposite was observed in Utility E, where most of the samples with NDMA concentrations over 20 ng/L occurred in ductile iron pipes.

Table D.13

Variations in NDMA concentrations (percent) between sampling locations with similar water age and pipe diameter, but different pipe material in Utility A (free chlorinated system)

Sampling date	Water temperature °C (°F)	Water age: 5.2 and 5.5 hours Pipe diameter: 8 in (20 cm) Pipe material: transite* and CI	Water age: 8.2 and 3.5 hours Pipe diameter: 16 in (40 cm) Pipe material: DI and CI
December 3, 2002	5.6°C (42°F)	+30 percent	+31 percent
January 7, 2003	6.1°C (43°F)	-46 percent	-30 percent
February 25, 2003	4.0°C (39°F)	-41 percent	-3.3 percent
August 12, 2003	20.1°C (68°F)	+100 percent	+125 percent
September 23, 2003	16.6°C (62°F)	+196 percent	+46 percent
Average		83 percent	47 percent

*Transite (lined asbestos cement)

Positive variations indicate that the NDMA concentration was greater in the cast iron pipe than in the transite or ductile iron pipe.

Negative variations indicate that the NDMA concentration was lower in the cast iron pipe than in the transite or ductile iron pipe

CI: Cast iron

DI: Ductile iron

Table D.14

Variations in NDMA concentrations (percent) between two sampling locations with similar water age and pipe diameter, but different pipe material in Utility E (chloraminated system that periodically switches to free chlorine)

Sampling date	Water temperature (°C, °F)	Water age: 7.1 and 8.9 hours Pipe diameter: 2 and 4 in (5 and 10 cm) Pipe material: DI and CI	Water age: 6.8 and 11 hours Pipe diameter: 8 in (20 cm) Pipe material: DI and CI
During chloramination:			
February 11, 2003	6.4°C (44°F) 9.0°C (48°F)	-36 percent	+124 percent
February 28, 2003	7.0°C (45°F) 8.0°C (46°F)	-155 percent	-128 percent
During free chlorination period:			
March 11, 2003	9.9°C (50°F) 10.2°C (50°F)	+135 percent	+49 percent
March 31, 2003	13.3°C (56°F)	+17 percent	+48 percent
Following return to chloramines:			
April 29, 2003	15.6°C (60°F) 15.5°C (60°F)	-43 percent	+53 percent
May 20, 2003	17.7°C (64°F) 17.5°C (64°F)	+23 percent	-31 percent
Average		68 percent	72 percent

Positive variations indicate that the NDMA concentration was greater in the cast iron pipe than in the ductile iron pipe.
Negative variations indicate that the NDMA concentration was lower in the cast iron pipe than in the ductile iron pipe
CI: Cast iron
DI: Ductile iron

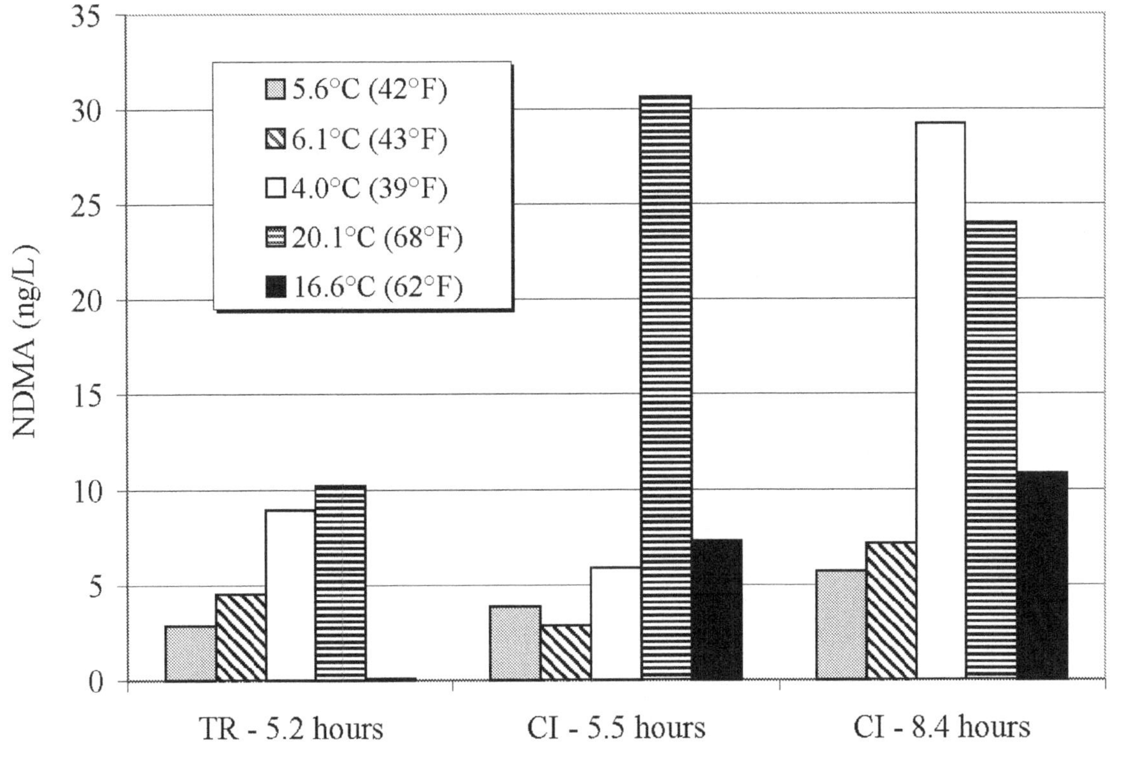

Figure D.3 Variation in NDMA concentration between sampling locations with similar water age and pipe diameter (8 or 10 in, 20 or 25 cm), but different pipe material (transite, TR [lined asbestos cement], and cast iron, CI) in Utility A (free chlorinated system)

Effect of Pipe Diameter

Trihalomethanes

The effect of pipe diameter was studied in only one of the free chlorinated systems, Utility A, and in both chloraminated systems, Utilities C and D. Tables D.15 through D.19 present the variations in THM4 concentrations between sampling locations with similar water age and pipe material, but different pipe diameter. Figure D.4 illustrates an example of representative results obtained. These results suggest that pipe diameter did not appear to affect THM4 concentrations in the systems studied as most variations were below 20 percent, except on isolated occasions.

Table D.15

Variations in THM4 concentrations (percent) between sampling locations with similar water age and pipe material, but different pipe diameter in Utility A (free chlorinated system)

Sampling date	Water temperature (°C, °F)	Water age: 5.2 and 8.2 hours Diameter: 8 and 16 in (20 and 40 cm) Pipe material: transite* and DI	Water age: 5.5 and 3.5 hours Diameter: 8 and 16 in (20 and 40 cm) Pipe material: CI
October 16, 2002	13.3°C (56°F)	+8.5 percent	+3.2 percent
December 3, 2002	5.6°C (42°F)	+60 percent	+9.3 percent
January 7, 2003	6.1°C (43°F)	+16 percent	-11 percent
February 25, 2003	4.0°C (39°F)	+4.5 percent	-58 percent
June 10, 2003	15.0°C (59°F)	-1.4 percent	+7.5 percent
August 12, 2003	20.1°C (68°F)	+34 percent	+53 percent
September 23, 2003	16.6°C (62°F)	+12 percent	-10 percent
Average		19 percent	22 percent

*Transite (lined asbestos cement)

Positive variations indicate that the THM4 concentration was greater in the larger diameter pipe than in the smaller diameter pipe.

Negative variations indicate that the THM4 concentration was lower in the larger diameter pipe than in the smaller diameter pipe

CI: Cast iron

DI: Ductile iron

Table D.16

Variations in THM4 concentrations (percent) between sampling locations with similar water age and pipe material, but different pipe diameter in Utility C (chloraminated system)

Sampling date	Water temperature (°C, °F)	Water age: 24 hours Diameter: 8 and 10 in (20 and 25 cm) Pipe material: CI	Water age: 6 and 8 hours Diameter: 6 and 14 in (15 and 35 cm) Pipe material: CI	Water age: 32 hours Diameter: 6 and 12 in (15 and 30 cm) Pipe material: CI
October 15, 2002	22.0°C (72°F)	-18 percent	+7.4 percent	+8.2 percent
November 18, 2002	16.0°C (61°F)	-12 percent	-10 percent	+19 percent
January 7, 2003	10.5°C (51°F)	-12 percent	0.0 percent	+4.7 percent
February 25, 2003	8.0°C (46°F)	+17 percent	-9.2 percent	-1.8 percent
April 29, 2003	20.0°C (68°F)	+2.5 percent	-13 percent	+20 percent
June 11, 2003	25.0°C (77°F)	-9.9 percent	+10 percent	+5.0 percent
August 12, 2003	27.0°C (81°F)	-7.1 percent	-16 percent	+3.4 percent
October 21, 2003	19.0°C (66°F)	+4.0 percent	+1.3 percent	+14 percent
Average		10 percent	8.4 percent	9.6 percent

Positive variations indicate that the THM4 concentration was greater in the larger diameter pipe than in the smaller diameter pipe.
Negative variations indicate that the THM4 concentration was lower in the larger diameter pipe than in the smaller diameter pipe
CI: Cast iron

Table D.17

Variations in THM4 concentrations (percent) between sampling locations with similar water age and pipe material, but different pipe diameter in Utility D (chloraminated system)

Sampling date	Water temperature °C (°F)	Water age: 1.6-5.2 hours Pipe diameter: 6 and 12 in (15 and 30 cm) Pipe material: AC
April 8, 2003	18.3°C (64°F)	0.0 percent
June 10, 2003	17.8°C (64°F)	+53 percent
July 22, 2003	21.1°C (70°F) 23.3°C (74°F)	-2.50 percent
September 23, 2003	21.1°C (70°F) 22.2°C (72°F)	-1.3 percent
Average		21 percent

Positive variations indicate that the THM4 concentration was greater in the larger diameter pipe than in the smaller diameter pipe. Negative variations indicate that the THM4 concentration was lower in the larger diameter pipe than in the smaller diameter pipe
AC: Asbestos cement

Table D.18

Variations in THM4 concentrations (percent) between sampling locations with low water age and similar pipe material, but different pipe diameter in Utility E (chloraminated system that periodically switches to free chlorine)

Sampling date	Water temperature (°C, °F)	Water age: 7.1 and 6.8 hours Diameter: 4 and 8 in (10 and 20 cm) Pipe material: DI	Water age: 7.1 and 5.6-6.0 hours Diameter: 4 and 16 in (10 and 40 cm) Pipe material: DI	Water age: 6.8 and 5.6-6.0 hours Diameter: 8 and 16 in (20 and40 cm) Pipe material: DI	Water age: 8.9 and 11.4 hours Diameter: 2 and 8 in (5 and 20 cm) Pipe material: CI
During chloramination:					
February 11, 2003	6.4°C (44°F) 9.0°C (48°F)	+6.7 percent	-5.4 percent	-12 percent	-4.2 percent
February 28, 2003	7.0°C (45°F) 8.0°C (46°F)	+0.21 percent	+2.1 percent	+1.9 percent	-3.9 percent
During free chlorination period:					
March 11, 2003	9.9°C (50°F) 10.2°C (50°F)	+5.2 percent	-9.9 percent	-15 percent	+2.7 percent
March 31, 2003	13.3°C (56°F)	-24 percent	-33 percent	-9.4 percent	-8.2 percent
Following return to chloramines:					
April 29, 2003	15.6°C (60°F) 15.5°C (60°F)	-9.0 percent	-12 percent	-2.6 percent	-2.3 percent
May 20, 2003	17.7°C (64°F) 17.5°C (64°F)	-10 percent	-14 percent	-3.9 percent	-11 percent
Average		9.1 percent	13 percent	7.5 percent	5.5 percent

Positive variations indicate that the THM4 concentration was greater in the larger diameter pipe than in the smaller diameter pipe.
Negative variations indicate that the THM4 concentration was lower in the larger diameter pipe than in the smaller diameter pipe
CI: Cast iron
DI: Ductile iron

Table D.19

Variations in THM4 concentrations (percent) between sampling locations with average and high water age and similar pipe material, but different pipe diameter in Utility E (chloraminated system that periodically switches to free chlorine)

Sampling date	Water temperature °C (°F)	Water age: 43 and 33-38 hours Diameter: 4 and 12 in (10 and 30 cm) Material: DI	Water age: 77 and 92 hours Diameter: 4 and 8 in (10 and 20 cm) Material: DI	Water age: 92 and 81 hours Diameter: 8 and 12 in (20 and 30 cm) Material: DI	Water age: 77 and 74 hours Diameter: 4 and 12 in (10 and 30 cm) Material: DI	Water age: 77 and 81 hours Diameter: 4 and 12 in (10 and 30 cm) Material: DI
During chloramination:						
February 11, 2003	6.4°C (44°F) 9.0°C (48°F)	-3.2 percent	+7.6 percent	-8.8 percent	+18 percent	-1.2 percent
February 28, 2003	7.0°C (45°F) 8.0°C (46°F)	+3.8 percent	+9.3 percent	-6.5 percent	+18 percent	+2.8 percent
During free chlorination period:						
March 11, 2003	9.9°C (50°F) 10.2°C (50°F)	-2.7 percent	+22 percent	-21 percent	+49 percent	+0.63 percent
March 31, 2003	13.3°C (56°F)	-22 percent	+17 percent	-18 percent	+40 percent	-0.60 percent
Following return to chloramines:						
April 29, 2003	15.6°C (60°F) 15.5°C (60°F)	+4.4 percent	-0.25 percent	-3.9 percent	+7.1 percent	-4.2 percent
May 20, 2003	17.7°C (64°F) 17.5°C (64°F)	-9.0 percent	+6.4 percent	-8.3 percent	+9.0 percent	-1.9 percent
Average		7.4 percent	10 percent	11 percent	24 percent	1.9 percent

Positive variations indicate that the THM4 concentration was greater in the larger diameter pipe than in the smaller diameter pipe.
Negative variations indicate that the THM4 concentration was lower in the larger diameter pipe than in the smaller diameter pipe
CI: Cast iron; DI: Ductile iron

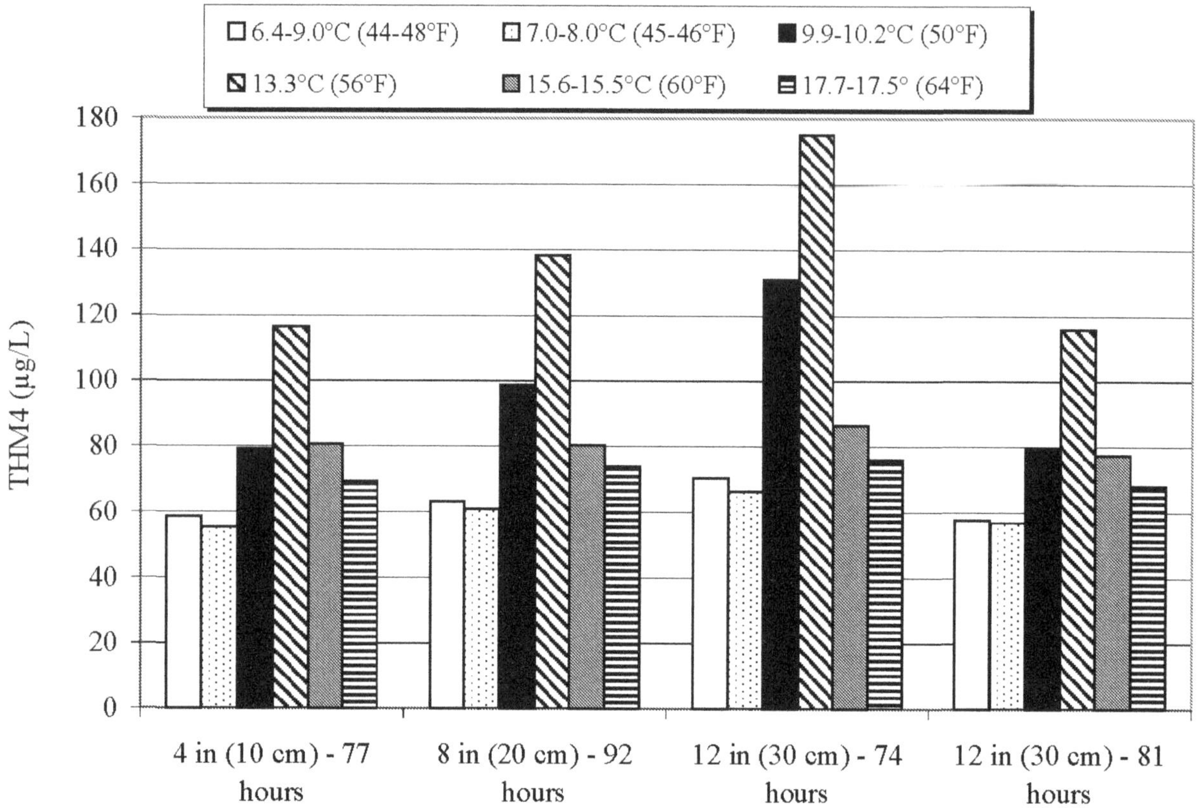

Figure D.4 Variation in THM4 concentration between sampling locations with similar water age and pipe material (ductile iron) but different pipe diameter in Utility E (chloraminated system that periodically switches to free chlorine)

Haloacetic Acids

Tables D.20 through D.24 present the variations in HAA9 concentrations between sampling locations with similar water age and pipe material but different pipe diameter at Utilities A, C, D and E, and Figure D.5 illustrates and example of representative results obtained. As observed for THM4, the variations in HAA9 concentrations were generally lower than 20 percent, suggesting no effect of pipe diameter.

Table D.20

Variations in HAA9 concentrations (percent) between sampling locations with similar water age and pipe material, but different pipe diameter in Utility A (free chlorinated system)

Sampling date	Water temperature (°C, °F)	Water age: 5.2 and 8.2 hours Diameter: 8 and 16 in (20 and 40 cm) Pipe material: transite* and DI	Water age: 5.5 and 3.5 hours Diameter: 8 and 16 in (20 and 40 cm) Pipe material: CI
October 16, 2002	13.3°C (56°F)	+6.5 percent	+18 percent
December 3, 2002	5.6°C (42°F)	+9.6 percent	-26 percent
January 7, 2003	6.1°C (43°F)	+25 percent	+0.39 percent
February 25, 2003	4.0°C (39°F)	-6.8 percent	+6.8 percent
June 10, 2003	15.0°C (59°F)	+3.3 percent	-8.1 percent
August 12, 2003	20.1°C (68°F)	+38 percent	-
September 23, 2003	16.6°C (62°F)	+4.5 percent	+6.8 percent
Average		13 percent	11 percent

*Transite (lined asbestos cement)

Positive variations indicate that the HAA9 concentration was greater in the larger diameter pipe than in the smaller diameter pipe.

Negative variations indicate that the HAA9 concentration was lower in the larger diameter pipe than in the smaller diameter pipe

CI: Cast iron

DI: Ductile iron

Table D.21

Variations in HAA9 concentrations (percent) between sampling locations with similar water age and pipe material, but different pipe diameter in Utility C (chloraminated system)

Sampling date	Water temperature °C (°F)	Water age: 24 hours Pipe diameter: 8 and 10 in (20 and 25 cm) Pipe material: CI	Water age: 6 and 8 hours Pipe diameter: 6 and 14 in (15 and 35 cm) Pipe material: CI	Water age: 32 hours Pipe diameter: 6 and 12 in (15 and 30 cm) Pipe material: CI
October 15, 2002	22.0°C (72°F)	-9.0 percent	0.0 percent	+6.9 percent
November 18, 2002	16.0°C (61°F)	+22 percent	-10 percent	-7.9 percent
January 7, 2003	10.5°C (51°F)	-26 percent	+2.0 percent	+24 percent
February 25, 2003	8.0°C (46°F)	12 percent	-23 percent	-19 percent
April 29, 2003	20.0°C (68°F)	-	+5.3 percent	+19 percent
June 11, 2003	25.0°C (77°F)	-7.0 percent	-2.9 percent	-20 percent
August 12, 2003	27.0°C (81°F)	+11 percent	-3.6 percent	0.0 percent
October 21, 2003	19.0°C (66°F)	+13 percent	-4.2 percent	+34 percent
Average		14 percent	6.4 percent	17 percent

Positive variations indicate that the HAA9 concentration was greater in the larger diameter pipe than in the smaller diameter pipe.
Negative variations indicate that the HAA9 concentration was lower in the larger diameter pipe than in the smaller diameter pipe
CI: Cast iron

Table D.22

Variations in HAA9 concentrations (percent) between sampling locations with similar water age and pipe material, but different pipe diameter in Utility D (chloraminated system)

Sampling date	Water temperature °C (°F)	Water age: 1.6-5.2 hours Pipe diameter: 6 and 12 in (15 and 30 cm) Pipe material: AC
April 8, 2003	18.3°C (64°F)	-0.92 percent
June 10, 2003	17.8°C (64°F)	+26 percent
July 22, 2003	21.1°C (70°F)	-6.0 percent
	23.3°C (74°F)	
September 23, 2003	21.1°C (70°F)	-7.2 percent
	22.2°C (72°F)	
Average		9.9 percent

Positive variations indicate that the HAA9 concentration was greater in the larger diameter pipe than in the smaller diameter pipe. Negative variations indicate that the HAA9 concentration was lower in the larger diameter pipe than in the smaller diameter pipe
AC: Asbestos cement

Table D.23
Variations in HAA9 concentrations (percent) between sampling locations with low water age and similar pipe material, but different pipe diameter in Utility E (chloraminated system that periodically switches to free chlorine)

Sampling date	Water temperature °C (°F)	Water age: 7.1 and 6.8 hours Diameter: 4 and 8 in (10 and 20 cm) Pipe material: DI	Water age: 7.1 and 5.6-6.0 hours Diameter: 4 and 16 in (10 and 40 cm) Pipe material: DI	Water age: 6.8 and 5.6-6.0 hours Diameter: 8 and 16 in (20 and 40 cm) Pipe material: DI	Water age: 8.9 and 11.4 hours Diameter: 2 and 8 in (5 and 20 cm) Pipe material: CI
During chloramination:					
February 11, 2003	6.4°C (44°F) 9.0°C (48°F)	+3.2 percent	-11 percent	-14 percent	-6.3 percent
February 28, 2003	7.0°C (45°F) 8.0°C (46°F)	+3.4 percent	+9.2 percent	+5.9 percent	+1.6 percent
During free chlorination period:					
March 11, 2003	9.9°C (50°F) 10.2°C (50°F)	+1.7 percent	-8.7 percent	-10 percent	-1.4 percent
March 31, 2003	13.3°C (56°F)	-2.1 percent	-7.9 percent	-5.8 percent	-0.64 percent
Following return to chloramines:					
April 29, 2003	15.6°C (60°F) 15.5°C (60°F)	-3.9 percent	-3.8 percent	+0.15 percent	+6.2 percent
May 20, 2003	17.7°C (64°F) 17.5°C (64°F)	+3.9 percent	+0.16 percent	-3.7 percent	+4.2 percent
Average		3.0 percent	6.8 percent	6.7 percent	3.4 percent

Positive variations indicate that the HAA9 concentration was greater in the larger diameter pipe than in the smaller diameter pipe.
Negative variations indicate that the HAA9 concentration was lower in the larger diameter pipe than in the smaller diameter pipe
CI: Cast iron
DI: Ductile iron

Table D.24

Variations in HAA9 concentrations (percent) between sampling locations with average and high water age and similar pipe material, but different pipe diameter in Utility E (chloraminated system that periodically switches to free chlorine)

Sampling date	Water temperature (°C, °F)	Water age: 43 and 33-38 hours Diameter: 4 and 12 in (10 and 30 cm) Pipe material: DI	Water age: 77 and 92 hours Diameter: 4 and 8 in (10 and 20 cm) Pipe material: DI	Water age: 92 and 81 hours Diameter: 8 and 12 in (20 and 30 cm) Pipe material: DI	Water age: 77 and 74 hours Diameter: 4 and 12 in (10 and 30 cm) Pipe material: DI	Water age: 77 and 81 hours Diameter: 4 and 12 in (10 and 30 cm) Pipe material: DI
During chloramination:						
February 11, 2003	6.4°C (44°F) 9.0°C (48°F)	-5.9 percent	+7.6 percent	-12 percent	+10 percent	-4.8 percent
February 28, 2003	7.0°C (45°F) 8.0°C (46°F)	-0.17 percent	+8.4 percent	-8.0 percent	-0.52 percent	+0.34 percent
During free chlorination period:						
March 11, 2003	9.9°C (50°F) 10.2°C (50°F)	+14 percent	-4.6 percent	+3.5 percent	+4.4 percent	-1.1 percent
March 31, 2003	13.3°C (56°F)	-9.0 percent	+5.7 percent	-1.0 percent	+12 percent	+4.7 percent
Following return to chloramines:						
April 29, 2003	15.6°C (60°F) 15.5°C (60°F)	+1.4 percent	+5.0 percent	-2.4 percent	+13 percent	+2.6 percent
May 20, 2003	17.7°C (64°F) 17.5°C (64°F)	-9.4 percent	+15 percent	-5.9 percent	+9.4 percent	+8.7 percent
Average		6.7 percent	7.6 percent	5.6 percent	8.2 percent	3.7 percent

Positive variations indicate that the HAA9 concentration was greater in the larger diameter pipe than in the smaller diameter pipe.
Negative variations indicate that the HAA9 concentration was lower in the larger diameter pipe than in the smaller diameter pipe
CI: Cast iron
DI: Ductile iron

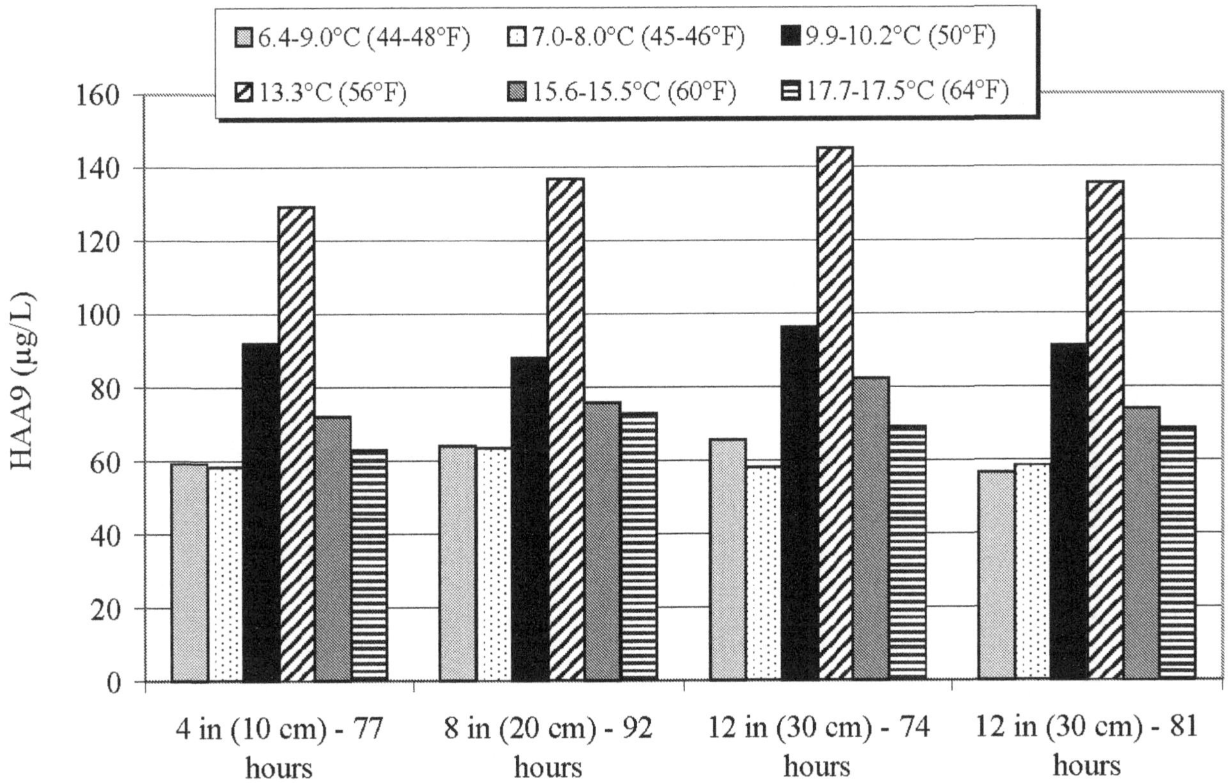

Figure D.5 Variation in HAA9 concentration between sampling locations with similar water age and pipe material (ductile iron) but different pipe diameter in Utility E (chloraminated system that periodically switches to free chlorine)

N-nitrosodimethylamine

Tables D.25 through D.28 present the variations in NDMA concentrations between sampling locations with similar water age and pipe material but different pipe diameter in Utilities A, C and E, and Figure D.6 illustrates and example of representative results obtained. Unlike THM4 and HAA9, high variations in NDMA concentrations were observed between sampling locations with different pipe diameter, but no specific trend could be distinguished. Even when the pipe diameters were relatively close to each other in Utility C (8 and 10 inch, 20 and 25 cm), NDMA concentrations showed high variations. Nonetheless, three of the four samples with NDMA concentrations over 20 ng/L occurred in small-diameter pipes in Utility A. In Utility C, the two highest NDMA concentrations measured, 25 and 46 ng/L, also occurred in small-diameter pipes. This trend (higher NDMA concentrations in smaller diameter pipes) was not observed in Utility E.

Table D.25
Variations in NDMA concentrations (percent) between sampling locations with similar water age and pipe material, but different pipe diameter in Utility A (free chlorinated system)

Sampling date	Water temperature (°C, °F)	Water age: 5.2 and 8.2 hours Pipe diameter: 8 and 16 in (20 and 40 cm) Pipe material: transite* and DI	Water age: 5.5 and 3.5 hours Pipe diameter: 8 and 16 in (20 and 40 cm) Pipe material: CI
December 3, 2002	5.6°C (42°F)	-55 percent	-54 percent
January 7, 2003	6.1°C (43°F)	-20 percent	-3.8 percent
February 25, 2003	4.0°C (39°F)	+6.3 percent	+44 percent
August 12, 2003	20.1°C (68°F)	-58 percent	-24 percent
September 23, 2003	16.6°C (62°F)	+196 percent	+53 percent
Average		67 percent	36 percent

*Transite (lined asbestos cement)
Positive variations indicate that the NDMA concentration was greater in the larger diameter pipe than in the smaller diameter pipe.
Negative variations indicate that the NDMA concentration was lower in the larger diameter pipe than in the smaller diameter pipe
CI: Cast iron
DI: Ductile iron

Table D.26
Variations in NDMA concentrations (percent) between sampling locations with similar water age and pipe material, but different pipe diameter in Utility C (chloraminated system)

Sampling date	Water temperature (°C, °F)	Water age: 24 hours Pipe diameter: 8 and 10 in (20 and 25 cm) Pipe material: CI	Water age: 6 and 8 hours Pipe diameter: 6 and 14 in (15 and 35 cm) Pipe material: CI	Water age: 32 hours Pipe diameter: 6 and 12 in (15 and 30 cm) Pipe material: CI
November 18, 2002	16.0°C (61°F)	-5.7 percent	+59 percent	-18 percent
January 7, 2003	10.5°C (51°F)	+59 percent	-68 percent	-82 percent
February 25, 2003	8.0°C (46°F)	-29 percent	-173 percent	-57 percent
June 11, 2003	25.0°C (77°F)	-165 percent	+153 percent	+151 percent
August 12, 2003	27.0°C (81°F)	+87 percent	+77 percent	-46 percent
October 21, 2003	19.0°C (66°F)	-0.56 percent	+55 percent	+13 percent
Average		58 percent	98 percent	61 percent

Positive variations indicate that the NDMA concentration was greater in the larger diameter pipe than in the smaller diameter pipe.
Negative variations indicate that the NDMA concentration was lower in the larger diameter pipe than in the smaller diameter pipe
CI: Cast iron

Table D.27

Variations in NDMA concentrations (percent) between sampling locations with low water age and similar pipe material, but different pipe diameter in Utility E (chloraminated system that periodically switches to free chlorine)

Sampling date	Water temperature °C (°F)	Water age: 7.1 and 6.8 hours Diameter: 4 and 8 in (15 and 20 cm) Pipe material: DI	Water age: 7.1 and 5.6-6.0 hours Diameter: 4 and 16 in (10 and 40 cm) Pipe material: DI	Water age: 6.8 and 5.6-6.0 hours Diameter: 8 and 16 in (20 and 40 cm) Pipe material: DI	Water age: 8.9 and 11.4 hours Diameter: 2 and 8 in (5 and 20 cm) Pipe material: CI
During chloramination:					
February 11, 2003	6.4°C (44°F) 9.0°C (48°F)	-44 percent	-124 percent	-93 percent	+118 percent
February 28, 2003	7.0°C (45°F) 8.0°C (46°F)	+72 percent	+44 percent	-30 percent	+114 percent
During free chlorination period:					
March 11, 2003	9.9°C (50°F) 10.2°C (50°F)	+54 percent	+26 percent	-30 percent	-57 percent
March 31, 2003	13.3°C (56°F)	-18 percent	-51 percent	-34 percent	+14 percent
Following return to chloramines:					
April 29, 2003	15.6°C (60°F) 15.5°C (60°F)	-69 percent	+4.7 percent	+73 percent	+25 percent
May 20, 2003	17.7°C (64°F) 17.5°C (64°F)	+28 percent	-54 percent	-79 percent	-26 percent
Average		47 percent	51 percent	56 percent	59 percent

Positive variations indicate that the NDMA concentration was greater in the larger diameter pipe than in the smaller diameter pipe.
Negative variations indicate that the NDMA concentration was lower in the larger diameter pipe than in the smaller diameter pipe.
CI: Cast iron
DI: Ductile iron

Table D.28

Variations in NDMA concentrations (percent) between sampling locations with average and high water age and similar pipe material, but different pipe diameter in Utility E (chloraminated system that periodically switches to free chlorine)

Sampling date	Water temperature °C (°F)	Water age: 43 and 33-38 hours Diameter: 4 and 12 in (10 and 30 cm) Pipe material: DI	Water age: 77 and 92 hours Diameter: 4 and 8 in (10 and 20 cm) Pipe material: DI	Water age: 92 and 81 hours Diameter: 8 and 12 in (20 and 30 cm) Pipe material: DI	Water age: 77 and 74 hours Diameter: 4 and 12 in (10 and 30 cm) Pipe material: DI	Water age: 77 and 81 hours Diameter: 4 and 12 in (10 and 30 cm) Pipe material: DI
During chloramination:						
February 11, 2003	6.4°C (44°F) 9.0°C (48°F)	-45 percent	+98 percent	-78 percent	+83 percent	+25 percent
February 28, 2003	7.0°C (45°F) 8.0°C (46°F)	+36 percent	+118 percent	+60 percent	+190 percent	+151 percent
During free chlorination period:						
March 11, 2003	9.9°C (50°F) 10.2°C (50°F)	+173 percent	+152 percent	-159 percent	+179 percent	-18 percent
March 31, 2003	13.3°C (56°F)	-3.1 percent	-	-18 percent	-	-
Following return to chloramines:						
April 29, 2003	15.6°C (60°F) 15.5°C (60°F)	-69 percent	-92 percent	+97 percent	-130 percent	+6.5 percent
May 20, 2003	17.7°C (64°F) 17.5°C (64°F)	+38 percent	+29 percent	+61 percent	+46 percent	+85 percent
Average		61 percent	98 percent	79 percent	126 percent	57 percent

Positive variations indicate that the NDMA concentration was greater in the larger diameter pipe than in the smaller diameter pipe.
Negative variations indicate that the NDMA concentration was lower in the larger diameter pipe than in the smaller diameter pipe.
CI: Cast iron
DI: Ductile iron

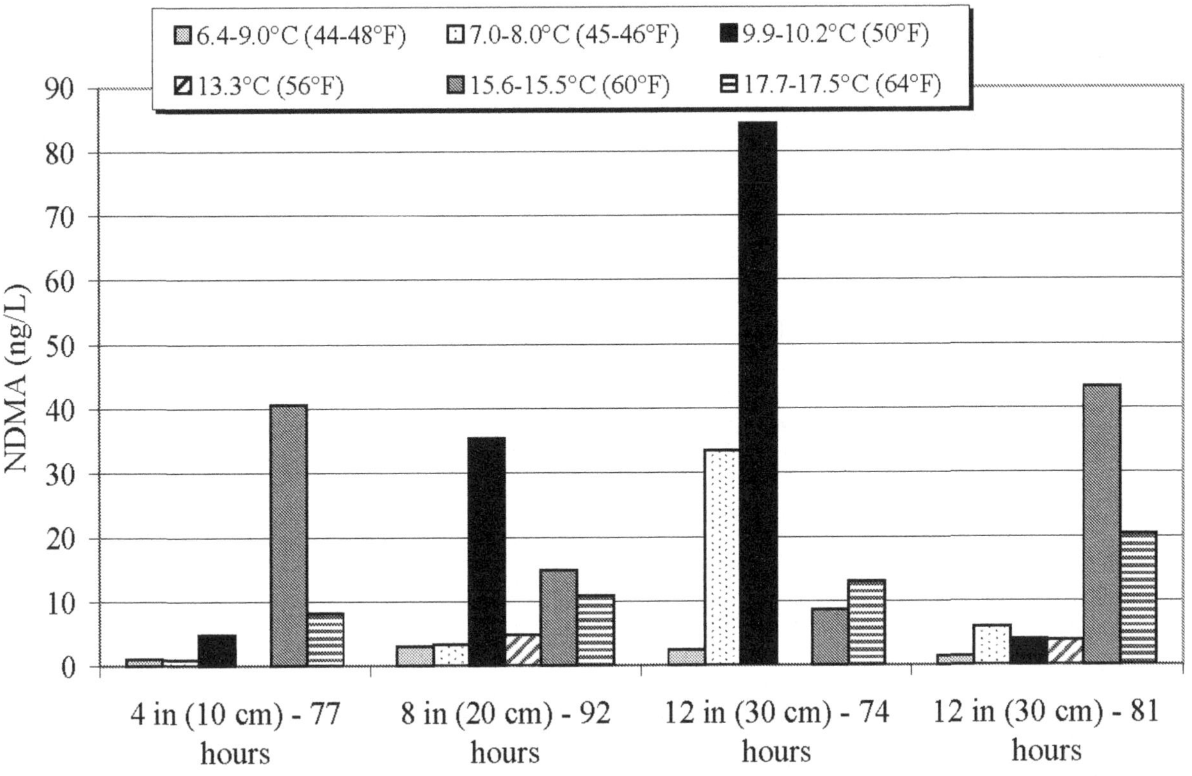

Figure D.6 Variation in NDMA concentration between sampling locations with similar water age and pipe material (ductile iron) but different pipe diameter in Utility E (chloraminated system that periodically switches to free chlorine)

Effect of a Booster Chlorination Station and a Storage Reservoir

The following section discusses the effect of a booster chlorination station and a storage reservoir in Utility A (free chlorinated system). Because water age in pipes around storage reservoirs may vary significantly depending on whether the tank is filling or emptying, the status of the reservoir during each sampling was documented and results are summarized in Table D.29. An example of reservoir water level curves is presented on Figure D.7; corresponding figures for the other samplings are presented on the enclosed CD-Rom.

The sampling sites located before and after the rechlorination station (RD028 and RD020) were very close to each other (approximately 6 feet, 1.8 m). As such, the sampling location after rechlorination station (RD020) was not considered representative because of the lack of adequate contact time, and a sampling site located a few miles downstream (RD025 with a water age of 323 hours) was used instead.

Table D.29
Status of the reservoir during samplings conducted at Utility A

Sampling date	Status
October 16, 2002 (13.3°C, 56°F)	Emptying
December 3, 2002 (5.6°C, 42°F)	Emptying (reservoir influent samples were collected at the beginning of the emptying cycle)
January 7, 2003 (6.1°C, 43°F)	Filling (reservoir influent samples were collected at the beginning of the fill cycle)
February 25, 2003 (4.0°C, 39°F)	Filling (reservoir influent samples were collected at the beginning of the fill cycle)
June 10, 2003 (15.0°C, 59°F)	Filling (reservoir influent samples were collected at the beginning of the fill cycle)
August 12, 2003 (20.1°C, 68°F)	Emptying
September 23, 2003 (16.6°C, 62°F)	Emptying

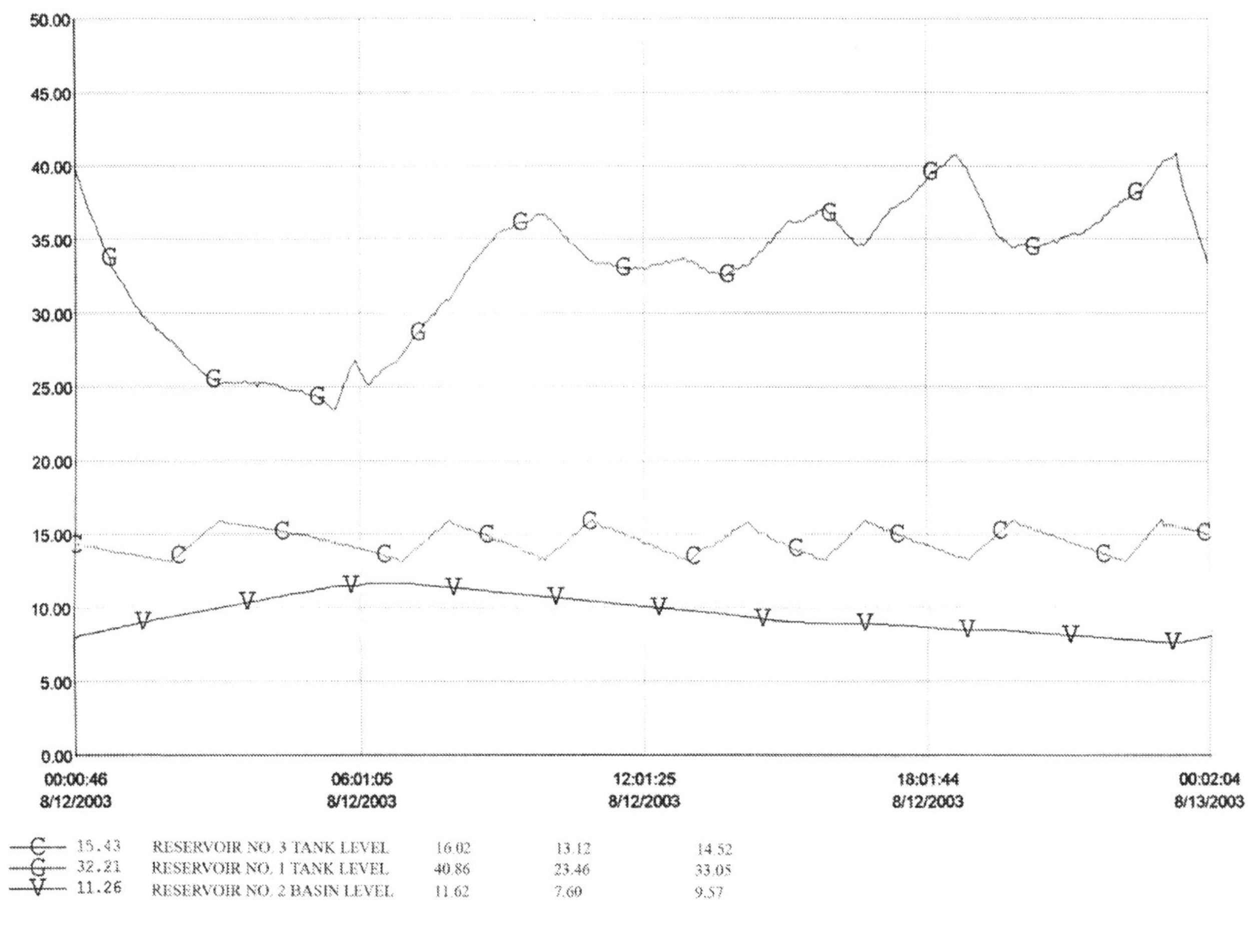

Figure D.7 Water level in the reservoirs of Utility A during the August 12, 2003 sampling (water temperature at the POE was 20.1°C, 68°F)

Figures D.8 through D.14 present graphs of THM4 concentration, HAA9 concentration and free chlorine residual around the reservoir and booster chlorination station, whereas Figure D.15 presents a corresponding graph for NDMA concentrations. Tables D.30 through D.32 summarize the changes in THM4, HAA9 and NDMA concentrations, respectively, in the reservoir and booster chlorination station. Positive variations indicate an increase in DBP concentration, whereas negative variations indicate a decrease in DBP concentration.

Results show that surprisingly, THM4 and HAA9 concentrations did not change significantly from the reservoir influent to effluent with variations of 20 percent or less most of the times. NDMA concentrations varied from the reservoirs' influent to effluent, but no particular trend could be distinguished. The status of the reservoir (filling or emptying) did not seem to affect the change in concentration of the DBPs studied. When interpreting these data, it should be kept in mind that the free chlorine residual data suggest that the sampling site located after the reservoir but before rechlorination may have been under the influence of the booster chlorination as chlorine residual was at times higher at that location than at the reservoir influent.

The booster chlorination itself increased THM4 concentrations by approximately 16 to 101 percent. The change in HAA9 concentrations associated with the booster chlorination station varied from sampling to sampling. Decreases in HAA9 concentrations were observed following booster chlorination during the first two samplings when the free chlorine residual decreased to low level at this location (<0.1 mg/L Cl_2). Increases in HAA9 concentrations were observed when the free chlorine residual remained high following booster chlorination (1.3 and 1.4 mg/L). When the free chlorine residual decreased but remained detectable (0.3 to 0.5 mg/L), HAA9 concentrations did not change following booster chlorination. The decrease in HAA9 concentrations in absence of free chlorine residual may have resulted from the biodegradation of the HAAs (van der Ploeg et al. 1991; Williams et al. 1996, 1997, 1998; Zhou and Xie 2002; Xie and Zhou 2002). When chlorine was present, the fate of HAAs resulted from a balance between their formation and degradation, which was related to the free chlorine residual present. Note that the sampling location after the rechlorination station was located a few miles downstream of this facility. As such, water age may have also affected the fate of HAAs. In the case of NDMA, changes in concentrations were observed during booster chlorination, but no particular trend could be distinguished.

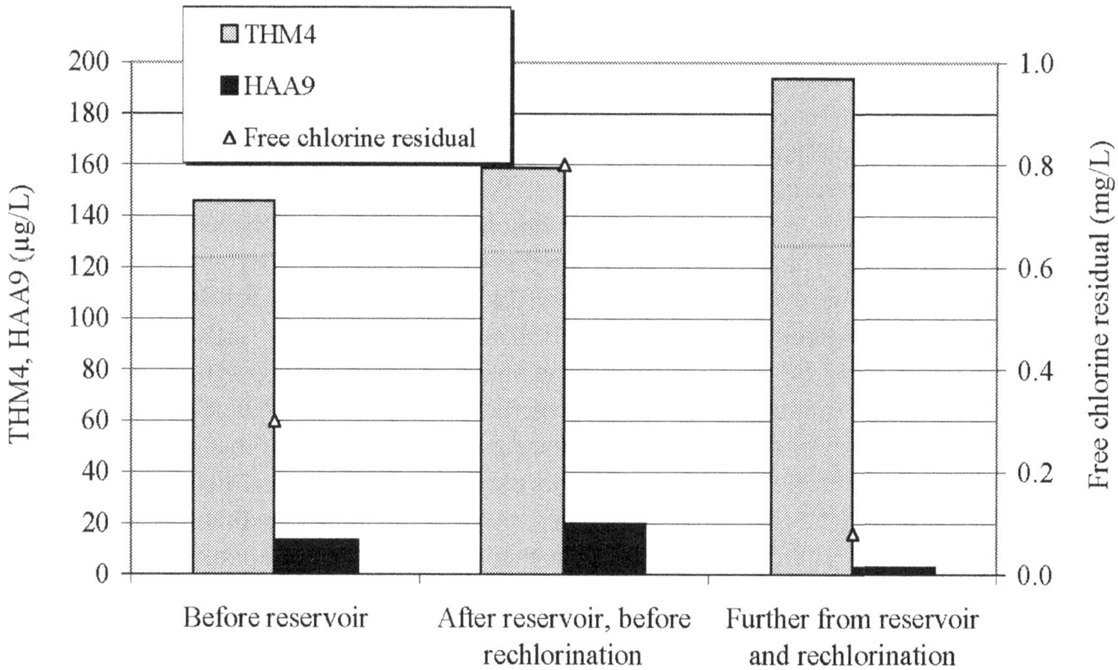

Figure D.8 THM4 concentrations, HAA9 concentrations and free chlorine residuals in the reservoir and booster chlorination station measured during the October 16, 2002 sampling in Utility A (water temperature at the POE was 13.3°C, 56°F)

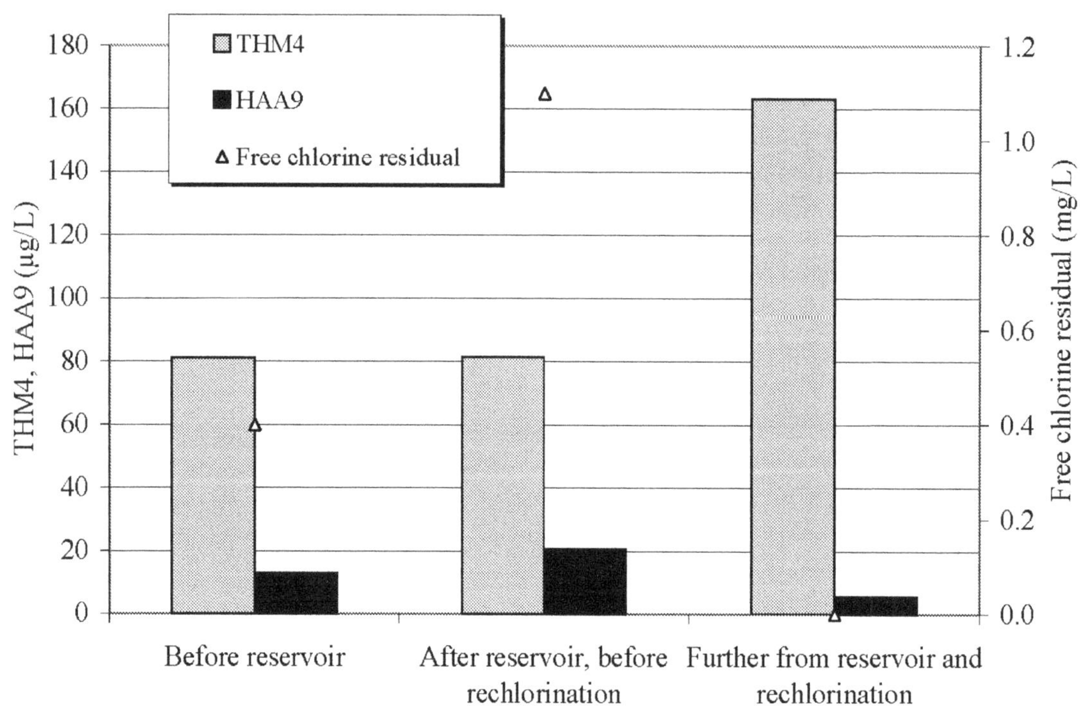

Figure D.9 THM4 concentrations, HAA9 concentrations and free chlorine residuals in the reservoir and booster chlorination station measured during the December 3, 2002 sampling in Utility A (water temperature at the POE was 5.6°C, 42°F)

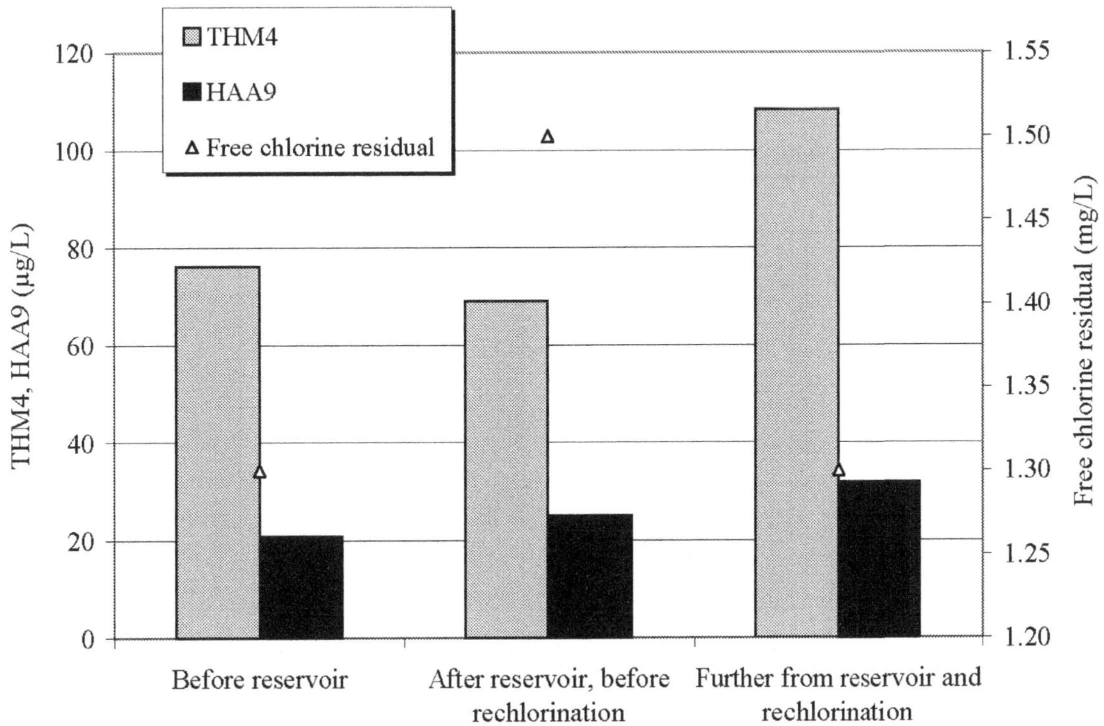

Figure D.10 THM4 concentrations, HAA9 concentrations and free chlorine residuals in the reservoir and booster chlorination station measured during the January 7, 2003 sampling in Utility A (water temperature at the POE was 6.1°C, 43°F)

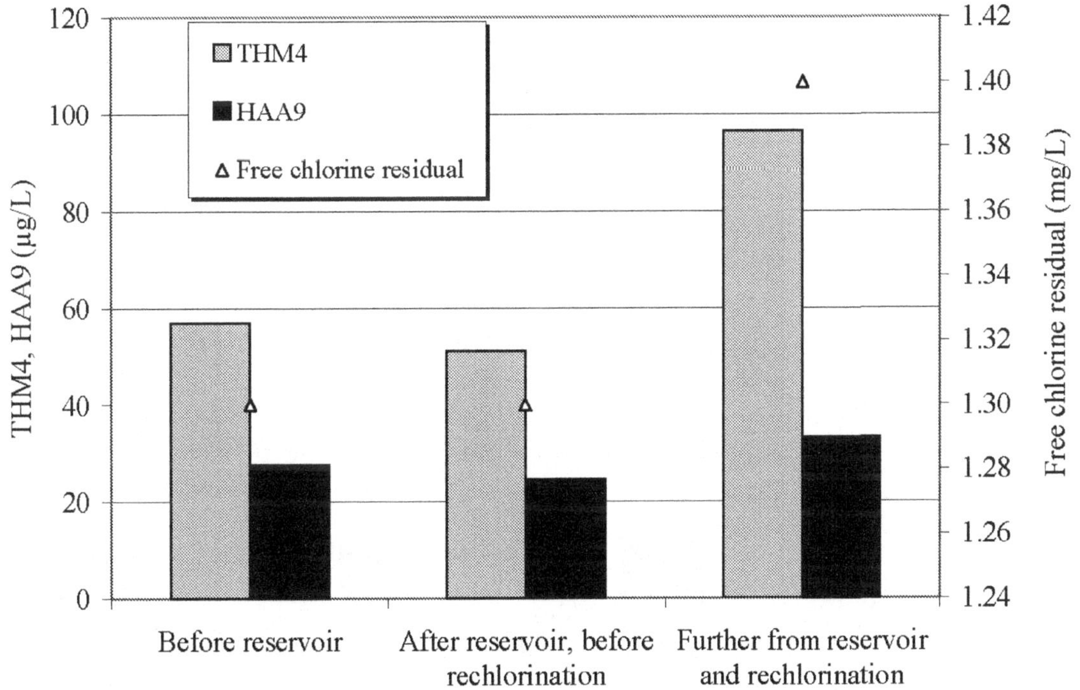

Figure D.11 THM4 concentrations, HAA9 concentrations and free chlorine residuals in the reservoir and booster chlorination station measured during the February 25, 2003 sampling in Utility A (water temperature at the POE was 4.0°C, 39°F)

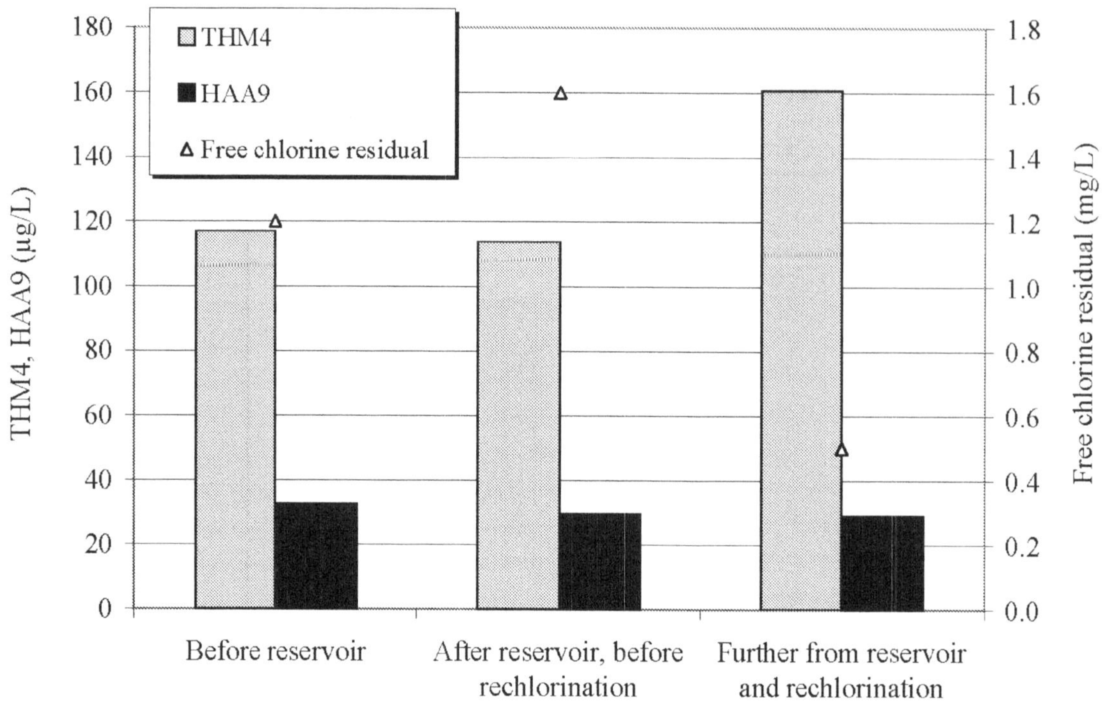

Figure D.12 THM4 concentrations, HAA9 concentrations and free chlorine residuals in the reservoirs and booster chlorination station measured during the June 10, 2003 sampling in Utility A (water temperature at the POE was 15.0°C, 59°F)

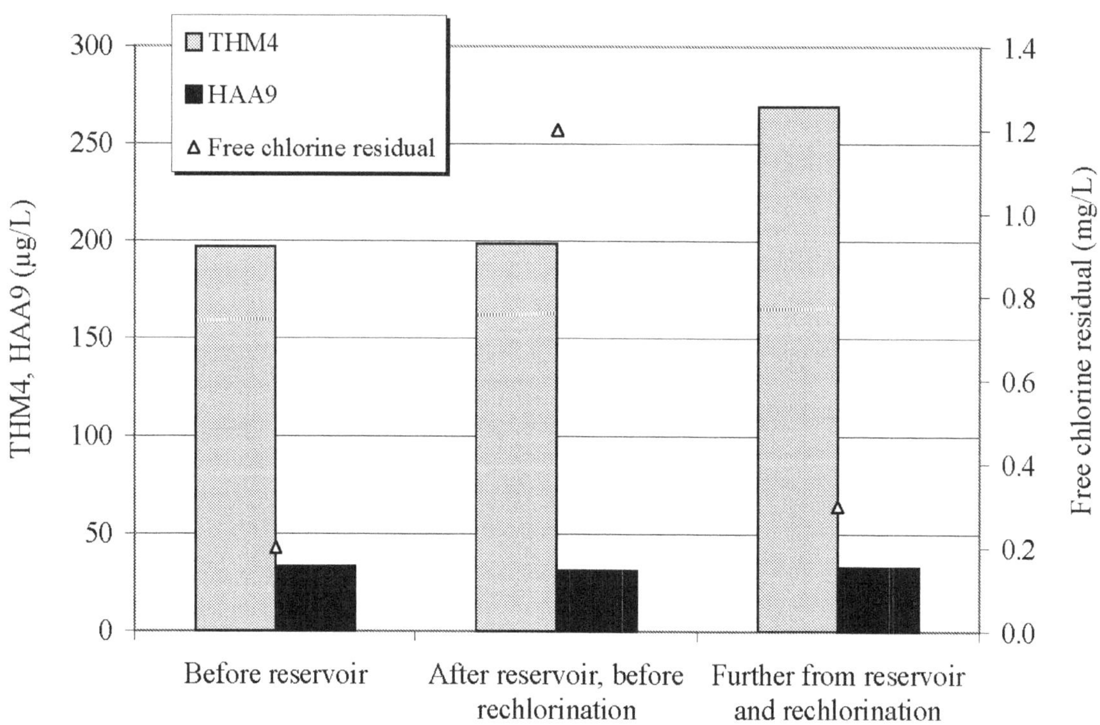

Figure D.13 THM4 concentrations, HAA9 concentrations and free chlorine residuals in the reservoir and booster chlorination station measured during the August 12, 2003 sampling in Utility A (water temperature at the POE was 20.1°C, 68°F)

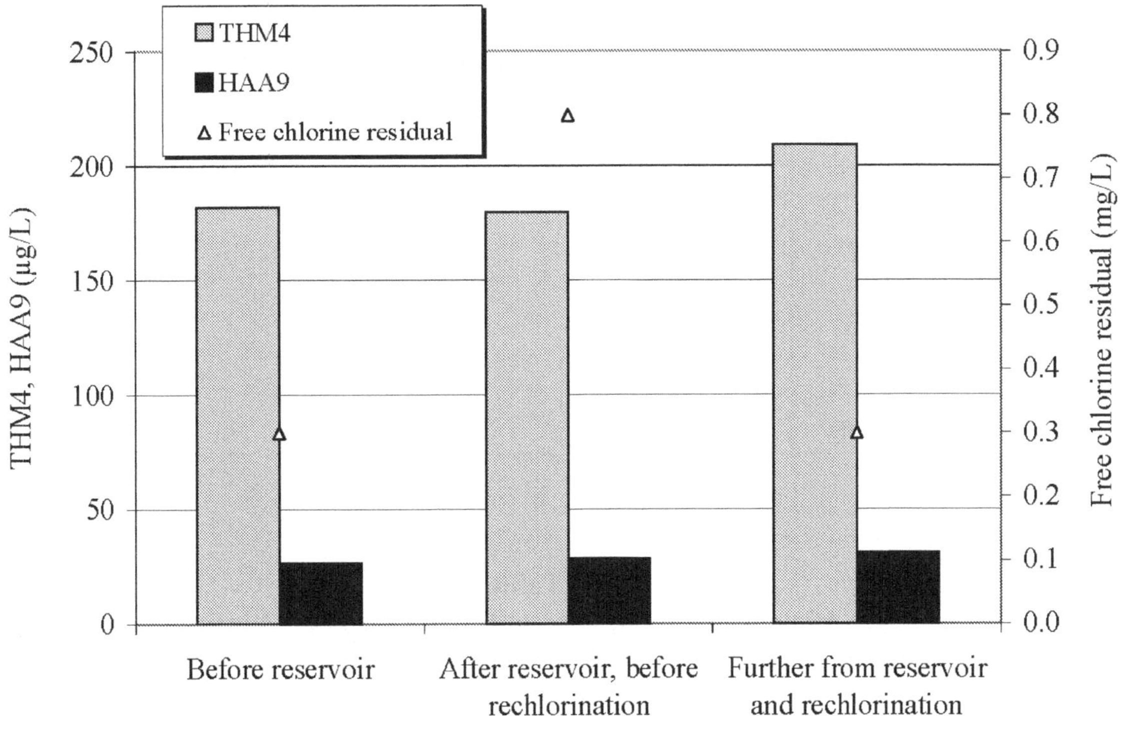

Figure D.14 THM4 concentrations, HAA9 concentrations and free chlorine residuals in the reservoir and booster chlorination station measured during the September 23, 2003 sampling in Utility A (water temperature at the POE was 16.6°C, 62°F)

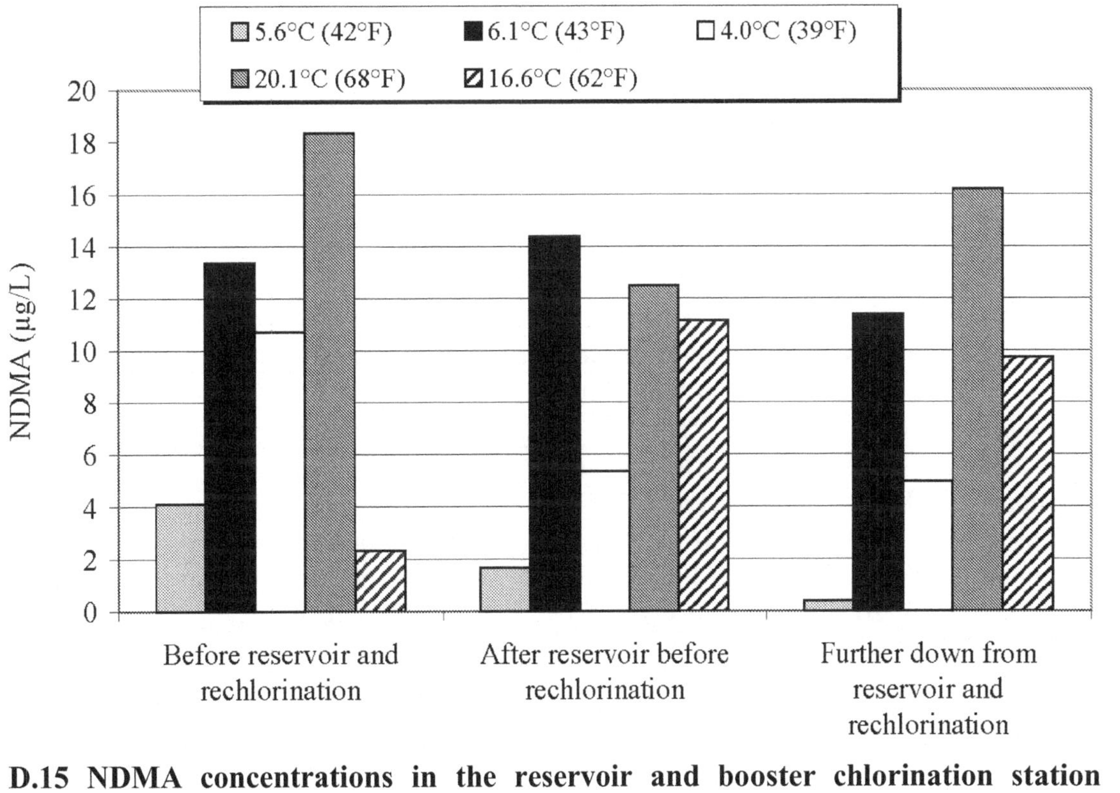

Figure D.15 NDMA concentrations in the reservoir and booster chlorination station measured during the samplings conducted in Utility A

Table D.30
Variations in THM4 concentrations (percent) from the reservoir influent to effluent, and from reservoir effluent to after booster chlorination station at Utility A

Sampling date	Water temperature °C (°F)	Effect of reservoir	Effect of booster chlorination station
October 16, 2002	13.3°C (56°F)	8.8 percent	22 percent
December 3, 2002	5.6°C (42°F(	0.26 percent	101 percent
January 7, 2003	6.1°C (43°F)	-9.4 percent	57 percent
February 25, 2003	4.0°C (39°F)	-10 percent	89 percent
June 10, 2003	15.0°C (59°F)	-2.7 percent	41 percent
August 12, 2003	20.1°C (68°F)	0.75 percent	36 percent
September 23, 2003	16.6°C (62°F)	-1.2 percent	16 percent

Table D.31
Variations in HAA9 concentrations (percent) from the reservoir influent to effluent, and from reservoir effluent to after booster chlorination station at Utility A

Sampling date	Water temperature °C (°F)	Effect of Reservoir	Effect of booster chlorination station
October 16, 2002	13.3°C (56°F)	49 percent	-85 percent
December 3, 2002	5.6°C (42°F(	60 percent	-73 percent
January 7, 2003	6.1°C (43°F)	20 percent	27 percent
February 25, 2003	4.0°C (39°F)	-11 percent	35 percent
June 10, 2003	15.0°C (59°F)	-8.2 percent	-1.9 percent
August 12, 2003	20.1°C (68°F)	-6.2 percent	5.9 percent
September 23, 2003	16.6°C (62°F)	7.5 percent	8.4 percent

Table D.32
Variations in NDMA concentrations (percent) from the reservoir influent to effluent, and from reservoir effluent to after booster chlorination station at Utility A

Sampling date	Water temperature °C (°F)	Effect of Reservoir	Effect of booster chlorination station
December 3, 2002	5.6°C (42°F(	-60 percent	-76 percent
January 7, 2003	6.1°C (43°F)	7.5 percent	-21 percent
February 25, 2003	4.0°C (39°F)	-50 percent	-7.5 percent
August 12, 2003	20.1°C (68°F)	-32 percent	30 percent
September 23, 2003	16.6°C (62°F)	379 percent	-13 percent

CONCLUSIONS

The following conclusions can be drawn from the evaluation of the effect of pipe material and diameter on THM4, HAA9 and NDMA concentrations:

- Pipe material and diameter did not have an appreciable effect on THM4 nor HAA9 concentrations in the distribution systems studied. Some important differences were observed at times, but these differences could result from variations in water age.
- In the case of NDMA, concentrations were higher in cast iron pipes at times, whereas at other times, concentrations were higher in transite or ductile iron pipes for samples with similar water ages. Overall, no specific trends could be drawn, suggesting that pipe material did not affect NDMA concentration in the distribution systems studied.
- Similarly, NDMA concentrations were greater in the small-diameter pipes at times, whereas at other times, concentrations were greater in the large-diameter pipes for samples with similar water ages. The majority of the samples with high NDMA concentrations occurred in small-diameter pipes in two of the systems studied, but not in the third system. Thus, pipe diameter did not seem to affect NDMA concentration.
- The type of distribution system residual disinfectant (free chlorine or chloramines) did not affect the trends observed.

The examination of the effect of a storage reservoir and a booster chlorination station on THM4, HAA9 and NDMA concentrations lead to the following conclusions:

- Surprisingly, storage reservoirs did not affect THM4 or HAA9 concentrations significantly, with variations of 20 percent of less most of the time.
- NDMA concentrations varied from the reservoirs' influent to effluent, but no particular trend could be distinguished.
- Booster chlorination increased THM4 concentrations by 16 to 101 percent.
- The fate of HAAs resulted from a balance between their formation and degradation, which was related to the free chlorine residual present. Decreases in concentrations were observed most likely due to HAA biodegradation when the free chlorine residual decreased to low levels (<0.1 mg/L Cl_2) downstream of the chlorination station. Increases in HAA9 concentrations were observed when the free chlorine residual remained high (1.3 and 1.4 mg/L) downstream of the booster chlorination station. When the free chlorine residual decreased but remained detectable (0.3 to 0.5 mg/L), HAA9 concentrations did not change following booster chlorination.
- Changes in NDMA concentrations were observed during booster chlorination, but no particular trend could be distinguished.

When examining water quality around a storage reservoir or a booster chlorination station, it is important to consider the hydraulic scenario present at the moment of sampling. In the case of a reservoir, depending on its piping configuration, aged water from the reservoir may flow back to the reservoir influent when the reservoir is emptying. It this study, it appeared that the status of the reservoirs did not affect the change in concentration of any of the DBPs studied, although further investigation is warranted to confirm this observation. The hydraulic scenario also deserves attention around booster chlorination stations. In this study, greater free chlorine residuals were observed at times upstream of the booster chlorination station, which suggest unexpected hydraulic conditions at this location and the possibility of data distortion.

APPENDIX E
INTER-LABORATORY COMPARISON EXPERIMENTS

Over the course of the study, interesting observations were made regarding various quality control (QC) aspects of the study, and are summarized in this chapter. Within each laboratory, strict internal QC procedures were followed and involved the use of field or travel control blanks, method and analysis control blanks, and matrix spikes and standards. Samples were collected and/or analyzed in replicate within the appropriate holding time. Because THM4 and HAA9 were analyzed at two different laboratories, Newport News Waterworks (NNWW) and University of North Carolina (UNC), inter-laboratory comparison experiments were conducted to ensure comparability of results. This appendix presents results of these inter-laboratory comparison experiments conducted for THM4 and HAA9.

MATERIALS AND METHODS

Inter-laboratory comparison experiments were conducted at the beginning of the project and then quarterly (approximately every three months) to ensure comparability of results:

- Inter-laboratory comparison test No.1: August 13, 2002, all utilities
- Inter-laboratory comparison test No.2: October 15, 2002, Utility C
- Inter-laboratory comparison test No.3: January 7, 2003, Utility D
- Inter-laboratory comparison test No.4: April 29, 2003, Utility E
- Inter-laboratory comparison test No.5: June 10, 2003, Utility B.

Following sample analysis, the relative percent difference (RPD) between the samples analyzed at both laboratories was calculated. It needed to be 20 percent or less for THMs (40 percent for low THM concentrations) and 40 percent or less for HAAs to ascertain that the data obtained from both laboratories were similar. If the RPD was above these percentages, the data obtained and related quality controls were reviewed.

First Inter-Laboratory Comparison Test

The first inter-laboratory comparison experiment involved sample collection at the POE of the five distribution systems studied, as well as spikes samples prepared in treatment plant effluent water by an external laboratory, as described below:

- Utility A: Samples were collected at the clearwell effluent (high-pressure pumps) from a tap in the laboratory.
- Utility B: Samples were collected in the Control Room, from a constant flowing tap that comes from one of the discharge mains of the high pressure pump station.
- Utility C: Samples were collected from the water treatment plant, at the closest location to the master meter.
- Utility D: Samples were collected at the main POE of the section of their distribution system that was studied during this project.

- Utility E: Samples were collected from WTP No.2, from a finished water tap in the laboratory, which provides water from the high-pressure pumps.
- External laboratory: Samples were collected from the effluent of a water treatment plant, from a sampling tap in the laboratory.

Two types of spiked samples were prepared as follows:

1. Spike samples were prepared in the collection containers of UNC and NNWW to compare the two different container preparation, preservative/quenching agents, and analytical methods used at these two laboratories. In these containers, two spike samples were prepared in duplicate.
2. Spike samples were prepared in the external laboratory collection containers to evaluate the potential differences between the analytical methods only. One spike sample for THM4 and one for HAA9 were prepared in these containers.

Six THM sample vials (three spike samples in duplicate) and a travel blank were sent to each laboratory (NNWW and UNC). All spiked samples were prepared in water collected at the effluent of a water treatment plant. Three THM4 samples were prepared by measuring three separate 500-mL volumes of plant effluent water, and were labeled THM-SP-1, THM-SP-2, and THM-SP-3. The spiking solution used was purchased from Supelco and contained 2,000 μg/L of each of the four THMs. Theoretical THM4 concentration in spike samples is presented in Table E.1. The THM-SP-3 sample consisted of plant effluent water without any additional THMs added.

Six HAA9 sample vials (three spike samples in duplicate) and a travel blank were also sent to each laboratory (NNWW and UNC). The three HAA9 samples were prepared by measuring three separate 500-mL volumes of treatment plant effluent water, and were labeled HAA-SP-1, HAA-SP-2, and HAA-SP-3. The spiking solution used was purchased from Supelco and contained a range of 100-300 μg/L of each analyte. Theoretical HAA concentration in spike samples is presented in Table E.2. The HAA-SP-3 sample consisted of plant effluent water without any additional HAA added.

The external laboratory also analyzed the THM4 spike samples prepared using USEPA Method 551.2 (USEPA 1989), and the HAA9 samples by USEPA Method 552.2 (USEPA 1992). At the time and location of sample collection, free and total chlorine residuals were also measured.

Table E.1
Theoretical THM concentration (in μg/L) in spike samples including spike and existing THM concentration in plant effluent water

THM	THM-SP-1	THM-SP-2
$CHCl_3$	82.5	66.5
$CHBrCl_2$	86.5	70.5
$CHBr_2Cl$	85.7	69.7
$CHBr_3$	75.9	59.9
THM4	330.6	266.6

Table E.2
Theoretical HAA concentration (in µg/L) in spike samples including spike and existing HAA concentration in plant effluent water

HAA	HAA-SP-1	HAA-SP-2
ClAA	60.0	3.0
BrAA	40.0	2.0
Cl_2AA	66.2	9.2
BrClAA	46.5	8.5
Br_2AA	24.4	5.4
Cl_3AA	25.5	6.5
$BrCl_2AA$	46.0	8.0
Br_2ClAA	43.6	5.6
Br_3AA	20.3	1.3
HAA9	372.5	49.5

Subsequent Inter-Laboratory Comparison Tests

The subsequent inter-laboratory comparison experiments were conducted as follows. During a sampling, additional containers were collected in duplicate and sent to both laboratories. Samples were collected at the POE and two additional locations in one of the distribution systems studied. Sampling locations were based on the concentration and speciation of DBPs present at the various locations. Different distribution systems were studied during the various inter-laboratory experiments conducted, as presented above. Sampling protocol for the inter-laboratory comparison test samples was similar to that used for the regular samples. All THM4 and HAA9 were analyzed during these inter-laboratory comparisons.

RESULTS

Five inter-laboratory comparison tests were conducted.

First Inter-Laboratory Comparison Test

The first test was conducted on August 13, 2002, and involved sample collection at the POE of the five distribution systems studied, as well as spikes samples prepared in treatment plant effluent water by an external laboratory. Multiple sample bottles were collected from each location and sent to both analytical laboratories, NNWW and UNC. The external laboratory also analyzed the spiked samples. Field blanks were collected at all utilities sampled.

Trihalomethanes

THM4 results are illustrated on Figure E.1. Results indicate that the methods used at NNWW and UNC provided similar trends. Due to a procedural error at UNC (a mistake was made by the new analytical team and an internal standard was not added to the samples), the RPDs between NNWW and UNC's samples were higher than expected.

THM4 results of the spike samples obtained from the external laboratory provided values between those obtained from NNWW and UNC. Between the external laboratory and NNWW, RPDs of 16 and 13 percent were calculated for THM-SP-1 and THM-SP-2, respectively, and 12 and 19 percent for THM-SP-1 and THM-SP-2 between the external laboratory and UNC. RPDs between NNWW and UNC were 25 and 26 percent for THM-SP-1 and THM-SP-2, respectively.

The difference between NNWW and UNC sample containers was determined by the analysis of THM-SP-1 and THM-SP-2 samples by the external laboratory. For THM4, RPDs were 3.3 and 5.5 percent for THM-SP-1 and THM-SP-2, respectively, which suggest that potential differences in vial washing procedures and preservative/dechlor(am)inating agents did not affect THM4 analytical results.

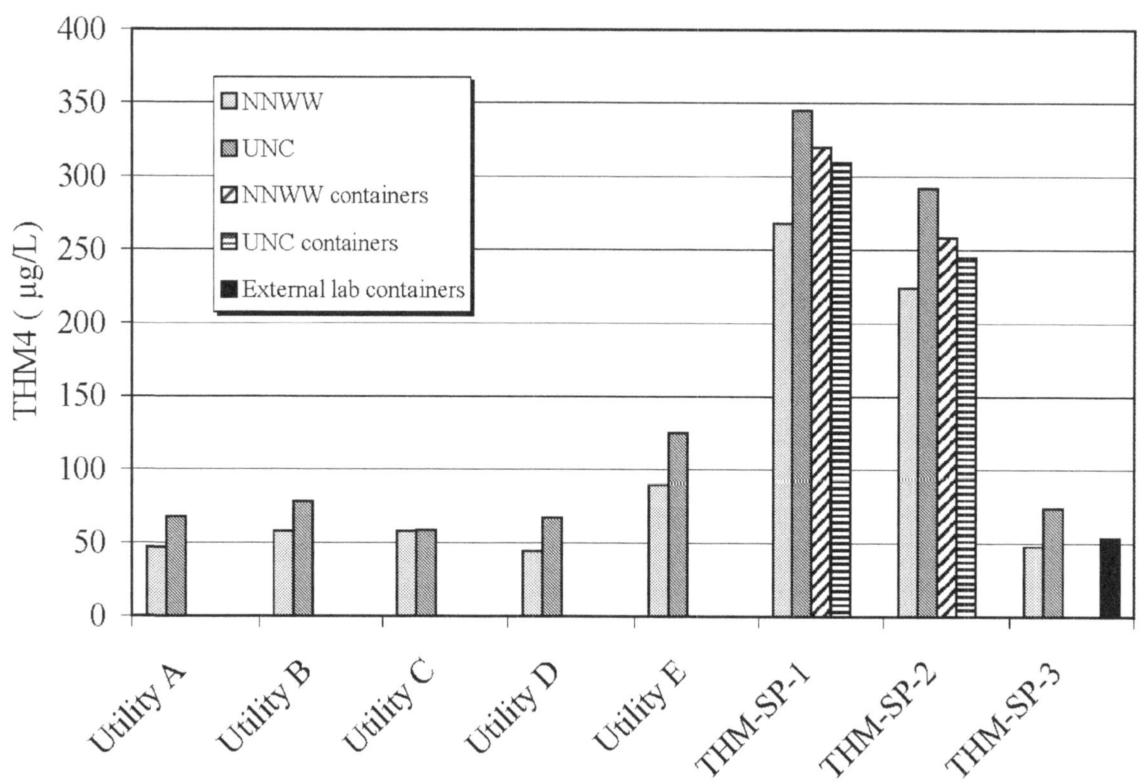

Figure E.1 THM4 concentrations measured during the inter-laboratory comparison test No.1 (all utilities were sampled)

Internal Quality Control Within Each Laboratory. At NNWW, RPDs of replicate samples varied between 0 and 22 percent for individual THM4 species (higher RPDs were calculated for $CHBr_3$ when concentrations were low), although the RPD of THM4 concentrations were 1.7 to 2.1 percent only and the differences in concentration between replicate samples were never more than 2 µg/L.

Procedural errors were encountered at UNC and some of the results obtained were outside the range of acceptability. RPDs of THM4 samples varied between 0.1 and 20 percent, with RPDs ranging from 0.3 to 27 percent for the individual THM4 species (one of the samples was not considered in this evaluation as it was clearly unacceptable with RPDs of 60 to 85 percent for individual THM4 species).

Haloacetic Acids

Results of HAA9 are presented on Figure E.2 and indicate that the methods used at NNWW and UNC provided similar trends. RPDs of individual HAA9 species and HAA9 indicated that the differences observed between both laboratories were generally within acceptable limits, except for Br_3AA. The high RPDs observed with Br_3AA were due to analyses of concentrations close to the MDL.

HAA9 analysis performed at the external laboratory provided lower results than those obtained by NNWW and UNC. Between the external laboratory and NNWW, RPDs were 14 and 37 percent for HAA-SP-1 and HAA-SP-2, respectively, and 21 and 33 percent for HAA-SP-1 and HAA-SP-2 between the external laboratory and UNC.

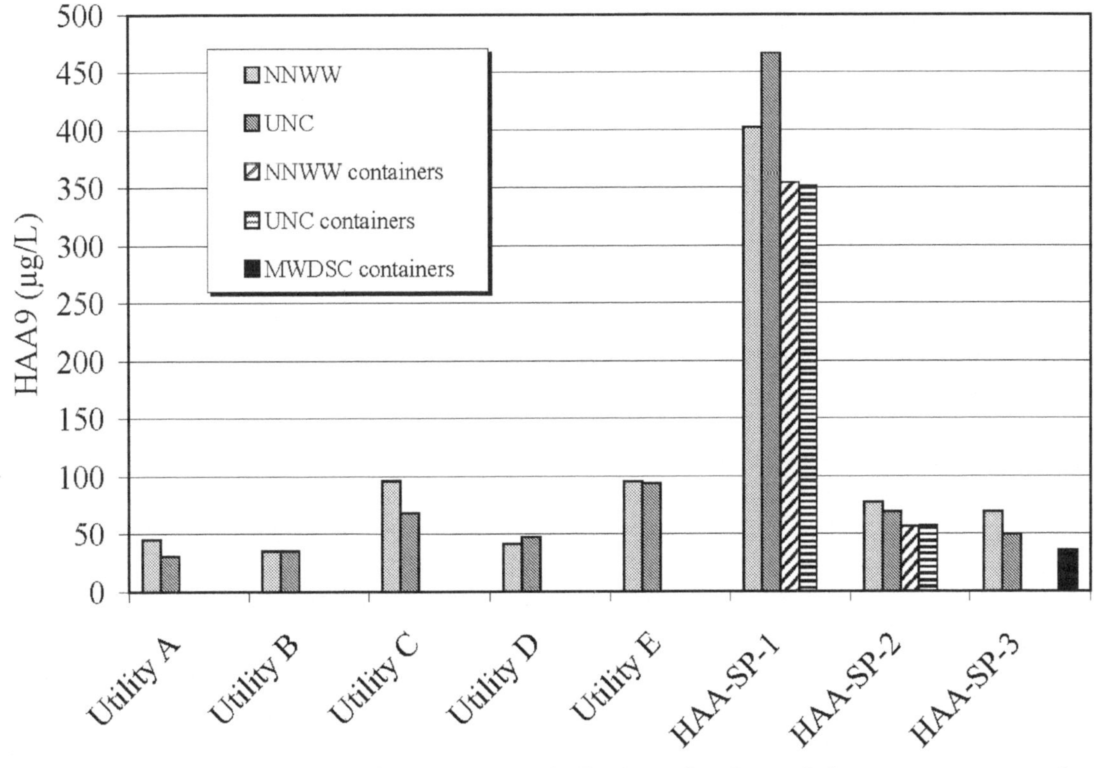

Figure E.2 HAA9 concentration measured during the inter-laboratory comparison test No.1 (all utilities were sampled)

The difference between NNWW and UNC containers was determined by the analysis of HAA-SP-1 and HAA-SP-2 samples by the external laboratory. For HAA9, the RPDs were 0.88 and 1.2 percent for HAA-SP-1 and HAA-SP-2, respectively. These results suggested that potential differences in vial washing procedures and preservative/dechlor(am)inating agents did not affect the analytical results.

Internal Quality Control Within Each Laboratory. At NNWW, RPDs of replicate samples were acceptable and varied between 0 and 20 percent for individual HAA9 species, with a 40 percent difference for ClAA, although the difference in concentration was only 2 µg/L between both duplicates. RPD of HAA5, HAA6 and HAA9 varied between 6.7 and 8.7 percent.

RPD of replicate HAA9 samples analyzed at UNC varied between 0.1 and 21 percent. Some RPDs for ClAA and Br_3AA were outside the range of acceptability, but these sample concentrations were close to the MDL.

Second Inter-Laboratory Comparison Test

The second test involved NNWW and UNC laboratories only and was conducted during the October 15, 2002 sampling. Three of the samples collected at Utility C (chloraminated system) were used: Nos.1, 8, and 15.

Trihalomethanes

THM4 results are presented on Figure E.3. Results showed inter-laboratory RPDs greater than the recommended limits.

The laboratory's individual QC indicated that at NNWW, THM concentrations in the travel blank were 1 µg/L for $CHCl_3$ and $CHBr_3$, and <1 µg/L for $CHBrCl_2$ and $CHBr_2Cl$. The RPDs between the replicates were within the acceptable limits, and ranged from 0 to 5.5 percent for the individual THM4 species. One of the QCSs lead to recoveries of 87 to 139 percent for the individual THM4 species, whereas the other lead to recoveries of 82 to 121 percent. Thus the QCSs were within the acceptable range except for $CHBr_3$, which had a low QCS concentration of 2.7 µg/L with a carry over of 1 µg/L leading to recoveries of 121 and 139 percent.

At UNC, RPDs between the replicates were within the acceptable limits and ranged from 0.3 to 16 percent. The matrix spikes performed at UNC lead to out-of-control recoveries ranging from 110 to 135 percent for the individual THM4 species.

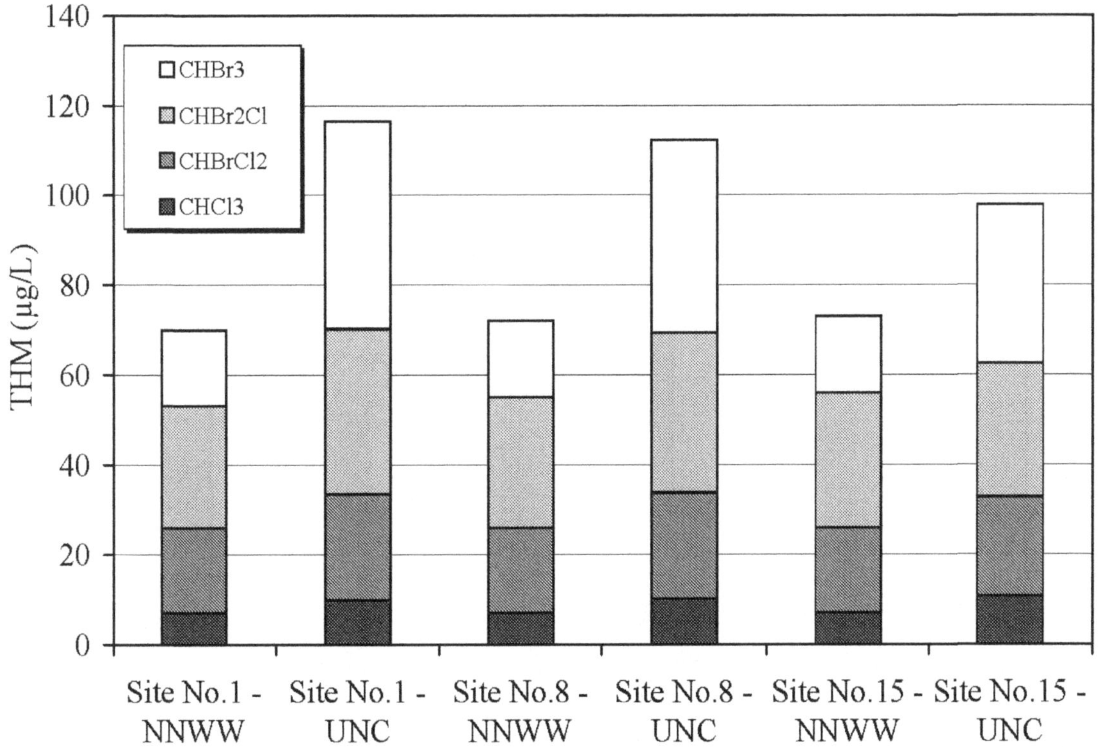

Figure E.3 THM4 concentrations measured during the inter-laboratory comparison test No.2 conducted at Utility C (chloraminated system) (water temperature at the POE was 22°C, 72°F)

Haloacetic Acids

HAA9 results obtained during the second test are presented on Figure E.4. Results showed RPDs that were greater than the recommended limits. In the case of ClAA, BrAA and to some extent Cl_3AA, results were close to or at the MDL, which explained the high RPDs calculated. Although NNWW detected some Br_2ClAA, it was below the MRL at UNC, which explained the high RPDs calculated.

At NNWW, RPDs between the replicate samples were 0 percent for the individual HAA9 species, with the exception of $BrCl_2AA$ for which it was 12 percent (within the acceptable limits), and ClAA which had a 67 percent RPD because one of the replicates showed a concentration of 4 µg/L while the other showed a level below the MRL of 2 µg/L. The QCSs showed acceptable recoveries ranging from 88 to 110 percent for the individual HAA9 species.

At UNC, the RPDs between replicate samples were acceptable and ranged from 0 to 18 percent. The matrix spike samples ranged from 98 to 116 percent for the individual HAA9 species, except for ClAA which showed a recovery of 43 percent.

Third Inter-Laboratory Comparison Test

The third test was conducted during the January 7, 2003 sampling, and used three of the samples collected from Utility D (chloraminated system): Nos.1, 5, and 11.

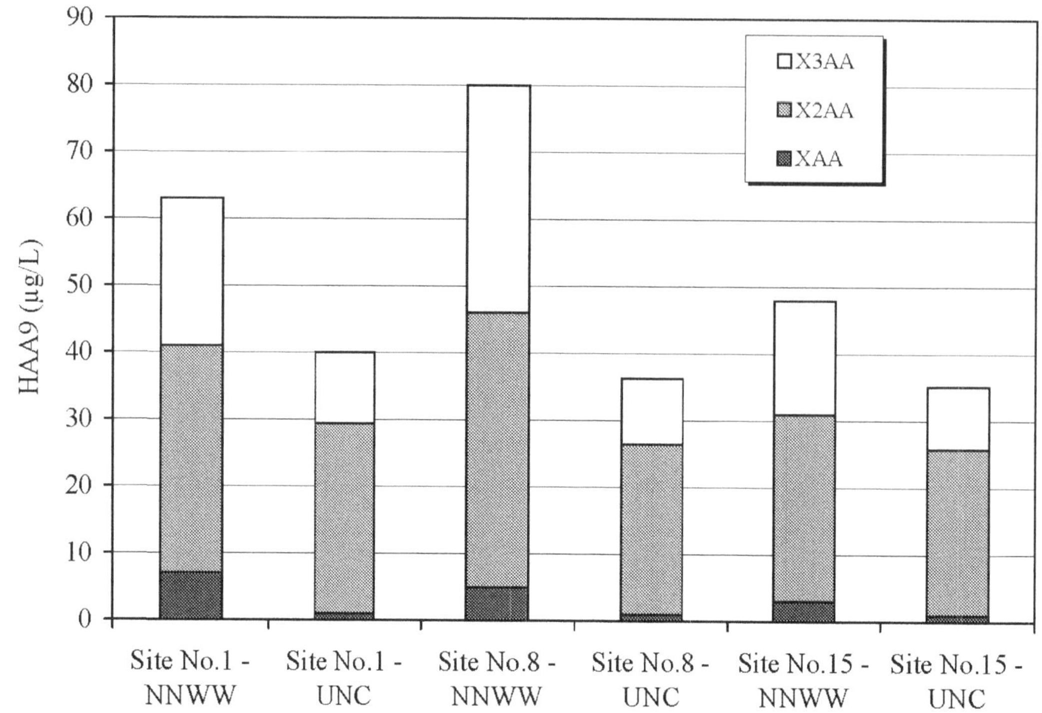

Figure E.4 HAA9 concentrations measured during the inter-laboratory comparison test No.2 conducted at Utility C (chloraminated system) (water temperature at the POE was 22°C, 72°F)

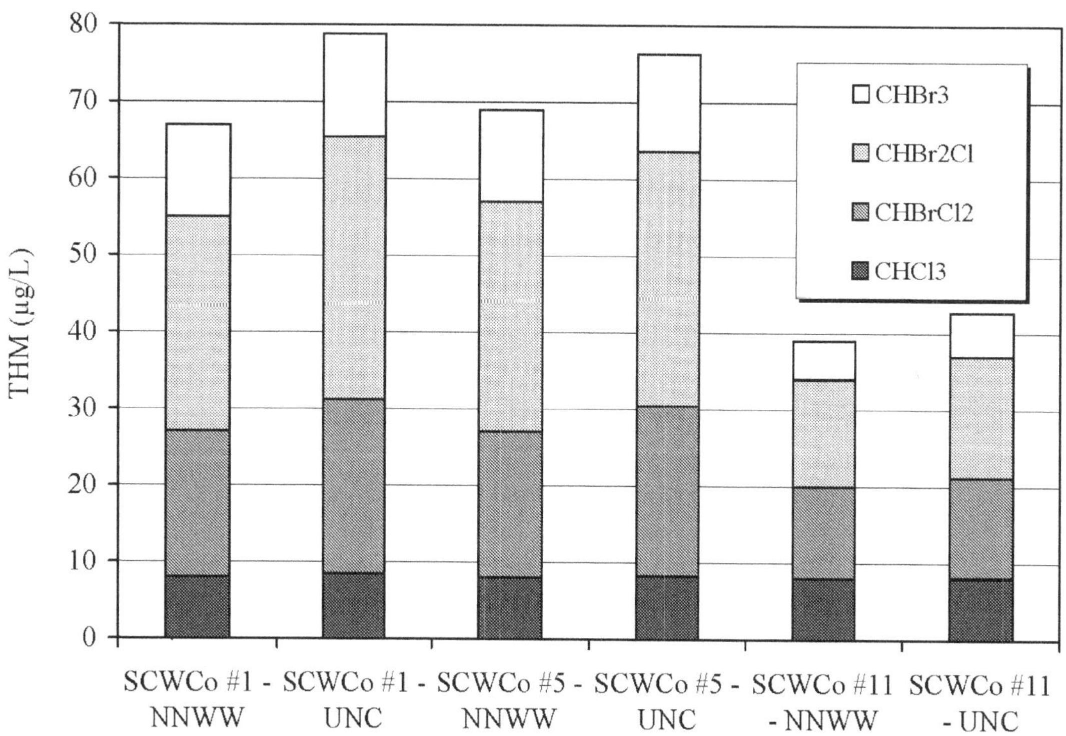

Figure E.5 THM4 concentrations measured during the inter-laboratory comparison test No.3 conducted at Utility D (chloraminated system) (water temperature at the POE was 15.6°C, 60°F)

Trihalomethanes

THM4 results are presented on Figure E.5. Results show that all inter-laboratory RPDs were within the recommended limits.

The laboratories' internal QC data showed that all data met acceptable requirements. At NNWW, travel blanks were <1 μg/L, LFBs ranged from 105 to 110 percent, and QCSs ranged from 86 to 110 percent. At UNC, field blanks were <1 μg/L, RPDs between replicate samples ranged from 0 to 6.2 percent, and matrix spike recoveries ranged from 81 to 88 percent.

Haloacetic Acids

HAA9 results are presented on Figure E.6. Results showed RPDs that were greater than the recommended limits. In general, HAA9 concentrations measured at NNWW were higher than those measured at UNC, which was responsible for the high RPDs calculated.

Both laboratories met their individual QC requirements. At NNWW, the LFBs ranged from 90 to 110 percent, and the QCSs from 88 to 125 percent (within the acceptable limits). The matrix spike samples showed recoveries ranging from 100 to 121 percent, with an out-of-control recovery of 57 percent for Br_3AA. At UNC, the field blanks were <2 μg/L. The RPDs between replicate samples were acceptable and ranged between 0 and 21 percent. The matrix spikes ranged from 101 to 132 percent, except for ClAA, which showed a recovery of 45 percent.

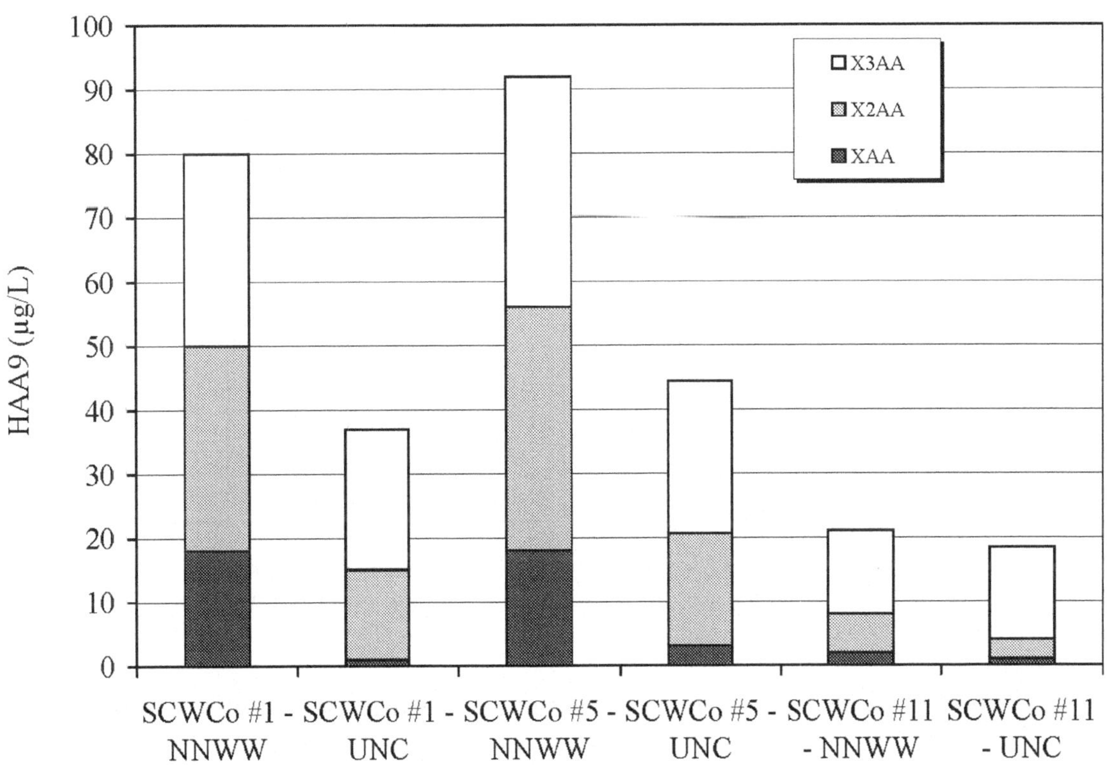

Figure E.6 HAA9 concentrations measured during the inter-laboratory comparison test No.3 conducted at Utility D (chloraminated system) (water temperature at the POE was 15.6°C, 60°F)

Fourth Inter-Laboratory Comparison Test

The fourth test was conducted during the April 29, 2003 sampling. Three of the samples collected from Utility E's distribution system (chloraminated system that periodically switches to free chlorine) were used: Sites Nos. 2, 4 and 14.

Trihalomethanes

THM4 results are presented on Figure E.7. Results showed that some of the inter-laboratory RPDs were outside the acceptable limits.

The individual laboratories met their respective QC requirements. At NNWW, the RPDs between replicate samples were 0 and 5.2 percent, the LFBs ranged from 95 to 106 percent, and the QCS ranged from 98 to 112 percent. At UNC, the field blanks were <1 µg/L, the RPDs between replicate samples ranged from 0 to 5.3 percent, and the matrix spike recoveries were 81 and 93 percent.

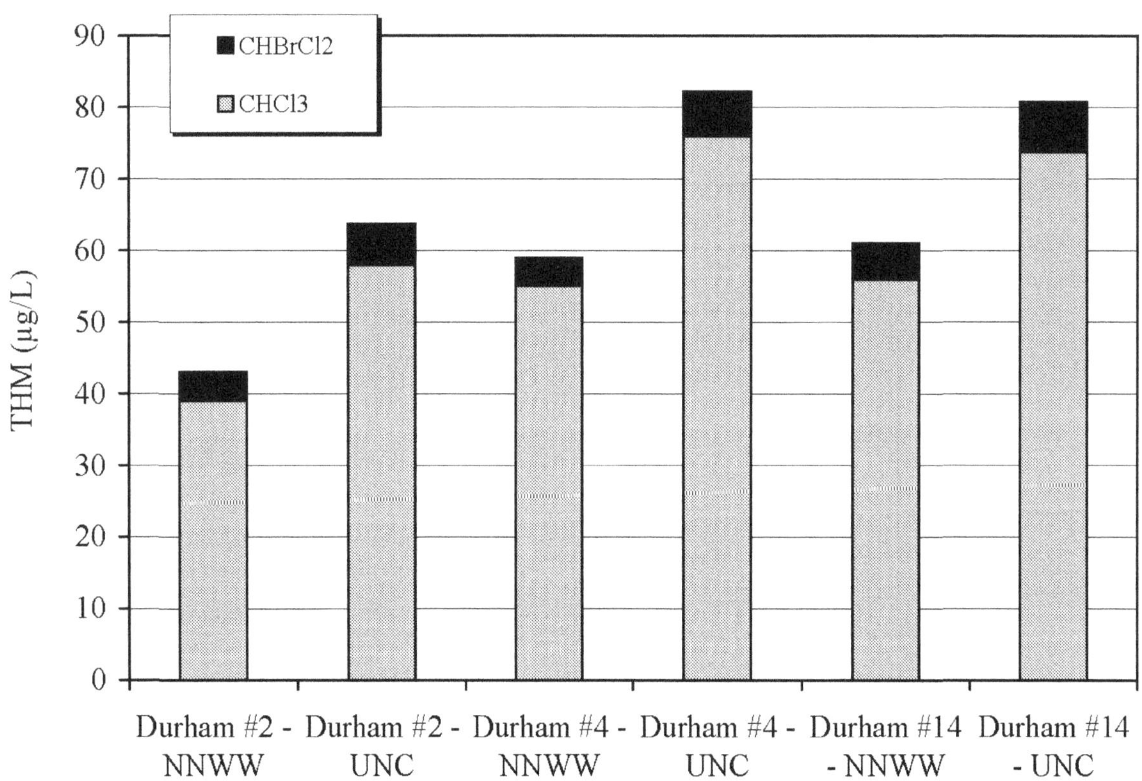

Figure E.7 THM4 concentrations measured during the inter-laboratory comparison test No.4 conducted at Utility E (chloraminated system that periodically switches to free chlorine) ($CHBr_2Cl$ and $CHBr_3$ were not detected) (water temperature at the POEs was 15.5-15.6°C, 60°F)

Haloacetic Acids

HAA9 results are presented on Figure E.8. Results showed acceptable RPDs between both laboratories, except when the concentrations measured were close to the MRL.

The individual laboratories met their respective QC requirements. At NNWW, the LFBs ranged between 90 and 97 percent at NNWW. The QCSs ranged from 66 to 105 percent, with an QCS of 10 percent for Br_3AA. The out-of-control QCSs corresponded to HAA9 species that were not detectable.

At UNC, the field blanks were <2 µg/L. The RPDs between replicate samples were acceptable and ranged from 0 and 21 percent, except for ClAA which showed an RPD of 33 percent because concentrations were close to the MRL. The matrix spike recoveries were acceptable and ranged from 88 to 98 percent.

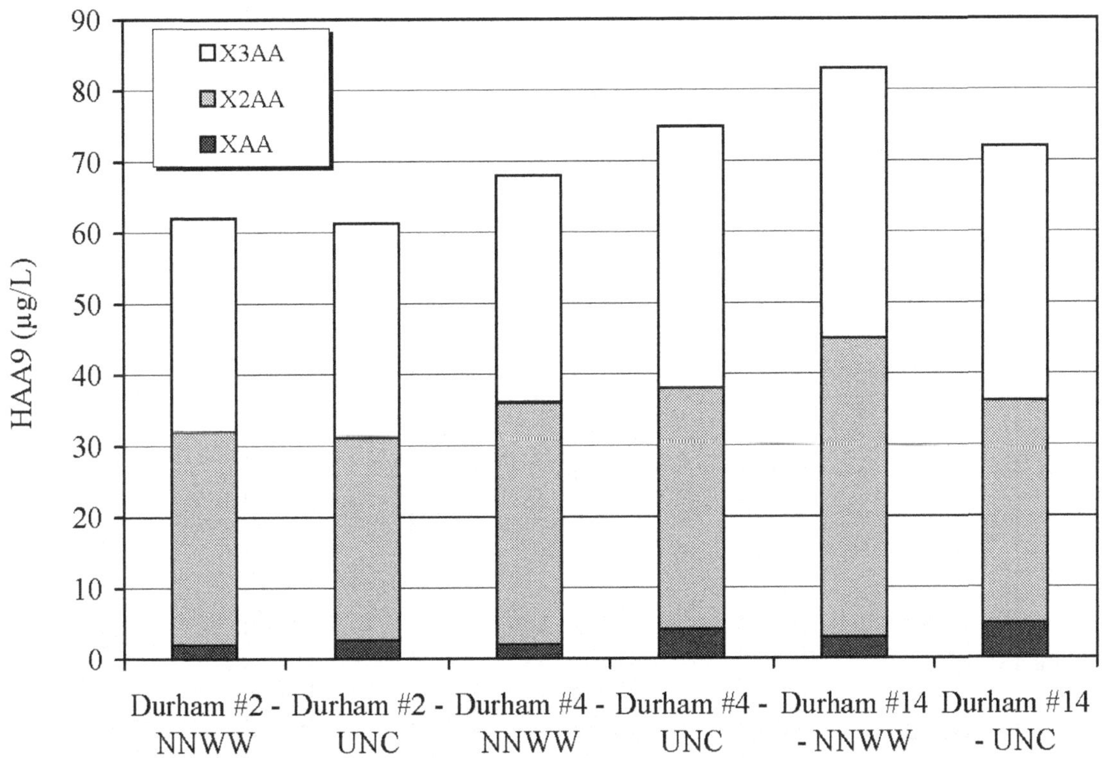

Figure E.8 HAA9 concentrations measured during the inter-laboratory comparison test No.4 conducted at Utility E (chloraminated system that periodically switches to free chlorine) (water temperature at the POEs was 15.5-15.6°C, 60°F)

Fifth Inter-Laboratory Comparison Test

The last test was conducted during the June 10, 2003 sampling. Three of the samples collected from Utility B's distribution system (free chlorinated system) were used: the POE, and Sites No. 883 and 251.

Trihalomethanes

THM4 results are presented on Figure E.9. Inter-laboratory RPD results were within the recommended limits, except for one of the $CHCl_2Br$ samples and one of the $CHClBr_2$ samples, although the out-of-control $CHClBr_2$ sample had a concentration very close to the MDL.

The individual laboratories' QC data indicate that at NNWW, the travel blanks were <1 μg/L, the LFBs ranged from 100 to 113 percent, and the QCSs ranged from 89 to 109 percent (within the acceptable limits). At UNC, the RPDs between replicates ranged from 0 to 9.7 percent, which is within the acceptable limits. Some of the matrix spikes were however out of control with recoveries ranging from 56 to 119 percent.

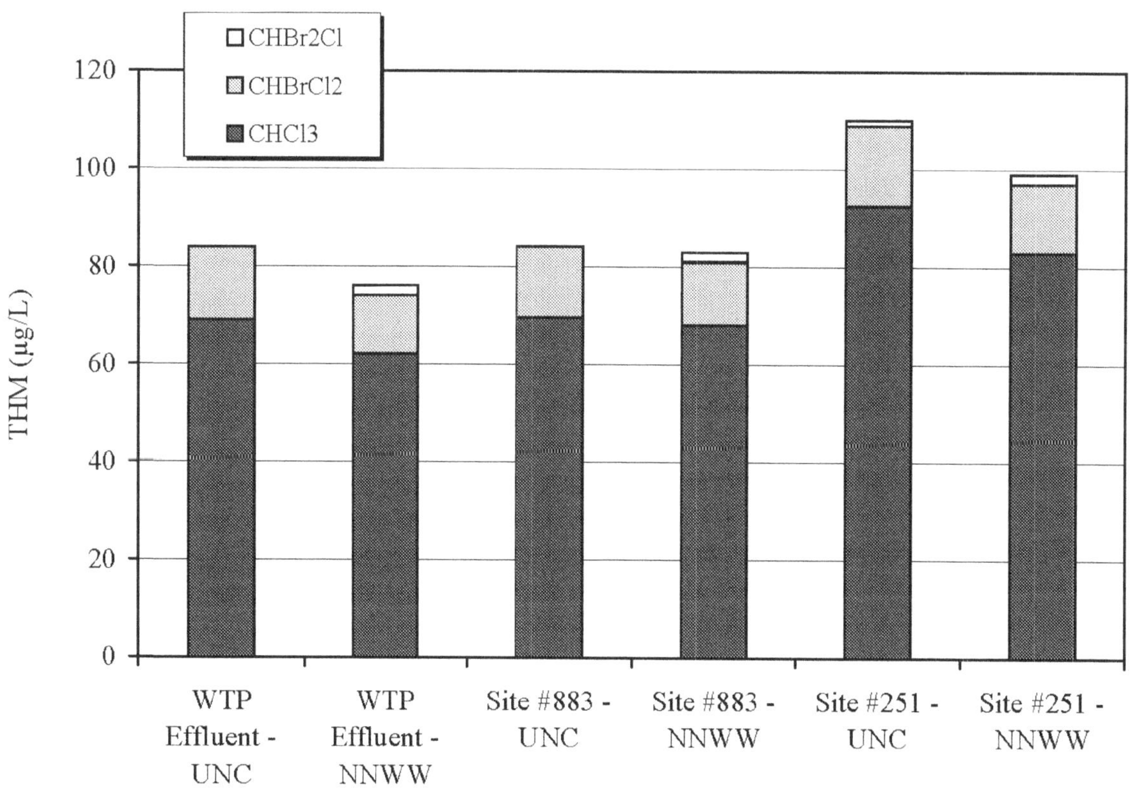

Figure E.9 THM4 concentrations measured during the inter-laboratory comparison test No.5 conducted at Utility B (free chlorinated system) (water temperature at the POE was 26°C, 79°F)

Haloacetic Acids

HAA9 results are presented on Figure E.10, and showed that some of the RPDs did not respect the acceptable limits.

The QC results of the individual laboratories can be summarized as follows. At NNWW, the LFBs were acceptable and ranged from 97 to 117 percent, except for Br_3AA which showed an LFB of 143 percent. The QCSs ranged from 79 to 110 percent at NNWW (within the acceptable limits).

At UNC, the RPDs between replicate samples were acceptable and ranged from 0 to 7.7 percent, except for ClAA which showed higher RPDs of 24 and 56 percent. The matrix spikes ranged from 79 to 126 percent, which was within the acceptable limits.

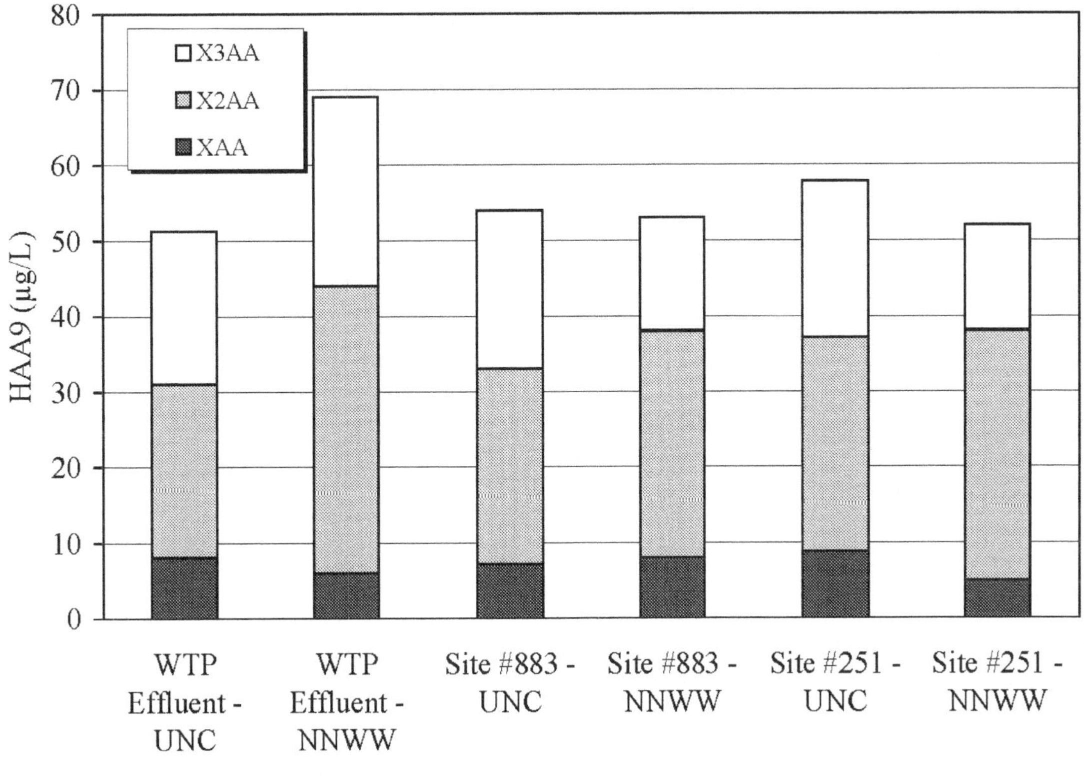

Figure E.10 HAA9 concentrations measured during the inter-laboratory comparison test No.5 conducted at Utility B (free chlorinated system) (water temperature at the POE was 26°C, 79°F)

SUMMARY AND CONCLUSIONS

This appendix presents results of five inter-laboratory comparison tests that were conducted to compare results obtained from the laboratories when analyzing THM4 and HAA9 for this study. Results show that differences were observed between both laboratories, although these differences were mainly observed when concentrations measured were low. Results suggest that these differences were not related to different collection vials, including washing procedures and preservative/dechlor(am)inating agents. The reasons for these differences are unknown, and this study was not designed to address them. To compensate for possible differences between laboratories, all samples collected from a utility were always sent to the same laboratory. Thus, consistency within each utility was respected. Also, considering the differences between the different full-scale systems studied (different source water, geographic locations, distribution system characteristics, etc.), absolute values were not compared across utilities; only the trends observed were compared.

These data emphasize the need to further examine the analytical methods that are suggested to comply with the Stage 2 DBPR. As observed during this study, even reputable laboratories that use strict QC procedures may show differences in THM or HAA concentrations, particularly when the concentrations are low. The source(s) of these differences need to be identified and examined in future studies.

REFERENCES

Adin, A., J. Katzhendler, D. Alkaslassy, and C. Rav-Acha. 1991. Trihalomethane Formation in Chlorinated Drinking Water: A Kinetic Model. *Wat. Res.*, 25(7):797-805.

Alarcon-Herrera, M.T., N. Biswas, and J.K. Bewtra. 1992. Seasonal Variations in Humic Substances and their Reduction Through Water Treatment Processes. AQTE - 4th Workshop on Drinking Water, Montréal, Québec, November 19, 353-362.

American Public Health Association (APHA). 1998. *Standard Methods for the Examination of Water and Wastewater*. Washington D.C.: 20th edition. APHA, AWWA, WPCF.

Amy, G.L., P.A. Chadik, and Z.K. Chowdhury. 1987. Developing Models for Predicting Trihalomethane Formation Potential and Kinetics. *Journal AWWA*, 79:7:89-97.

Amy, G.L., T. Tan, and M. Davis. 1991. The Effects of Ozonation and Activated Carbon Adsorption on Trihalomethane Speciation. *Water Research*, 25(2):191-202.

Andrews, S.A., and V.Y. Taguchi. 2000. NDMA - Canadian Issues. *In proceedings of the Water Quality Technology Conference*. Salt Lake City, Utah.

Andrews, S.A., P.M. Huck, and R.T. Coutts. 1992. Ozonation and Post-chlor(am)ination By-products of Fractionated Natural Organic Matter Obtained Using an Improved Isolation Procedure. AWWA Water Quality Technology Conference. Toronto, Ontario.

Appel, R.D., A. Bairoch, and D.F. Hochstrasser, 1994. A New Generation of Information Retrieval Tools for Biologists: The Example of the ExPASy WWW Server. *Trends Biochem. Sci.*, 19:258-260.

Baribeau, H., S.W. Krasner, R. Chinn, and P.C. Singer. 2000. Impact of Biomass on the Stability of Haloacetic Acids and Trihalomethanes in a Simulated Distribution System. *In proceedings of the AWWA Water Quality Technology Conference*, Denver, Colo.

Baribeau, H., M. Prévost, R. Desjardins, and P. Lafrance. 2001. Changes in Chlorine and DOX Concentrations in Distribution Systems. *Jour. AWWA*, 93(12):102-114.

Barrett, S., C. Hwang, Y. Guo, S.A. Andrews, and R.L. Valentine. 2003. Ocurrence of NDMA in Drinking Water: A North American Survey, 2001-2002. AWWA Annual Conference and Exhibition, Anaheim, CA.

Barthe, C., N. Brassard, F. Gamache, and A. Riopel. 1991. Phénomène de recroissance bactérienne dans les systèmes de distribution d'eau potable. AQTE 3rd Workshop on Drinking Water, Montréal, Québec, November 14.

Ben Amor, H. 1988. Le bioxyde de chlore dans la filière de production des eaux de consummation: Contribution à l'étude de son mécanisme d'action sur des molecules modèles et sur les substances humiques aquatiques. Ph.D. thesis, Order No.175, University of Poitiers, France, October.

Bergmann, J.G., and J. Sanik, 1957. Determination of Trace Amounts of Chlorine in Naphtha. *J. Anal. Chem.*, 29:241-242.

Biswas, P., C. Lu, and R.M. Clark, 1993. A Model for Chlorine Concentration Decay in Drinking Water Distribution Pipes. *Wat. Res.*, 27(12):1715.

Blank, V., H.M. Shukairy, and J. McLain. 2002. Unregulated Organic DBPs in ICR Finished Water and Distribution Systems. In: *Information Collection Rule Data Analysis*. Prepared by M.J. McGuire, J.L. McLain, and A. Obolenski. Denver, Colo.: AwwaRF and AWWA, 277-297.

Blum, H., H. Beier, and H.J. Gross. 1987. Improved Silver Staining of Plant Proteins, RNA, and DNA in Polyacrylamide Gels. *Electrophoresis*, 8:93-99.

Bontrager, S.L., Z. Chowdhury, and J. Hayes. 1992. Optimization of Conventional Treatment for Disinfection By-product Control. AWWA Annual Conference, Vancouver, British Columbia, June 18-22.

Bove, F.J., M.C. Fulcomer, J.B. Klotz, J. Esmart, E.M. Dufficy, and J.E. Savrin. 1995. Public Drinking Water Contamination and Birth Outcomes. *American Journal of Epidemiology*, 141:850-862.

Boyce, S.D., and J.F. Hornig. 1983. Reaction Pathways of Trihalomethanes Formation from the Halogenation of Dihydroxyaromatic Model Compounds for Humic Acid. *Environmental Science Technology*, 17(4):202-211.

Brett, R.W., and R.A. Calverley. 1979. A One-Year Survey of Trihalomethane Concentration Changes Within a Distribution System. *Jour. AWWA*, 71(9):515-520.

Brophy, K.S., H.S. Weinberg, and P.C. Singer. 2000. Quantification of Nine Halogenated Acetic Acids Using Gas Chromatography with Electron Capture Detection. In: *Natural Organic Matter and Disinfection By-Products: Characterization and Control in Drinking Water, ACS Symposium Series 761*. Edited by S.E. Barrett, S.W. Krasner, and G.L. Amy. Washington, D.C.: American Chemical Society, 343-355.

Bryant, E.A., G.P. Fulton, and G.C. Budd. 1992. *Disinfection Alternatives for Safe Drinking Water*. Ed. Hazen and Sawyer, Van Nostrand Reinhold.

Bull, R.J., and F.C. Kopfler. 1991. *Health Effects of Disinfectants and Disinfection By-products*. Denver, Colo.: AWWARF and AWWA.

Bull, R.J., S.W. Krasner, P.A. Daniel, R.D. and Bull. 2001. *Health Effects and Occurrence of Disinfection By-products*. Denver, Colo.: AWWARF.

Bull, R.J., I.M. Sanchez, M.A. Larsen, and A.J. Lansing, 1990. Liver Tumor Induction in B6C3F1 Mice by Dichloroacetate and Trichloroacetate. *Toxicol.*, 63:341-359.

California Code of Regulation (CCR). 2000. Title 22, Section 12705 (b)(1).

California Department of Health Services (DHS). 2002a. Studies on the Occurrence of NDMA in Drinking Water. http://www.dhs.ca.gov/ps/ddwem/chemicals/NDMA/studies.html. March 15.

California Department of Health Service (DHS). 2002b. California Drinking Water: NDMA Related Activities. http://www.dhs.ca.gov/ps/ddwem/NDMA/NDMAindex.htm

Campbell, W.H. 1999. BL483 Biochemistry Techniques. Houghton: Michigan Technological University. http://www.bio.mtu.edu/campbell/bl482/lectures/lec3/482ek2.htm [Accessed 28 November 2000].

Cantor, K.P., C.F. Lynch, M.E. Hildesheim, M. Dosemeci, J. Lubin, M. Alavanja, and G. Craun. 1998. Drinking Water Source and Chlorination By-products. I. Risk of Bladder Cancer. *Epidemiology*, 9:211-228.

Capellier, M., C. Picoche, and A. Deguin. 1992. Devenir du CODB dans les réseaux de distribution – Étude de deux cas. *Revue des Sciences de l'Eau*, 5:51-67.

Caprette, D.R. 1997. Absorbance Assay (280 nm). Houston: Rice University. http://www.ruf.rice.edu/~bioslabs/methods/protein/abs280.html [Accessed 3 March 2004].

Carlson, M. and D. Hardy. 1998. Controlling DBPs with Monochloramine. *Journal AWWA*, 90(2):95-106.

Cattley, R.C., J. DeLuca, C. Elcombe, P. Fenner-Crisp, B.G. Lake, D.S. Marsman, T.A. Pastoor, J.A. Popp, D.E. Robinson, B. Schwetz, J. Tugwood, and W. Wahli. 1998. Do Peroxisome Proliferating Compounds Pose a Hepatocarcinogenic Hazard to Humans. *Regulatory Toxicology and Pharmacology*, 27:47-60.

Chang, C.-H., J.F. Schinder, C.F. Unkerfer, L.A. Vanderberg, J.R. Brainard, and T.C. Terwilliger. 1999. In Vivo Screening of Haloalkane Dehalogenase Mutants. *Bioorg. Med. Chem.*, 7:2175-2181.

Chen, W.J., and C.P. Weisel. 1998. Halogenated DBP Concentrations in a Distribution System. *Jour. AWWA*, 90(4):151-163.

Child, P., G. Kaa, D. Benitz, P. Fowlie, and R. Hong-You. 1996. Reaction Between Chlorine and a Dimethylamine Containing Polyelectrolyte Leading to the Formation of N-nitrosodimethylamine. http://www.pollutech.com/papers/p11.htm.

Choi, J., and R.L. Valentine. 2001. Studies on the Formation of N-Nitrosodimethylamine (NDMA) in Drinking Water: A New Chloramination Disinfection By-product. AWWA Annual Conference, Washington, D.C., June 17-21.

Chowdhury, K.K., and G.L. Amy. 1999. Modeling Disinfection By-product Formation. In: *Formation and Control of Disinfection By-products in Drinking Water*. Singer, P.C. ed., AWWA Publishers.

Chu, I., D.C. Villeneuve, V.E. Secours, G.C. Becking, and V.E. Valli. 1982a. Toxicity of Trihalomethane: I The Acute and Subacute Toxicity of Chloroform, Bromodichloromethane, Chlorodibromomethane, and Bromoform in Rats. *Jour. Environmental Science and Health*, B17(3):205-224.

Chu, I., D.C. Villeneuve, V.E. Secours, G.C. Becking, and V.E. Valli. 1982b. Toxicity of Trihalomethane: II Reversibility of Toxicological Changes Produced by Chloroform, Bromodichloromethane, Chlorodibromomethane, and Bromoform in Rats. *Jour. Environmental Science and Health*, B17(3):225-240.

Coallier, J., J. Lavoie, D. Duchesne, M. Prévost, and R. Desjardins. 1991. Comparaison de la qualité de l'eau dans les réseaux de distribution de Ville de Laval. *In proceedings of AQTE – 3rd Workshop on Drinking Water*, Montréal, Québec, November 14, 395-441.

Colbourne, J.S., P.J. Dennis, A.J. Rachwal, W. Keevil, and C. Mackerness, 1991. The Operational Impact of Growth on Coliforms in London's Distribution System. *In proceedings of the AWWA – Water Quality Technology Conference*, Orlando, Flo.

Condie, L.W., C.L. Smallwood, and R.D. Laurie. 1983. Comparative Renal and Hepatoxicity of Halomethanes: Bromodichloromethane, Bromoform, Chloroform, Dibromochloromethane, and Methylene Chloride. *Drug and Chemical Toxicology*, 6(6):563-578.

Cowman, G.A., and P.C. Singer. 1994. Effects of Bromide Ion on Haloacetic Acid Speciation Resulting from Chlorination and Ozonation of Aquatic Humic Substances. AWWA Water Quality Technology Conference, San Francisco, CA, November 6-10.

Cowman, G.A., and P.C. Singer. 1996. Effect of Bromide Ion on Haloacetic Acid Speciation Resulting from Chlorination and Chloramination of Aquatic Humic Substances. *Environ. Sci. & Tech.*, 30(1):16-24.

Croué, J.P. 1987. Contribution à l'étude de l'oxydation par le chlore et l'ozone d'acides fulviques naturels extaits d'eau de surface. Ph.D. thesis, Order No.89, École Supérieure d'Ingénieurs de Poitiers, France, July.

Damborsky, J., E. Rorije, A. Jesenska, Y. Nagata, G. Klopman, and W. Peijenburg. 2000. Structure-Specificity Relationships for Haloalkane Dehalogenases. *Environ. Toxicol. Chem.*, 20:2681-2689.

Damborsky, J., S. Luengo, F. Gago, A. Ramirez, and R. Wade. 2001. Comparative Binding Energy Analysis of the Substrate Specificity of Haloalkane Dehalogenase from *Xanthobacter autotrophicus* GJ-10. *Biochemistry*, 40:8905-8917.

De Angelo, A.B. and L.P. McMillan. 1989. Carcinogenicity of Chlorinated Acetic Acids. In: *Water Chlorination, Vol.6*, Lewis Publishers, Inc., 193-199.

De Angelo, A.B., F.B. Daniel, B.M. Most, and G.R. Olson. 1996. The Carcinogenicity of Dichloroacetic Acid Administered in the Drinking Water to Produce Liver Cancer in Male F344/N Rats. *Jour. of Toxicology and Environmental Health*, 52(5):425-445.

De Laat, J., N. Merlet, and M. Doré. 1982. Chloration de composés organiques: Demande en chlore et réactivité vis-à-vis de la formation des trihalométhanes. *Wat. Res.*, 16:1437-1450.

Desjardins, R., J. Lavoie, P. Lafrance, and M. Prévost. 1991. Comparaison de l'évolution de la qualité de l'eau dans deux réseaux de distribution. *Sciences et techniques de l'eau*, 24(4):321-331.

DiGiano, F.A. 2003. Personal communication.

Dodds, L., W. King, C. Woolcott, and J. Pole. 1999. Trihalomethanes in Public Water Supplies and Adverse Birth Outcomes. *Epidemiology*, 10:233-237.

Doré, M. 1989. Chimie des oxydants et traitement des eaux. *Technique et Documentation.* Lavoisier, Editors.

Doyle, T.J., W. Zheng, J.R. Cerhan, C.-P. Hong, T.A. Sellers, L.H. Kushi, and A.R. Folsom. 1997. The Association of Drinking Water Source and Chlorination By-products with Cancer Incidence among Postmenopausal Women in Iowa: A Prospective Cohort Study. *American Journal of Public Health*. 87:1168-1176.

Edzwald, J.K., and J.E. Van Benschoten. 1990. Aluminum Coagulation of Natural Organic Matter. In: *Fourth International Gothenburg Symposium on Chemical Treatment.* Madrid, Spain.

Falkinham, J.O., III, C.D. Norton, and M.W. LeChevallier. 2001. Factors Influencing Numbers of *Mycobacterium avium*, *Mycobacterium intracellular*, and Other *Mycobacteria* in Drinking Water Distribution Systems. *Appl. Environ. Microbiol.*, 67(3):1225-1231.

Foster, A.M., A.G. Prentice, J.A. Copplestone, R.A. Cartwright, and C. Ricketts. 1997. The Distribution of Leukemia in Association with Domestic Water Quality in Southwest England. *European Journal of Cancer Prevention*, 6:11-19.

Geldreich, E.E. 1996. *Microbial Quality of Water Supply in Distribution Systems*. New York: Lewis Publishers.

Gill, S.C., and P.H. von Hippel. 1989. Calculation of Protein Extinction Coefficients from Amino Acid Sequence Data. *Anal. Biochem.*, 182:319-326.

Graham, J.E., O. Meresz, G.J. Farquhar, and S.A. Andrews. 1996. Thiram as an NDMA Precursor in Drinking Water Treatment. *In proceedings of AWWA Annual Conference.* Toronto, Ontario.

Grulois, P., G. Matejka, and B. Parinet. 1989. Les précurseurs de THM dans les eaux de surface du Togo: Cas de substances humiques extraites des eaux de retenue de Notse. *Environmental Technology Letters*, 10:221-230.

Haas, C.N., M.A. Meyer, and M.S. Paller. 1983. Microbial Alterations in Water Distribution Systems and their Relationship to Physical-Chemical Characteristics. *Jour. AWWA*, 75(9):475-481.

Hardman, D.J. 1991. Biotransformation of Halogenated Compounds. *Crit. Rev. Biotechnol.*, 11:1-40.

Harms, L.L., and R.W. Loogenga. 1977. Chlorination Adjustment to Reduce Chloroform Formation. *Jour. AWWA*, 69(5):258-263.

Harrington, G.W., A. Bruchet, D. Rybacki, and P.C. Singer. 1996. Characterization of Natural Organic Matter and its Reactivity with Chlorine. In: *Water Disinfection and Natural Organic Matter.* Edited by Minear R.A. and Amy G.L. Washington D.C.: American Chemical Society, 138-158.

Harris, D.C. 2003. *Quantitative Chemical Analysis*. New York: W.H. Freeman and Company.

Harvey, S., and R. Harvey. 1998. An Introduction to Artificial Intelligence. *Appita Journal*, 51(1):20-24.

Herren-Freund, S.L., M.A. Pereira, M.D. Khoury, and G. Olson. 1987. The Carcinogenicity of Trichloroethylene and its Metabolites, Trichloroacetic Acid and Dichloroacetic, Acid in Mouse Liver. *Toxicol. Appl. Pharm.*, 90:183-189.

Hoigné, J., and H. Bader. 1988. The Formation of Trichloronitromethane (Chloropicrin) and Chloroform in a Combined Ozonation/Chlorination Treatment of Drinking Water. *Wat. Res.*, 22(3):313-319.

Holloway, P., J.T. Trevors, and H. Lee. 1998. A Colorimetric Assay for Detecting Haloalkane Dehalogenase Activity. *Jour. Microbiol. Methods,* 32:31-36.

Hooper, S.M., R.S. Summers, H.M. Shukairy, and D.M. Owen. 1994. Development of a New Test for the Assessment of Disinfection By-product Formation: Uniform Formation Conditions. AWWA Water Quality Technology Conference, San Francisco, CA, November 6-10.

Hozalski, R., L. Zhang, and W. Arnold. 2001. Reduction of Haloacetic Acids by Fe°: Implications for Treatment and Fate. *Environ. Sci. Technol.*, 35(11):2258-2263.

Huck, P.M., W.B. Anderson, E. Savage, R.C. von Borstel, S.A. Daignault, D.W. Rector, G.A. Irvine, and D.T. Williams. 1987. Pilot Scale Evaluation of Ozone and Other Drinking Water Disinfectants Using Mutagenicity Testing. 8th Ozone World Congress, Zurich, Switzerland, September 15-18.

Ibrahim, E.A., M.L. Temmer, and R.G. Lee. 1994. Seasonal Variations in Disinfection By-products Precursors and their Removal by Existing Treatment Practices. AWWA Annual Conference, New York City, NY, June 19-23.

Jacangelo, J.G., N.L. Patania, K.M. Reagan, E.M. Aieta, S.W. Krasner, and M.J. McGuire. 1989. Ozonation: Assessing its Role in the Formation and Control of Disinfection By-products. *Jour. AWWA*, 81(8):74-84.

Jadas-Hécart, A. 1989. Contribution à l'étude de la demande en chlore à long terme d'une eau potable – Modélisation et identification de précurseurs organiques. Ph.D. thesis, Order No.282, University of Poitiers, France, November.

Jammes, C., C. Hochereau, and A. Bruchet. 1994. Formation et évolution de quelques sous-produits d'oxydation dans les usines de traitement d'eau potable comportant une étape d'ozonation. *Congrès du Groupe de Recherche Universitaire sur les Techniques de Traitement et d'Épuration des Eaux (GRUTTEE) and Laboratoire de Chimie des Eaux et des Nuisances*, Poitiers, France, September 29-30.

Janssen, D.B., F. Pries, and J.R. Van der Ploeg. 1994. Genetics and Biochemistry of Dehalogenating Enzymes. *Annu. Rev. Microbiol.*, 48:163-191.

Janssen, D.B., A. Scheper, and B. Witholt. 1984. Biodegradation of 2-Chloroethanol and 1,2-Dicholoroethane by Pure Bacterial Cultures. *Innovations in Biotechnology*, 20:169-178.

Jobb, B., R.B. Hunsinger, O. Meresz, and V. Tagushi. 1993. *Removal of N-Nitrosodimethylamine (NDMA) from the Ohsweken (Six Nations) Water Supply, Interim Report*. Ontario Ministry of Environment and Energy, Toronto, Ontario.

Jobb, B., R.B. Hunsinger, O. Meresz, and V. Taguchi. 1994a. Ultraviolet Degradation of N-nitrosodimenthylamine (NDMA) in the Ohsweken (Six Nations) Water Supply. *In proc. of AWWA Annual Conference*, New York, N.Y. (Poster).

Jobb, B., R.B. Hunsinger, O. Meresz, and V. Taguchi. 1994b. Removal of N-Nitrosodimethylamine (NDMA) from the Oshweken (Six Nations) Water Supply. *In proc. of, AWWA Annual Conference.* New York, N.Y. (Poster).

Johnson, J.D., and J.N. Jensen. 1986. THM and TOX Formation: Routes, Rates and Precursors. *Jour. AWWA*, 78(4):156-162.

Johnson, J.D., and W. Sun. 1975. Bromine Disinfection of Wastewater. In: *Disinfection Water and Wastewater*. Edited by Johnson, J.D. Ann Arbor, Michigan, Ann Arbor Science Publishers.

Kaplan, L.A., and M.W. LeChevallier. 1993. *Assimilable Organic Carbon Measurement Techniques*. Denver, Colo.: AwwaRF and AWWA.

Kavanaugh, M.C., A.R. Trussell, J. Cromer, and R.R. Trussell. 1980. An Empirical Kinetic Model of Trihalomethane Formation: Applications to Meet the Proposed THM Standard. *Journal AWWA*, 72(10):578-582.

Kawasaki, H., K. Tsuda, I. Matsushita, and K. Tonomura. 1992. Lack of Homology Between Two Haloacetate Dehalogenase Genes Encoded on a Plasmid from Moraxella Species Strain B. *Jour. Gen. Microbiol.*, 138:1317-1323.

Keuning, S., D.B. Janssen, and B. Witholt. 1985. Purification and Characterization of Hydrolytic Haloalkane Dehalogenase from *Xanthobacter autotrophicus* GJ-10. *Jour. Bacteriol.*, 163:635-639.

Kiéné, L., W. Lu, and Y. Lévi. 1998. Relative Importance of the Phenomena Responsible for Chlorine Decay in Drinking Water Distribution Systems. *Wat. Sci. Tech.*, 38(6):219-227.

Kiéné, L., 1994. Can Chlorine Decay in Distribution Systems be Predicted by Corrosion Monitoring? *In proceedings of the AWWA – Water Quality Technology Conference*, San Francisco, Calif.

King, W.D., and L.D. Marrett. 1996. Case-control Study of Water Source and Bladder Cancer. *Cancer Causes Control*, 7:596-604.

Kirschenbaum, D.M. 1975. Molar Absorptivity and A1%/1cm Values for Proteins at Selected Wavelengths of the Ultraviolet and Visible Regions. *Anal. Biochem.*, 68:465-484.

Klinefelter, G.R., E.S. Hunter, and M.G. Narotsky. 2001. Reproductive and Developmental Toxicity Associated with Disinfection By-Products of Drinking Water. In: *Microbial Pathogens and Disinfection By-products in Drinking Water: Health Effects and Management of Risks*. Edited by Craun G.F., Hauchman F.S., and Robinson D.E. Washington, D.C.: ILSI Press, 309-323.

Klotz, J.B., and L.A. Pyrch. 1999. Neural Tube Defects and Drinking Water Disinfection By-Products. *Epidemiology*, 10(4):383-390.

Koch, B., E.W. Crofts, W.K. Schimpff, and M.K. Davis. 1988. Analysis of Halogenated Disinfection By-products by Capillary Chromatography. AWWA Water Quality Technology Conference, St. Louis, MO, November 13-17.

Koch, B., S.W. Krasner, M.J. Sclimenti, and W.K. Schimpff. 1991. Predicting the Formation of DBPs by the Simulated Distribution System. *Jour. AWWA*, 83(10):62-70.

Kohut, K.D., and S.A. Andrews. 2002. N-Nitrosodimethylamine Formation in Drinking Water due to Amine-Based Polyelectrolytes. AWWA Water Quality Technology Conference. Seattle, WA, November.

Krasner, S.W. 1999. Chemistry of Disinfection By-product Formation. In: *Formation and Control of Disinfection By-products in Drinking Water*. Edited by Singer, P.C. AWWA Publishers.

Krasner, S.W., J.-P. Croué, J. Buffle, and E.M. Perdue. 1996. Three Approaches for Characterizing NOM. *Jour. AWWA*, 88(6):66-79.

Krasner, S.W., W.H. Glaze, H.S. Weinberg, P.A. Daniel, and I.N. Najm. 1993. Formation and Control of Bromate During Ozonation of Waters Containing Bromide. *Jour. AWWA*, 85(1):73-81.

Krasner, S.W., M.J. McGuire, J.G. Jacangelo, N.L. Patania, K.M. Reagan, and E.M. Aieta. 1989. The Occurrence of Disinfection By-products in US Drinking Water. *Jour. AWWA*, 81(8):41-53.

Kurihara, T., E. Nobuyoshi, and K. Soda. 2000. Bacterial 2-haloacid Dehalogenases: Structures and Reaction Mechanisms. *Jour. Mol. Catal. B: Enzyme*, 10(1-3):57-65.

Kurihara, T., T. Yamauchi, S. Ichiyama, H. Takahata, and N. Esaki. 2003. Purification, Characterization, and Gene Cloning of a Novel Fluroacetate Dehalogenase from *Burkholderia* sp. FA1. *Jour. Mol. Catal. B: Enzyme*, 23:347-355.

Laemmli, U.K. 1970. Cleavage of Structural Proteins during the Assembly of the Head of Bacteriophage T4. *Nature*, 227:680-685.

Lagoa, A.L., H.P. Diogo, M.P. Dias, M.E. Minas da Piedade, L.M. Amaral, M.A. Ribeiro da Silva, J.A. Martinho Simoes, R.C. Guedes, B.J. Costa Cabral, K. Schwarz, and M. Epple, 2001. Energetic of C-Cl, C-Br, and C-I Bonds in Haloacetic Acids: Enthalpies of Formation of XCH_2COOH (X =Cl, Br, I) Compounds and the Carboxymethyl Radical. *Chemistry*. 7:483-489.

Lavoie, J. 1991. Étude des parameters influençant l'évolution de la qualité de l'eau dans un réseau de distribution d'eau potable. Master thesis, École Polytechnique of Montréal, Québec, December.

LeBel, G.L., F.M. Benoit, and D.T. Williams. 1995. Variations of Chlorinated Disinfection By-Products in Water from Treatment Plants Using Three Different Disinfection Processes. *In proc. of AWWA Water Quality Technology Conference, November 12-16*. New Orleans, La., 1339-1356.

LeBel, G.L., F.M. Benoit, and D.T. Williams. 1996. *A One-Year Survey of Halogenated Disinfection By-Products in the Distribution System of Treatment Plants Using Three Different Disinfection Processes.* Minister of Public Works and Government Services, Canada.

LeChevallier, M.W., C. Cawthon, and R.G. Lee. 1988. Inactivation of Biofilm Bacteria. *Appl. Environm. Microbiol.*, 54(10):2492.

LeChevallier, M.W., C.D. Lowry, and R.G. Lee. 1990. Disinfecting Biofilms in a Model Distribution System. *Jour. AWWA*, 82(7):87-99.

LeChevallier, M.W., N.E. Shaw, L.A. Kaplan, and T.L. Bott. 1993. Development of a Rapid Assimilable Organic Carbon Method for Water. *Appl. Environm. Microbiol.*, 59(5):1526-1531.

Levine, A.D., E.R. Baumann, and L.B. Lind. 1987. Curbing THMs in Small Water Systems. *Jour. AWWA*, 79(5):52-56.

Liang, S., J.H. Min, M.K. Davis, J.F. Green, and D.S. Remer. 2000. Treatment of MDNA by Pulsed-Ultraviolet Irradiation and Pulsed-UV/Hydrogen Peroxide Processes. AWWA Water Quality Technology Conference, Salt Lake City, UT.

Loos, M.A. 1975. Indicator Media for Microorganisms Degrading Chlorinated Pesticides. *Can. J. Microbiol.*, 21:104-107.

Loveland, J.P., E.G. Means III, and M. Rasouli. 2001. Pilot Testing to Inhibit N-Nitrosodimethylamine (NDMA) Formation During Anion Exchange Treatment. AWWA Water Quality Technology Conference, Nashville, TN, November.

Lykins, B.W., J.A. Goodrich, W.E. Koffskey, and M.H. Griese. 1991. Controlling Disinfection By-products with Alternative Disinfectants. AWWA Annual Conference, Philadelphia, PA, June 23-27.

Madigan, M.T., J.M. Martinko, and J. Parker. 2000. Brock Biology of Microorganisms. New Jersey: Prentice Hall.

Mathieu, L., J.L. Paquin, J.C. Block, P. Hartemann, and F. Colin. 1992. Paramètres gouvernant la prolifération bactérienne dans les réseaux de distribution. *Sciences de l'eau*, 5:91-112.

Matsuoka, A, K. Yamakage, H. Kusakabe, S. Wakuri, M. Asakura, T. Noguchi, T. Sugiyama, H. Shimada, S. Nakayama, Y. Kasahara, Y. Takahashi, K.F. Miura, M. Hatanaka, M. Ishidate Jr., T. Morita, K. Watanabe, M. Hara, K. Odawara, N. Tanaka, M. Hayashi, and T. Sofuni. 1996. Re-evaluation of Chromosomal Aberration Induction on Nine Mouse Lymphoma Assay "Unique Positive" NTP Carcinogens. *Mutation Research*, 369(3-4):243-252.

Marcus, P.M., D.A. Savitz, R.C. Millikan, and H. Morgenstern. 1998. Female Breast Cancer and Trihalomethanes in Drinking Water in North Carolina, *Epidemiology*, 9:156-160,

McClellan, J.N., D.A. Reckhow, J.E. Tobiason, and J.K. Edzwald. 1999. Modeling chlorine and DBPs in Distribution Systems: A Mechanistic Approach. AWWA Annual Conference, Chicago, IL, June 20-24.

McGeehin, J.S., G.R. Lankas, and R.E. Morrissey. 1994. Toxicokinetic and Mechanistic Considerations in the Interpretation of the Rodent Bioassay. *Toxicologic Pathology*, 22:124-140.

McGuire, M.J., and N. Graziano. 2002. Trihalomethanes in U.S. Drinking Water - NORS to ICR. In: *Information Collection Rule Data Analysis*. Prepared by M.J. McGuire, J.L. McLain, and A. Obolenski. Denver, Colo.: AwwaRF and AWWA, 81-110.

McGuire, M.J., and M.L. Hotaling. 2002. Relationships Between Source Water Quality and Choices of Primary and Secondary Disinfectants. In: *Information Collection Rule Data Analysis*. Prepared by M.J. McGuire, J.L. McLain, and A. Obolenski. Denver, Colo.: AwwaRF and AWWA, 299-315.

McGuire, M.J., J.L. McLain, and A. Obolensky. 2002. *Information Collection Rule Data Analysis*. Denver, Colo.: AwwaRF and USEPA.

McLain, J.L., A. Obolenski, and H.M. Shukairy. 2002. Disinfection By-Product Speciation. In: *Information Collection Rule Data Analysis*. Prepared by M.J. McGuire, J.L. McLain, and A. Obolenski. Denver, Colo.: AwwaRF and AWWA, 141-168.

Merlet, N. 1986. Contribution à l'étude du mécanisme de formation des trihalométhanes et des composés organo-halogénés non volatils lors de la chloration de molecules modèles. Ph.D. thesis, Order No.426, University of Poitiers, France, February.

Meusel, M. and H.J. Rehm. 1993. Biodegradation of Dichloroacetic acid by Freely Suspended and Adsorptive Immobilized *Xanthobacter autotrophicus* GJ10 in Soil. *Appl. Microbiol. and Biotechnol.*, 40:165-171.

Meyer, I. 1992. Suivi des sous-produits de la chloration et de leurs précurseurs de l'eau brute à l'eau distribuée. Ph.D. thesis, University of Nancy 1, France, September.

Meyer, I., T. Français, B. Lykins, M. Dutang, and P. Hartemann. 1993. Influence of Biofilm on Disinfection By-products in a Distribution Network. *Water Supply*, 11(3-4):355-364.

Miller, J.W., and P.C. Uden. 1983. Characterization of Nonvolatile Aqueous Chlorination Products of Humic Substances. *Environmental Science & Technology*, 17(3):150-157.

Mitch, W.A., and D.L. Sedlak. 2002. Formation of N-Nitrosodimethylamine (NDMA) from Dimethylamine during Chlorination. *Environmental Science & Technology*, 36(4):588-595.

Mitch, W.A., J.O. Sharp, R.R. Trussell, R.L. Valentine, L. Alvarez-Cohen, and D.L. Sedlak. 2003. N-nitrosodimethylamine (NDMA) as a Drinking Water Contaminant: A Review. *Environmental Engineering Science*, 20(5):389-404.

MOE (Ontario Ministry of Environment). 1998. *Drinking Water Surveillance Program, 1996-1997 Executive Summary Report.* http://www.ene.gov.on.ca/envision/dwsp/index96_97.htm.

MOE (Ontario Ministry of Environment). 2000. *Regulation Made Under the Ontario Water Resources Act.* O. Regulation 459/00. http://www.ene.gov.on.ca/envision/WaterReg/Reg-final.pdf

Morris, R.D., A.M. Audet, I.F. Angelillo, T.C. Chalmers, and F. Mosteller. 1993. Chlorination, Chlorination By-products, and Cancer: A Meta-analysis 1992. *American Journal of Public Health*, 82:955-963.

Najm, I.N., N.L. Patania, J.G. Jacangelo, and S.W. Krasner. 1994. Evaluating Surrogates for Disinfection By-products. *Jour. AWWA*, 86(6):98-106.

Najm, I.N., and R.R. Trussell. 2001. NDMA Formation in Water and Wastewater. *Jour. AWWA*, 93(2):92-99.

Narayanan, L., A. Moghaddam, A. Taylor, G. Sudberry, and J. Fisher. 1999. Sensitive High-Performance Liquid Chromatography Method for the Simultaneous Determination of Low levels of Dichloroacetic Acid and Its Metabolites in Blood and Urine, *Jour. of Chromatography*, 729:271-277.

National Cancer Institute (NCI). 1976. *Report on Carcinogenesis Bio-Assay of Chloroform.* Springfield, VA.: National Technical Information Service, PB-264018.

National Center for Biotechnology Information (NCBI). 2004. S)-2-haloacid dehalogenase (2-haloalkanoic acid dehalogenase) (L-2-haloacid dehalogenase) (Halocarboxylic acid halidohydrolase) ACCESSION Q60099 GI:3122178. Bethesda: NCBI. http://www.ncbi.nlm.nih.gov/entrez/viewer.fcgi?db=protein&val=3122178 [Accessed 18 July 2004].

National Toxicology Program (NTP). 1986. *Toxicological and Carcinogenesis Studies of Bromodichloromethane in F344/N Rats and B6C3F1 Mice Technical Report No. 321.* NIH publication No. 88-2537; CAS No. 75-25-2, Washington D.C.

National Toxicology Program (NTP). 2000. N-Nitrosodimethylamine CAS No. 62-75-9, 9th Report on Carcinogens, Public Health Service, U.S. Department of Health and Human Services.

Newman, J., T.S. Peat, R. Richard, L. Kan, P.E. Swanson, J.A. Affholter, I.H. Holmes, J.F. Schindler, C.J. Unkefer, and T.C. Terwilliger. 1999. Haloalkane Dehalogenases: Structure of a Rhodococcus Enzyme. *Biochemistry*, 38:16105-16114.

Nguyen, M., B.A. Lawrence, and P. Westerhoff. 2002. DOC and DBP Precursors in Western U.S. Watersheds and Reservoirs. *Jour. AWWA,* 94(5):98-112.

Nichols, C. 2004. Factors Affecting the Formation and Decay of Trihalomethanes and Haloacetic Acids in a Full-Scale Drinking Water Distribution System. Master's thesis. University of North Carolina at Chapel Hill, NC.

Nieminski, E.C., S. Chaudhuri, and T. Lamoreaux. 1993. The Occurrence of DBPs in Utah Drinking Waters. *Jour. AWWA*, 85(9):98-105.

Noack, M.G., and R.L. Doerr. 1978. Reactions of Chlorine, Chlorine Dioxide, and Mixtures Thereof with Humic Acid. In: *Water Chlorination: Environmental Impact and Health Effects*, Vol.2. Jolley et al Editors. Ann Arbor Science Publishers, 49-58.

Nokes, C.J., E. Fenton, and C.J. Randall. 1999. Modelling the Formation of Brominated Trihalomethanes in Chlorinated Drinking Waters. *Wat. Res*., 33(17):3557-3568.

Obolensky, A. 2004. *Historical DBP Data from a Chloramine System IDSE Standard Monitoring Plan Design and Expectations*. AwwaRF Technology Transfer Conference, Washington D.C., April 22.

Obolensky, A., and M. Frey. 2002. Distribution System DBP Results and SDS Performance. In: *Information Collection Rule Data Analysis*. Prepared by M.J. McGuire, J.L. McLain, and A. Obolensky. Denver, Colo.: AwwaRF and AWWA, 169-198.

Obolensky, A., H.M. Shukairy, and V. Blank. 2002. Occurrence of Haloacetic Acids in ICR Finished Water and Distribution Systems. In: *Information Collection Rule Data Analysis*. Perpared by M.J. McGuire, J.L. McLain, and A. Obolensky. Denver, Colo.: AwwaRF and AWWA, 111-140.

Olaniran, A.O., G.O. Babalola, and A.I. Okoh. 2001. Aerobic Dehalogenation Potentials of Four Bacterial Species Isolated from Soil and Sewage Sludge. *Chemosphere*, 45:45-50.

Oliver, B.G. 1978. Chlorinated Non-volatile Organics Produced by the Reaction of Chlorine with Humic Materials. *Canadian Research*, 11(6):21-22.

Oliver, B.G. 1980. Effect of Temperature, pH and Bromide Concentration on the Trihalomethane Reaction of Chlorine with Aquatic Humic Material. In: *Water Chlorination: Environmental Impact and Health Effects, Vol.3.* R.L. Jolley et al. Editors. Ann Arbor: Science Publishers, 141-149.

Oliver, B.G., and J. Lawrence. 1979. Haloforms in Drinking Water: A Study of Precursors and Precursor Removal. *Jour. AWWA*, 71(3):161-163.

Olivieri, V.P., M.C. Snead, C.W. Krusé, and K. Kawata. 1986. Stability and Effectiveness of Chlorine Disinfectants in Water Distribution Systems. *Environmental Health Perspectives*, 69:15-29.

Owen, D.M., G.L. Amy, and Z.K. Choudhury. 1993. *Characterization of Natural Organic Matter and its Relationship to Treatability*. Denver, Colo.: AwwaRF and AWWA.

Pereira, M.A. 1996. Carcinogenic Activity of Dichloroacetic Acid and Trichloroacetic Acid in the Liver of Female B6C3F1 Mice. *Fundamental and Applied Toxicology*, 31(2):192-199.

Pereira, V., H.S. Weinberg, and P.C. Singer. 2001. Development of a Sampling Approach to Assess Patterns of Exposure to Disinfection By-Products in a Chloraminated Drinking Water System. *In proc. of AWWA Water Quality Technology Conference*. Nashville, TN, November 11-15 (poster).

Poelarends, G.J., J.E.T. Van Hylckama Vlieg, J.R. Marchesi, L.M. Freitas Dos Santos, and D.B. Janssen. 1999. Degradation of 1,2-Dibromoethane by *Mycobacterium* sp. Strain GP1. *Jour. Bacteriol.*, 181:2050-2058.

Powell, J.C., N.B. Hallam, J.R. West, C.F. Forster, and J. Simms. 2000. Factors Which Control Bulk Chlorine Decay Rates. *Wat. Res.*, 34(1):117-126.

Rebhun, M., J. Manka, and A. Zilberman. 1988. Trihalomethane Formation in High-bromide Lake Galilée Water. *Jour. AWWA*, 80(6):84-89.

Reckhow D.A. 1984. Organic Halide Formation and the Use of Pre-Ozonation and Alum Coagulation to Control Organic Halide Precursors. Ph.D. Thesis, University of North Carolina at Chapel Hill, Department of Environmental Sciences and Engineering, North Carolina.

Reckhow D.A., and P.C. Singer. 1985. Mechanisms of Organic Halide Formation During Fulvic Acid Chlorination and Implications with Respect to Preozonation. In: *Water Chlorination: Chemistry, Environmental Impact and Health Effects, Vol.5.* Lewis Publishers, Inc.

Reckhow, D.A., P.C. Singer, and R.L. Malcolm. 1990. Chlorination of Humic Materials: Byproduct Formation and Chemical Interpretations. *Environ. Sci. & Tech.*, 24(11):1655.

Rehm, H.J., and U. Heinze. 1993. Biodegradation of Dichloroacetic Acid by Entrapped and Adsorptive Immobilized *Xanthobacter autotrophicus* GJ-10. *Appl Microbiol Biotechnol.*, 40:158-164.

Reif, J.S., M.C. Hatch, M. Bracken, L.B. Holmes, B.A. Schwetz, and P.C. Singer. 1996. Reproductive and Developmental Effects of Disinfection By-products in Drinking Water. *Environmental Health Perspectives*, 104(10):1056-1061.

Reif, J., A. Bachand, and M. Andersen. 2000. *Reproductive and Developmental Effects of Disinfection By-Products*. Prepared for Bureau of Reproductive and Child Health, Health Canada.

Rizet, M., J. Blanc, M. Cognet, M. Derons, N. Dorizon, and G. Filippi. 1984. Qualité de l'eau dans les réseaux de distribution, son évolution biologique et chimique. *Journal Français d'Hydrologie*, 15(3):207-224.

Robert, M.G., P.C. Singer, and A. Obolensky. 2002. Comparing Total HAA and Total THM Concentrations Using ICR Data. *Jour. AWWA*, 94(1):103-114.

Rook, J.J. 1974. Formation of Haloforms During Chlorination of Natural Waters. *Water Treatment Exam.*, 23(2):234-243.

Rook, J.J. 1976. Haloforms in Drinking Water. *Jour. AWWA*, 68:168-172.

Rook, J.J. 1980. *Water Chlorination: Environmental Impact and Health Effects*, Vol.3. R.L. Jolley et al, Editors, Ann Arbour Science Publishers.

Rossman, L.A., R.A. Brown, P.C. Singer, and J.R. Nuckols. 2001. DBP Formation Kinetics in a Simulated Distribution System. *Wat. Res.*, 35(14):3483-3489.

Rossman, L.A., R.M. Clark, and W.M. Garyman. 1994. Modeling Chlorine Residuals in Drinking Water Distribution Systems. *Jour. Environmental Engineering ASCE*, 120(4):803.

Saber, D.L., and R.L. Crawford. 1985. Isolation and Characterization of *Flavobacterium* Strains That Degrade Pentachlorophenol. *Appl. Environ. Microbiol.*, 50(6):1512-1518.
Savitz, D.A., P.C. Singer, K.E. Hartmann, A.H. Herring, H.S. Weinberg, C. Makarushka, C. Hoffman, R. Chan, and R. Maclehose. 2005. *Drinking Water Disinfection By-Products and Pregnancy Outcome*. Denver, Colo.: AwwaRF.
Schwarze, R., A. Brokamp, and F.R.J. Schmidt. 1997. Isolation and Characterization of Dehalogenases from 2,2-Dichloropropionate-Degrading Soil Bacteria. *Curr. Microbiol.*, 34:103-109.
Sedlak, D.L., and W. Mitch. 2001. Factors Controlling the Formation of N-Nitrosodimethylamine (NDMA) During Chlorination. Water Reuse Foundation – 2001 Annual Research Conference, Monterey, California, June 4-5.
Siddiqui, M.S., and G.L. Amy. 1993. Factors Affecting DBP Formation During Ozone-Bromide Reactions. *Jour. AWWA*, 85(1):63-72.
Sim, M., 2002. Case Studies in the Use of Toxicological Measures in Epidemiological Studies. *Toxicology*, 181-182:405-409.
Singer, P.C. 1990. Assessing Ozonation Research Needs in Water Treatment. *Jour. AWWA*, 82(10):78-88.
Singer, P.C. 1999a. *Formation and Control of Disinfection By-Products in Drinking Water*. Denver, Colo.: AWWA, Library of Congress.
Singer, P.C. 1999b. Variability and Assessment of Disinfection By-Product Concentrations in Water Distribution Systems. 2nd International Conference on the Safety of Water Disinfection. Washington D.C.: International Life Sciences Inst, November.
Singer, P.C. 2001. Variability and Assessment of Disinfection By-Product Concentrations in Water Distribution Systems. In: *Microbial Pathogens and Disinfection By-Products in Drinking Water: Health Effects and Management Risks.* International Life Sciences, 211-223.
Singer P.C., J.J. Barry III, G.M. Palen, and A.E. Scrivner. 1981. Trihalomethane Formation in North Carolina Drinking Waters. *Jour. AWWA*, 73(8):392-401.
Singer P.C., and S.D. Chang. 1989. Correlations Between Trihalomethanes and Total Organic Halides Formed During Water Treatment. *Jour. AWWA*, 81(8):61-65.
Singer P.C., V. Pereira, H.S. Weinberg, D. Radiloff. 2001. Spatial and Temporal Variability of Disinfection By-Products in a Water Distribution System. International Society for Exposure Assessment Conference. Charleston, S.C., November.
Singer, P.C., R.D.G. Pyne, M. Avs, C.T. Miller, and C. Mojonnier. 1993. Examining the Impact of Aquifer Storage and Recovery on DBPs. *Jour. AWWA*, 85(11):85-94.
Singer, P.C., H.S. Weinberg, K. Brophy, L. Liang, M. Roberts, I. Grisstede, S.W. Krasner, H. Baribeau, H. Aurora, and I. Najm. 2002. *Relative Dominance of Haloacetic Acids and Trihalomethanes in Treated Drinking Water*. Denver, Colo.: AwwaRF and AWWA.
Slater, J.H., A.T. Bull, and D.J. Hardman. 1997. Microbial Dehalogenation of Halogenated Alkanoic Acids, Alcohols, and Alkanes. In: *Advances in Microbial Physiology, Vol. 38.* Poole, R.K., (Ed.). San Diego, Calif.: Academic Press, 133-176.
Slater, J.H., and Weightman A. 1980. Selection of *Pseudomonas putida* Strains with Elevated Dehalogenase Activities by Continuous Culture Growth on Chlorinated Alkanoic Acids. *Jour. of General Microbiology*, 121:187-193.
Slater, J.H., A. Weightman, and A. Bull. 1979. The Partial Purification of Two Dehalogenases From *Pseudomonas putida* PP3. *FEMS Microbiology Letters*, 6:231-234.

Smith, J., K. Harrison, and J. Colby. 1990. Purification and Characterization of D-2-Haloacid Dehalogenase from *Pseudomonas putida*. *Jour. General Microbiology,* 136:881-886.

Smith, D.B., A.F. Hess, and D. Opheim. 1989a. Control of Distribution System Coliform Regrowth. *In proc. of AWWA Water Quality Technology Conference.* Philadelphia, PA, November 12-16, 1009-1029.

Snoeyink, V.L. 1994. Water Quality Impacts on Corrosion. AWWA Annual Conference, New York, NY, June 19-23.

Soda, K., T. Kurihara, and N. Esaki. 2000. Bacterial 2-Haloacid Dehalogenases: Structures and Reaction Mechanisms. *Jour. of Molecular Catalysis B:Enzymatic*, 10:57-65.

Soroushian, F., Y. Shen, M. Patel, and M. Wehner. 2001. Evaluation and Pilot Testing of Advanced Treatment Processes for NDMA Removal and Reformation Prevention. AWWA Water Quality Technology Conference, Salt Lake City, Utah.

Speitel, G.E. 1999. Control of Disinfection By-product Formation Using Chloramines. In: *Formation and Control of Disinfection By-products in Drinking Water*. Singer, P.C. ed., Denver, Colo.: AWWA Publishers.

Stauber, A.J., and R.J. Bull. 1997. C-Jun and c-Fos immunoreactivity Differentiates Modes of Action Among Haloacetates, Abstract 1129. *Toxicologist*, 36:222.

Stevens, A.A., L.A. Moore, and R.J. Miltner. 1989. Formation and Control of Non-trihalomethane Disinfection By-products. *Jour. AWWA*, 81(8):54-60.

Stevens, A.A., C.J. Slocum, D.R. Seeger, and G.G. Robeck. 1976. Chlorination of Organics in Drinking Water. *Jour. AWWA*, 68(11):675.

Strotmann, U., and R. Roschenthaler. 1987. A Method for Screening Bacteria: Aerobically Degrading Chlorinated Short-chain Hydrocarbons. *Curr. Microbiol.*, 15:159-163.

Summers, R.S., S.M. Hooper, H.M. Shukairy, G. Solarik, and D. Owen. 1996. Assessing DBP Yield: Uniform Formation Conditions. *Jour. AWWA*, 88(6):80-93.

Swan, S.H., K. Waller, B. Hopkins, G. Windham, L. Fenster, C. Schaefer, and R.R. Neutra. 1998. A Prospective Study of Spontaneous Abortion: Relation to Amount and Source of Drinking Water Consumed in Early Pregnancy. *Epidemiology*, 9(2):126-133.

Swiss Institute of Bioinformatics (SIB). 2004. Expert Protein Analysis System. www.au.expasy.org.

Symons, J.M., P.L.K. Fu, R.C. Dressman, and A.A. Stevens. 1987. The Influence of Bromide Ion on Organic Bromine Formation During Free Chlorination. *Jour. AWWA*, 79(9):114-118.

Symons, J.M., S.W. Krasner, L.A. Simms, and M.J. Sclimenti. 1993. Measurement of THM and Precursor Concentrations Revisited: The Effect of Bromide Ion. *Jour. AWWA*, 85(1):51.

Symons, J.M., G.E. Speitel Jr., C.J. Hwang, S.W. Krasner, and S.E. Barrett. 1998. *Factors Affecting Disinfection By-Product Formation During Chloramination*. Denver, Colo.: AwwaRF and AWWA.

Taguchi, V.Y., S.W.D. Jenkins, D.W. Wang, J.P. Palmentier, E.J. Reiner. 1994. Determination of N-Nitrosodimethylamine by Isotope Dilution High Resolution Mass Spectrometry. *Can. J. Appl. Spec.*, 39(3):87-93.

Thai, S.F., J.W. Allen, A.B. DeAngelo, M.H. George, and J.C. Fuscoe. 2003. Altered Gene Expression in Mouse Livers After Dichloroacetic Acid Exposure. *Mutat. Res.*, 543:167-180.

Trun, N., and J. Trempy. 2003. Fundamental Bacterial Genetics. Blackwell Publishing Inc.

Trussell, R.R. 1992. Impact of Corrosion on Chlorine Decay is Distribution Systems. Memorandum Presented at the Project Advisory Committee for AWWARF Project 815-92: Characterization and Modeling of Chlorine Decay in Distribution Systems, Toronto.

Trussell, R.R., and M.D. Umphres. 1978. The Formation of Trihalomethanes. *Jour. AWWA*, 70(11):604-612.

Tsang, J.S.H., and B.C.M. Pang. 2000. Identification of the Dimerization Domain of Dehalogenase IVa of *Burkholderia cepacia* MBA4. *Appl. Environ. Microbiol.*, 66:3180-3186.

Tsang, J.S.H., P.J. Sallis, A.T. Bull, and D.J. Hardman. 1988. A Monobromoacetate Dehalogenase from *Pseudomonas cepacia* MBA4. *Arch. Microbiol.*, 150:441-446.

Tsang, J.S.H., and J. Sze. 2002. Sec-dependent and Sec-independent Translocation of Haloacid Dehalogenase Chd1 of *Burkholderia cepacia* MBA4 in *Escherichia coli. FEMS Microbiol. Letter*, 211:259-264.

Tuthill, R.W., and G.S. Moore. 1980. Drinking Water Chlorination: A Practice Unrelated to Cancer Mortality. *Jour. AWWA*, 72(10):570-573.

Uber, J.G., M. Webster, M. Tryby, E. Bissonnette, and R.S. Summers. 2000. Observational Study of Population Exposure to Regulated Disinfection By-Products. AWWA Water Quality Technology Conference, Salt Lake City, Utah, November 5-9.

Uchiyama, H., T. Nakajima, O. Yagi, and T. Nakahara. 1992. Role of Heterotrophic Bacteria in Complete Mineralization of Trichloroethylene by *Methylocystis* sp. Strain M. *Appl. Environm. Microbiol.*, 58(9):3067-3071.

United States Environmental Protection Agency (USEPA) 1978a. Method 353.2 – Nitrogen, Nitrate-Nitrite. Office of Research and Development, Cincinnati, Ohio.

USEPA. 1978b. Method 350.1 – Nitrogen, Ammonia. Office of Research and Development, Cincinnati, Ohio.

USEPA. 1979. *Federal Register*, 44:85:25505, May 1.

USEPA. 1979b. Methods for Chemical Analysis of Water and Wastes. EPA/600/4-79/020. Cincinatti, OH.

USEPA. 1990. Revised External Review Draft for the Drinking Water Criteria Document on Trihalomethanes. Government Printing Office, TR-1242-34B, Washington, D.C.

USEPA. 1992. Methods for the Determination of Organic Compounds in Drinking Water, Supplement 2, EPA/600/R-92/129, April, 143-172.

USEPA. 1993. Method 300.0 – Methods for the Determination of Inorganic Substances in Environmental Samples. EPA/600/R-93/100.

USEPA. 1995. Method 502.2: Methods for the Determination of Organic Compounds in Drinking Water - Supplement III, USEPA, EPA/600/R-95/131, August.

USEPA. 1996. National Primary Drinking Water Regulation: Monitoring Requirements for Public Drinking Water Supplies; Final Rule. Federal Register, 61:94:24353-24388, May 14.

USEPA. 1997a. N-Nitrosodimethylamine; CASRN 62-75-9 (04/01/97), Integrated Risk Information Service (IRIS) Substance File, http://www.epa.gov/ngisogm3/iris/subst/0042.htm.

USEPA. 1997b. Method 300.1 - Determination of Inorganic Anions in Drinking Water by Ion Chromatography, Revision 1.0. EPA/600/R-98/118. USEPA National Exposure Research Laboratory, Office of Research and Development, Cincinnati, Ohio.

USEPA. 1997c. Manual for the Certification of Laboratories Analyzing Drinking Water: Criteria and Procedures Quality Assurance. EPA 815-B-97-001.

USEPA. 2000a. ICR Auxiliary Database Version 5.0, Query Tool Version 2.0 [CD-ROM]. EPA 815-C-00-002, Cincinnati, OH.

USEPA. 2000b. Conducting a Risk Assessment of Mixtures of Disinfection By-Products (DBPs) for Drinking Water Treatment Systems. USEPA, National Center for Environmental Assessment, Cincinnati, OH.

USEPA. 2000c. Stage 2 Microbial and Disinfection Byproducts Federal Advisory Committee Agreement in Principle. *Fed. Reg.*, 65:251:83015-83024. (December 29).

USEPA. 2003a. Draft Stage 2 Disinfectant and Disinfection Byproduct Rule – Initial Distribution System Evaluation. No. TBD. USEPA Office of Water, Washington D.C., July.

USEPA. 2003b. The Stage 2 Disinfectants and Disinfection Byproducts Rule (Stage 2 DBPR) Implementation Guidance. Office of Water (4606M), EPA 816-D-03-002.

Valentine, R.L. 1986. Bromochloramine Oxidation of N,N-Diethyl-p-phenylenediamine in the Presence of Monochloramine. *Environ. Sci. & Tech.*, 20(2):166-170.

Valentine, R.L., J. Choi, Z. Chen, S.E. Barrett, C. Hwang, Y. Guo., M. Wehner, S. Fitzsimmons, and S.A. Andrews. 2004. *Factors Affecting the Formation of NDMA in Water and Occurrence*. Draft Final Report submitted to AwwaRF, Denver, Colo.

van den Wijngaard, A.J., K.W. van der Kamp, J. van der Ploeg, F. Pries, B. Kazemier, and D.B. Janssen. 1992. Degradation of 1,2-Dichloroethane by *Ancylobacter aquaticus* and Other Facultative Methylotrophs. *Appl. Environ. Microbiol.*, 58(3):976-983.

van der Kooij, D. 1990. Assimilable Organic Carbon (AOC) in Drinking Water. In: *Drinking Water Microbiology*. G.A. McFeters, Editor, Springer-Verlag, Publishers, New York, 57-65.

van der Kooij, D. 1992. Assimilable Organic Carbon as an Indicator of Bacterial Regrowth. *Jour. AWWA*, 84:57-65.

van der Ploeg, J.; G. van Hall, and D.B. Janssen. 1991. Characterization of the Haloacid Dehalogenase from *Xanthobacter autotrophicus* GJ10 and Sequencing of the dhlB Gene. *Jour. Bacteriol.*, 173(24):7925-7933.

Vasconcelos, J., P. Boulos, W. Grayman, L. Kiene, O. Wable, P. Biswas, A. Bhari, L. Rossman, R. Clark, and J. Goodrich. 1996. Characterization and Modeling of Chlorine Decay in Distribution Systems. Denver, Colo.: AwwaRF and AWWA.

Vasconcelos J.J., L.A. Roseman, W.M. Grayman, P.F. Boulos, and R.M. Mark. 1997. Kinetics of Chlorine Decay. *Journal AWWA*, 89(7):54-65.

Vena J.E., S. Graham, J. Dreudenheim, J. Marshall, M. Zielezny, M. Swanson, and G. Sufrin. 1993. Drinking Water, Fluid Intake, and Bladder Cancer in Western New York. *Archives of Environmental Health*, 48:191-198

Wable O. et al. 1991. Modeling chlorine Concentrations in a Network and Application to Paris Distribution Network. *In proc. of 1991 AWWARF/USEPA Conference on Water Quality Modeling in Distribution System*. Cincinnati, Ohio.

Wahman, D.G., A.E. Henry, and G.E. Speitel. 2003. Trihalomethane Cometabolism by Nitrifying Bacteria: Kinetics and Biofilter Performance. AWWA Water Quality Technology Conference, Philadelphia, Penn., November 2-6.

Waller K., S.H. Swan, G. DeLorenze, and B. Hopkins. 1998. Trihalomethanes in Drinking Water and Spontaneous Abortion. *Epidemiology*, 9(2):134-140.

Weightman, A.J., A.L. Weightman, and Slater, J.H. 1982. Stereospecificity of 2-Monochloropropionate Dehalogenation by Two Dehalogenases of *Pseudomonas putida* PP3: Evidence for Two Different Dehalogenation Mechanisms. *Jour. General Microbiology,* 128(8):1755-1762.

Welander, T., and P. Yu, 1995. Growth of an Aerobic Bacterium with Trichloroacetic Acid as the Sole Source of Energy and Carbon. *Appl. Microbiology Biotechnology.*, 42(5):769-774.

Wempe, M.F., W.B. Anderson, H.F. Tzeng, P.G. Board, M.W. Anders. 1999. Glutathione Transferase Zeta-Catalyzed Biotransformation of Deuterated Dihaloacetic Acids. *Biochem. Biophys. Res. Commun.*, 261:779-783.

Wilczak A., A. Assadi-Rad, C.Y. Wong, L.L. Hoover, R. Berger, R.B. Hunsinger, J.F. Smith, F. Rodigari, L.J. Lazzelle, C.T. Heaney, H.H. Lai, A. Sheikholeslami. 2002. Screening of Treatment Processes for NDMA Control. AWWA Annual Conference, New Orleans, Louisiana.

Wilkes C.R., A.D. Mason, L.L. Niang, H.E. Rector, F.W. Power, A.M. Tsang, P.M. Stephan, H.S. Lyn, J.N. Blancato, and C.H. Stephen. 2002. Developing Individual Human Exposure Estimates for Individual DBPs - Developing Exposure Estimates, Final Report. In: *The Feasibility of Performing Cumulative Risk Assessments for Mixtures of Disinfection By-Products in Drinking Water*. USEPA, National Center for Environmental Assessment, EPA/600/R-03/051, Cincinnati, Ohio, September.

Williams S.L., D.F. Rindfleisch, and R.L. Williams. 1994. Dead-end on Haloacetic Acids (HAA). *In proc. of AWWA Water Quality Technology Conference.* San Francisco, CA, November 6-10, 1053-1058.

Williams S.L., R.L. Williams, and A.S. Gordon, 1995. Degradation of Haloacetic Acids (HAA) at Maximum Residence Time Locations (MRTLs). *In proc. of AWWA Water Quality Technology Conference*. New Orleans, LA, November 12-16, 1353-1366.

Williams S.L., R.L. Williams, and A.S. Gordon, 1996. The Impact of Bacterial Degradation of Haloacetic Acids (HAA) in the Distribution System. AWWA Water Quality Technology Conference, Boston, Massachusetts, November 17-21.

Williams, S.L., R.L. Williams, A. Gordon. 1997. Degradation of Haloacetic Acids (HAAs) in Drinking Water. *In procof 213th Annual ACS Meeting, Div. of Environmental Chemistry.* 37:1285-1287.

Williams S.L., R.L. Williams, and J. Yuan. 1998. Bacterial Degradation of Haloacetic Acids in the Distribution System. AWWA Water Quality Technology Conference, San Diego, California, November 1-4.

Xie, Y.F. and H. Zhou. 2002. Use of BAC for HAA Removal - Part 2, Column Study. *Journal AWWA*, 94(5):126 – 134.

Yang C.Y., H.F. Chiu, M.F. Cheng, and S.S. Tsai. 1998. Chlorination of Drinking Water and Cancer Mortality in Taiwan. *Environmental Research*, 78:1-6.

Yuan, J. 2001. *Enzymatic Degradation of Various DBPs in Drinking Water*. Hall, E. (Ed.). Virginia Section of AWWA, pp.8-9.

Zavaleta, J.O., F.S. Hauchman, and M.W. Cox, 1999. Epidemiology and Toxicology of Disinfection By-Products. In: *Formation and Control of Disinfection By-Products in Drinking Water*. Singer, P.C., ed. AWWA Publishers.

Zhang, X., S. Echigo, R.A. Minear, and M.J. Plewa. 2000. Characterization and Comparison of Disinfection By-Products of Four Major Disinfectants. In: *Natural Organic Matter and Disinfection By-Products: Characterization and Control in Drinking Water*. S.E. Barrett, S.W. Krasner, and G.L. Amy Editors. Washington D.C., American Chemical Society.

Zhang, X., and R.A. Minear. 2002. Decomposition of Trihaloacetic Acids and Formation of the Corresponding Trihalomethanes in Drinking Water. *Water Research*, 36:3665-3673.

Zhang, Q., and S.J. Stanley. 1997. Forecasting Raw-Water Quality Parameters for the North Saskatchewan River by Neural Network Modeling. *Water Research*, 31(9):2340-2350.

Zhou, H., and Y.F. Xie. 2002. Use of BAC for HAA Removal - Part 1, Column Study. *Journal AWWA*, 94(4):194 – 200.

Zumstein, J., and J. Buffle. 1989. Circulation of Pedogenic and Aquagenic Organic Matter in an Eutrophic Lake. *Water Research*, 23(2):229.

ABBREVIATIONS

ANOVA	analysis of variance
ANN	artificial neural networks
AOC	assimilable organic carbon
APHA	American Public Health Association
ASR	aquifer storage recovery
AwwaRF	Awwa Research Foundation
BDOC	biodegradable dissolved organic carbon
BrAA	monobromoacetic acid
Br_2AA	dibromoacetic acid
Br_3AA	tribromoacetic acid
BrClAA	bromochloroacetic acid
$BrCl_2AA$	bromodichloroacetic acid
Br_2ClAA	dibromochloroacetic acid
BrO_3^-	bromate ion
BSA	bovine serum albumin
°C	degrees Celsius
$CaCO_3$	calcium carbonate
CCC	continuing calibration check
CCR	California Code of Regulation
CFS	coagulation, flocculation, sedimentation
CFU	colony forming unit
$CHBr_2Cl$	dibromochloromethane
$CHBrCl_2$	bromodichloromethane
$CHBr_3$	bromoform
$CHCl_3$	chloroform
CI	cast iron
ClAA	monochloroacetic acid
Cl_2	chlorine
Cl_2AA	dichloroacetic acid
Cl_3AA	trichloroacetic acid
CLM	chloramines as secondary disinfectant
ClO_2^-	chlorite ion
cm	centimeter
C. parvum	*Cryptosporidium parvum*
CWSs	community water systems
DCAN	dichloroacetonitrile
DBP	disinfection by-products
DBPfp	DBP formation potential
DBPR	Disinfectants/Disinfection By-Products Rule
DCAN	dichloroacetonitrile

DHS	Department of Health Services
DI	ductile iron
DMA	di-methylamine
DOC	dissolved organic carbon
DOX	dissolved organic halides
DVEM	discrete volume element method
DWSP	Drinking Water Surveillance Program
E. coli	*Escherichia coli*
ECD	electron capture detection
EPS	extended period simulations
°F	degrees Fahrenheit
FACA	Federal Advisory Committee Act
ft^3	cubic feet
G. duodenalis	*Giardia duodenalis*
G. muris	*Giardia muris*
GAC	granular activated carbon
GC	gas chromatography
GSTZ1	glutathione S-transferaze zeta-1
GW	groundwater
h	hour
HAA5	sum of five HAAs: ClAA, Cl_2AA, Cl_3AA, BrAA and Br_2AA
HAA5fp	HAA5 formation potential
HAA9	sum of nine HAAs: ClAA, Cl_2AA, Cl_3AA, BrAA, Br_2AA, Br_3AA, BrClAA, $BrCl_2AA$ and Br_2ClAA
HAAs	haloacetic acids
HANs	haloacetonitriles
HCl	hydrochloric acid
HKs	haloketones
HPC	heterotrophic plate count
HOBr	hypobromous acid
HOC1	hypochlorous acid
ICR	Information Collection Rule
IDSE	Initial Distribution System Evaluation
IESWTR	Interim Enhanced Surface Water Treatment Rule
in	inch
IS	internal standard
km	kilometers

L	liter
LFB	Laboratory Fortified Blank
LRAA	locational running annual average
LRB	Laboratory Reagent Blank
M	molar
m	meters
m^3/d	cubic meters per day
MAC	maximum acceptable concentration
MCAN	monochloroacetonitrile
MCL	maximum contaminant level
MDL	method detection limit
MG	million gallons
$MgCl_2$	magnesium chloride
mgd	million gallons per day
mg/L	milligrams per liter
min	minute
mL	milliliter
mM	millimolar
MOE	Ministry of the Environment
MOA	mode of action
MRL	minimum reportable limit
MRTL	maximum residence time location
MS	mass spectrometry
MtBE	methyl-*tert*-butyl ether
mg	milligram
μg	microgram
μL	microliter
μM	micromolar
NaCl	sodium chloride
NaOH	sodium hydrohyde
Na_2S_2O	sodium thiosulfate
NCI	National Cancer Institute
NDMA	N-nitrosodiumethylamine
ng	nanogram
ng/L	nanogram per liter
nm	nanometer
NH_4Cl	ammonium chloride
NOM	natural organic matter
NNWW	Newport News Waterworks
NTNCWSs	Non-transient non-community water systems
NTP	National Toxicology Program

OBr^-	hypobromite ion
OEHHA	Office of Environmental Health Hazard Assessment
P25	25th percentile
P50	50th percentile
P75	75th percentile
P90	90th percentile
PAC	powdered activated carbon
PCR	polymerase chain reaction
PID	photo ionization detector
POE	point of entry
PRV	pressure reducing valve
PVC	polyvinyl chloride
QC	quality control
QCS	quality control samples
RAA	running annual average
RPD	relative percent difference
SD	standard deviation
SDS	simulated distribution system
SDWA	Safe Drinking Water Act
sec	second
SIM	selected ion monitoring
SM	standard method
SOP	standard operating procedure
sp.	Species
SMP	standard monitoring program
SSS	system-specific study
Stage 1 DBPR	Stage 1 Disinfectants/Disinfection By-Products Rule
Stage 2 DBPR	Stage 2 Disinfectants/Disinfection By-Products Rule
SUVA	specific UV absorbance
SW	surface water
TCAN	trichloroacetonitrile
TCR	Total Coliform Rule
TEM	total exposure model
THMfp	THM formation potential
THMs	trihalomethanes
THM4	four THMs
TOC	total organic carbon
TOX	total organic halides
TR	transite
TSB	tryptic soy broth

U	unit
UDMH	1,1-dimethyhydrazine
UNC	University of North Carolina
U.S.	United States
USEPA	United States Environmental Protection Agency
UV	ultraviolet
v/v	volume/volume
w/v	weight/volume
WIDB	water industry database
WTP	water treatment plant
XAAs	monohalogenated acetic acids (i.e., ClAA and BrAA)
X_2AA	dihalogenated acetic acids (i.e., Cl_2AA, Br_2AA and BrClAA
X_3AAS	trihalogenated acetic acids (i.e., Cl_3AA, Br_3AA, $BrCl_2AA$ and Br_2ClAA)

IWA Publishing's authorised EU representative for General Product Safety Regulations is Diane D'Arras, 15 rue Duret, 75116 Paris, France, e-mail: safety@iwap.co.uk.

Printed and bound by CPI Group (UK) Ltd, Croydon, CR0 4YY
17/09/2026
02218811-0003